T0313815

From LED to Solid State Lighting

Principles, Materials, Packaging, Characterization, and Applications

Shi-Wei Ricky Lee
Chair Professor, Department of Mechanical and Aerospace Engineering
Director, HKUST Foshan Research Institute for Smart Manufacturing
Chairman, HKUST SZ-HK Collaborative Innovation Research Institute
Acting Dean, Systems Hub, HKUST (GZ)
The Hong Kong University of Science and Technology
Kowloon, HKSAR, China

Jeffery C. C. Lo
Assistant Director, HKUST Foshan Research Institute for Smart Manufacturing
Program Manager, Electronic Packaging Laboratory
The Hong Kong University of Science and Technology
Kowloon, HKSAR, China

Mian Tao
Lead Engineer, Technology Division, Group of Integrated Circuits and Systems
Hong Kong Applied Science and Technology Research Institute
New Territories, HKSAR, China

Huaiyu Ye
Associate Professor, School of Microelectronics
South University of Science and Technology
Shenzhen, Guangdong, China

This edition first published 2022

Registered Offices
John Wiley & Sons, Inc., 111 River Street, Hoboken, NJ 07030, USA
John Wiley & Sons Singapore Pte. Ltd, 1 Fusionopolis Walk, #07-01 Solaris South Tower, Singapore 138628

Editorial Office
1 Fusionopolis Walk, #07-01 Solaris South Tower, Singapore 138628

For details of our global editorial offices, customer services, and more information about Wiley products visit us at www.wiley.com.

Wiley also publishes its books in a variety of electronic formats and by print-on-demand. Some content that appears in standard print versions of this book may not be available in other formats.

Library of Congress Cataloging-in-Publication Data

Names: Lee, Shi-Wei Ricky, author. | Lo, Jeffery C. C., author. | Tao, Mian, author. | Ye, Huaiyu, author.
Title: From LED to solid state lighting : principles, materials, packaging, characterization, and applications / Shi-Wei Ricky Lee, Jeffery C. C. Lo, Mian Tao, Huaiyu Ye.
Description: Hoboken, NJ : Wiley, 2022.
Identifiers: LCCN 2021003853 (print) | LCCN 2021003854 (ebook) | ISBN 9781118881477 (hardback) | ISBN 9781118881583 (adobe pdf) | ISBN 9781118881552 (epub)
Subjects: LCSH: Light emitting diodes.
Classification: LCC TK7871.89.L53 L4365 2022 (print) | LCC TK7871.89.L53 (ebook) | DDC 621.32/8–dc23
LC record available at https://lccn.loc.gov/2021003853
LC ebook record available at https://lccn.loc.gov/2021003854

Cover Design: Wiley
Cover Image: © ModernewWorld/DigitalVision/Getty Images

Set in 9.5/12.5pt STIXTwoText by Straive, Chennai, India

Printed and bound by CPI Group (UK) Ltd, Croydon, CR0 4YY

C091074_310821

Contents

Preface

The light-emitting diode (LED) is a semiconductor device based on the effect of electroluminescence, which is a form of energy conversion from electrons directly to photons. Although the fundamental principle of physics was discovered more than 100 years ago, devices for actual applications were not invented until the 1960s. It took another 30 years for researchers to complete the visible light spectrum of LEDs with various recipes of semiconductor compounds. By the turn of the century, owing to the development of high-brightness and high-power LEDs with white light illumination, the era of solid-state lighting (SSL) began. LED evolved to be the light source of the fourth generation of lighting systems since the turn of the millennium.

Although diodes seem to be the simplest solid-state device, the manufacturing of LED light sources actually involves many fabrication processes and supply chains. Starting from the preparation of substrate wafers and the growth of epitaxial layers, sophisticated semiconductor processing control must be enforced in order to ensure good internal quantum efficiency. LED chip design and fabrication are essential elements that can substantially affect light extraction efficiency. The subsequent packaging will define the luminous efficiency, and the power supply at the system level will determine the overall lighting efficiency. There are a number of books on LEDs, but most of them are on sciences and technologies before packaging. This book intends to cover the mid- to downstream topics related to LEDs and so is mainly based on the research outcomes and teaching materials of authors over the past decade. The selected subjects are for postgraduate students in universities and professional engineers in the industries involved in the design, material development, packaging and assembly processes, reliability testing, and applications of LEDs for SSL. This book is aimed at the introductory to intermediate levels. A bachelor's degree or equivalent in a relevant engineering or science discipline greatly help the reader's understanding of its contents.

Chapter 1 of this book is a concise review of LEDs and lighting systems for readers who have not had former exposure to these areas. Chapter 2 introduces the fundamentals of packaging processes and materials for conventional LEDs. Chapter 3 covers more advanced topics on chip scale and wafer level packaging for LEDs, which are mainly based on the authors' research outcomes. Chapter 4 provides information on board level assembly including chip-on-board and LED modules. Chapter 5 reviews the approaches to evaluating the optical, electrical, and thermal performance of LEDs. Chapter 6 covers the topic of reliability engineering for LED packaging, with detailed test methods and failure analyses. Chapter 7 introduces emerging applications of LEDs, such as automotive lighting, micro/mini-LED displays, and visible light communications. Chapter 8 discusses LEDs beyond visible light, including ultraviolet and infrared applications, and relevant technology trends.

As mentioned in the first chapter of this book, the gap between the onset of generations of lighting systems is approximately 60 years. It is the consensus that the dawn of SSL occurred in the late 1990s. Twenty years have gone by and there is no hint of new sciences that may lead to the fifth generation of lighting systems in the future. Some people tend to believe LEDs may be the last form of light source for general lighting applications for human societies. Nevertheless, there are always surprises in scientific and technological development. We will see whether there is to be a new form of light source for general lighting in four decades' time. In the meantime, the authors hope this book will contribute to the education and training of students and professionals for the improvement of LED packaging and its relevant applications in terms of reliability and performance. The 2014 Nobel Prize in Physics highlighted that the award was conferred "for the invention of efficient blue light-emitting diodes which has enabled bright and energy-saving white light sources." The most significant word in this award citation is "energy-saving." Lighting accounts for 20% of electricity consumption on earth. Any saving on energy in lighting is meaningful. Let's work together harder and further to achieve more efficient lighting through the design, materials, and packaging of LEDs.

March 2021

Shi-Wei Ricky Lee
Jeffery C. C. Lo
Mian Tao
Huaiyu Ye

About the Authors

Shi-Wei Ricky Lee, PhD, FIEEE, LFASME, LFIMAPS, FInstP
Chair Professor, Department of Mechanical & Aerospace Engineering
Director, HKUST Foshan Research Institute for Smart Manufacturing
Director, Electronic Packaging Laboratory
Acting Dean, Systems Hub, HKUST (GZ)
The Hong Kong University of Science and Technology
Kowloon, HKSAR, China

Ricky Lee received his PhD degree in Aeronautical and Astronautical Engineering from Purdue University in 1992. Currently, he is Chair Professor of Mechanical and Aerospace Engineering and Director of Foshan Research Institute for Smart Manufacturing, and Director of Electronic Packaging Laboratory, at the Hong Kong University of Science and Technology (HKUST). He also has concurrent appointments as Acting Dean of Systems Hub of the HKUST Guangzhou Campus and Director of HKUST LED-FPD Technology R&D Center at Foshan, Guangdong, China. Dr. Lee has been focusing his research on the development of packaging and assembly technologies for electronics and optoelectronics. His R&D activities include wafer level packaging and 3D IC integration, additive manufacturing for microsystems packaging, LED packaging for solid-state lighting and applications beyond lighting, lead-free soldering, and reliability analysis. Dr. Lee is a Fellow of IEEE, ASME, IMAPS, and Institute of Physics (UK). He is also Editor-in-Chief of *ASME Journal of Electronic Packaging*.

Jeffery C. C. Lo, PhD
Assistant Director, HKUST Foshan Research Institute for Smart Manufacturing
Program Manager, Electronic Packaging Laboratory
The Hong Kong University of Science and Technology
Kowloon, HKSAR, China

Jeffery Lo received his bachelor (first class honors) and MPhil degrees in Mechanical Engineering Department from the Hong Kong University of Science and Technology (HKUST) in 2002 and 2004, respectively. He then joined Electronic Packaging

Laboratory of HKUST as a Senior Technical Officer and provided professional technical support to various users, including undergraduate and postgraduate students, from different departments and industrial partners around the world. To further develop his R&D career, he completed his PhD degree at HKUST in 2008 and continued offering support to the lab at the same time. He is now the Program Manager of Electronic Packaging Laboratory and Assistant Director of Foshan Research Institute for Smart Manufacturing, focusing on R&D projects with various international companies. The topics of his research interests include flip-chip technologies, wafer level packaging, and LED packaging. He was granted the 2004 ECTC Best Poster Paper Award (May 2005), Young Award in IEEE 9th VLSI Packaging Workshop in Japan (December 2008), and IEEE-CPMT Outstanding Young Engineer Award in 2015. He was the IEEE-CPMT Hong Kong Chapter Chairman in 2015/16.

Mian Tao, PhD
Lead Engineer, Technology Division, Group of Integrated Circuits and Systems
Hong Kong Applied Science and Technology Research Institute
New Territories, HKSAR, China

Mian Tao received his MSc degree from the Hong Kong University of Science and Technology (HKUST) in Mechanical Engineering in 2010. He then worked in HKUST LED-FPD Technology R&D Centre at Foshan from 2011 to 2013 focusing on the optical, thermal, and electrical characterization of light-emitting diode (LED) chips and devices. He received his PhD degree in 2016 from HKUST. His research topic focused on the effect of nonuniformity junction temperature on the performance of an LED device. Afterward, he worked as a research associate in HKUST's Electronic Packaging Laboratory and later joined the Hong Kong Applied Science and Technology Research Institute (ASTRI) as a lead engineer. His research interests include thermal characterization and the management of high-power electronics and the development of advanced microelectronic packages.

Huaiyu Ye, PhD
Associate Professor, School of Microelectronics
South University of Science and Technology
Shenzhen, Guangdong, China

Huaiyu Ye received his PhD degree from the Department of Microelectronics of Delft University of Technology in 2014. He is Associate Professor of Southern University of Science and Technology (SUSTech). He won the Outstanding Innovation Youth Award in 2017 and Alliance Contribution Award in 2018 of the China Advanced Semiconductor Industry Innovation Alliance and was selected as the Shenzhen Overseas High-Caliber Personnel in 2019. He mainly conducts his research on advanced packaging of solid-state lighting and power electronic devices, focusing on materials, technology, structure, testing, thermal management, and reliability. He is also responsible for the construction of advanced packaging platforms, R&D, and industrialization. He chaired and participated in 16 projects in China and overseas. He has served as an academic committee member of several international conferences and participated in Technology Roadmap for Wide Band Gap Power Semiconductor 2018.

1

LEDs for Solid-State Lighting

1.1 Introduction

A light-emitting diode (LED) is one kind of semiconductor that can emit light. A diode is the simplest solid-state device in electronics [1]. It consists of a p-doped and an n-doped semiconductor, as illustrated in Figure 1.1. The two types of semiconductors form a junction which has the current–voltage (I–V) characteristics given in Figure 1.2. An LED is a diode that can emit photons when it is subject to the forward bias, as shown in Figure 1.3 [2]. In the twentieth century, LEDs were mainly used as signal indicators on instrument panels or simple displays for commercial signages. Owing to the development of white light illumination and the improvement in power rating/efficiency, LEDs have been widely used for general lighting since the 1990s [3].

As an opening, in this chapter the classification of light sources is defined first and the four generations of artificial lighting systems introduced. Subsequently, the historical development of LEDs will be reviewed. Afterward, the implementation of white light illumination with LEDs is given as background information for the subsequent chapters. Finally, certain examples are given to illustrate the applications of LEDs for general lighting.

1.2 Evolution of Light Sources and Lighting Systems

In principle, the light sources on earth may be classified into two main categories: hot light sources and cold light sources. Hot light sources include combustion (e.g. candles) and thermal radiation (e.g. incandescent lamps). The former is a chemical reaction, while the latter is a physical phenomenon. Cold light sources include chemiluminescence (e.g. fireflies), electrical discharge (e.g. fluorescent tubes), and electroluminescence (e.g. LEDs) [4]. Electroluminescence is direct energy conversion from electrons to photons [5]. As illustrated in Figure 1.4, when mobile electrons and holes meet at the p–n junction, photons are emitted due to radiative recombination [6]. This is the lighting mechanism of the LED.

Human history has witnessed four generations of lighting systems, as shown in Figure 1.5. Owing to the development of gas pipelines in the early nineteenth century, gas lamps were considered the first generation of lighting systems. In 1879, Thomas Edison

From LED to Solid State Lighting: Principles, Materials, Packaging, Characterization, and Applications, First Edition.
Shi-Wei Ricky Lee, Jeffery C. C. Lo, Mian Tao, and Huaiyu Ye.

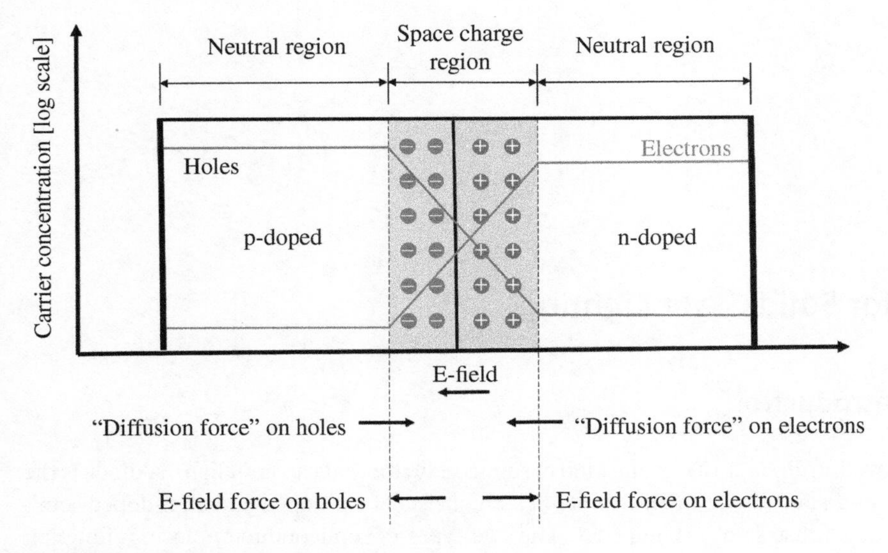

Figure 1.1 Structure of a typical diode.

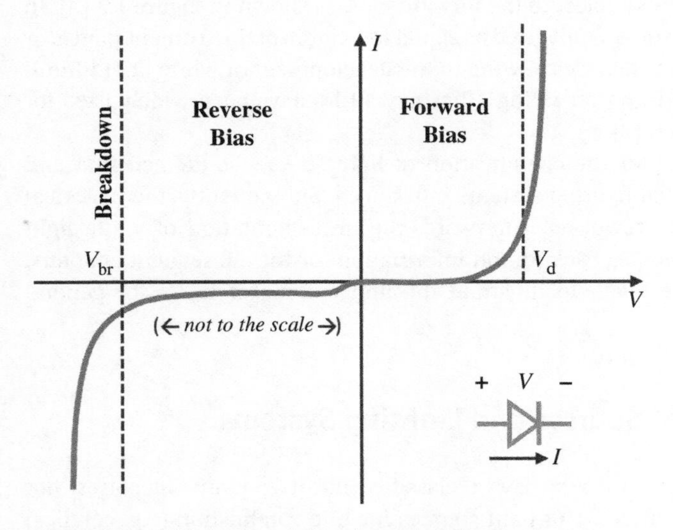

Figure 1.2 *I*–*V* characteristics of a typical diode.

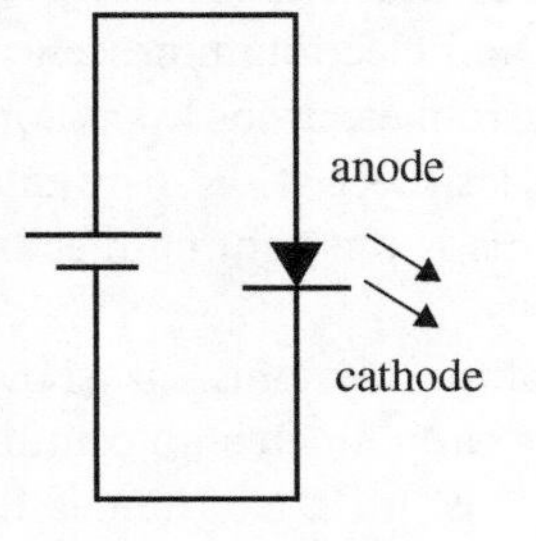

Figure 1.3 LED under forward bias.

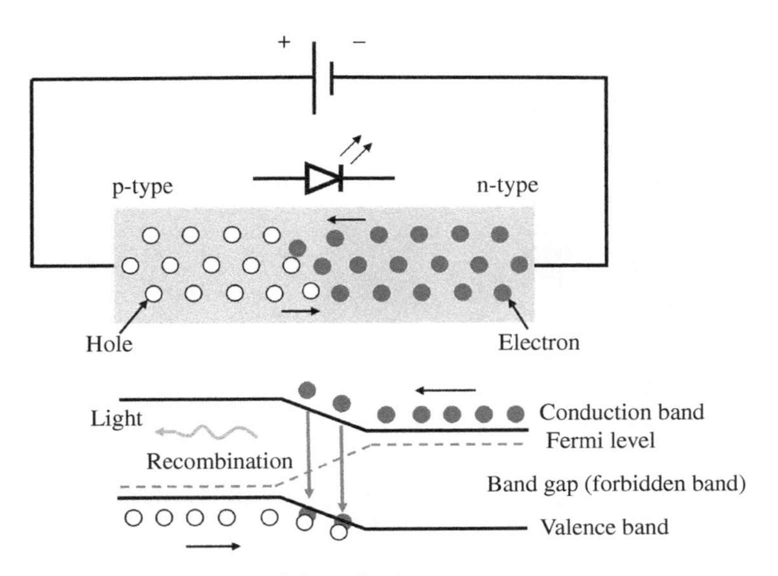

Figure 1.4 Mechanism of electroluminescence.

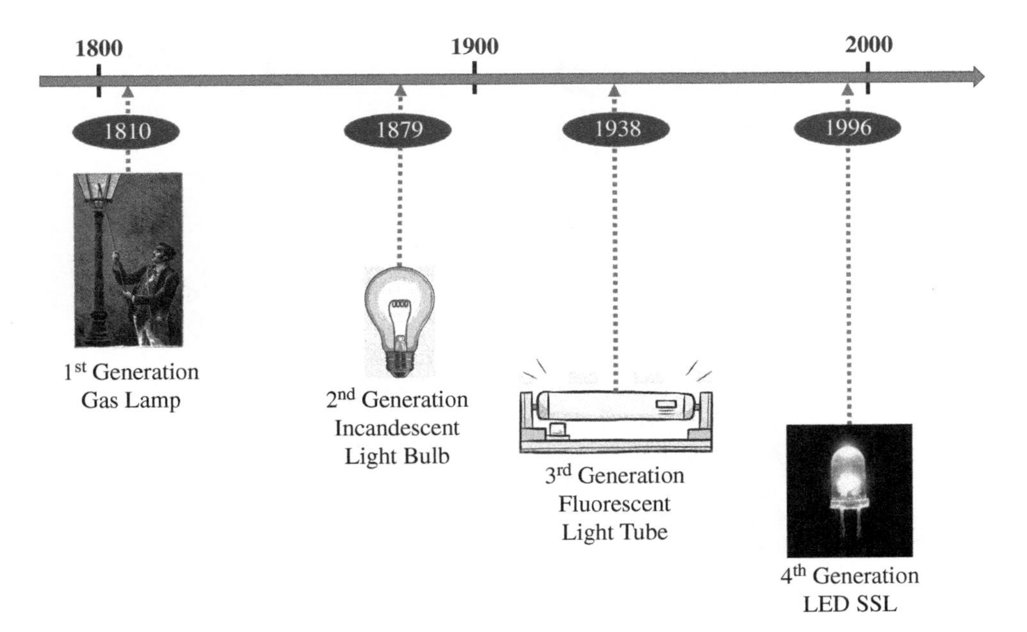

Figure 1.5 Four generations of lighting systems. Source: Chronicle/Alamy Stock Photo.

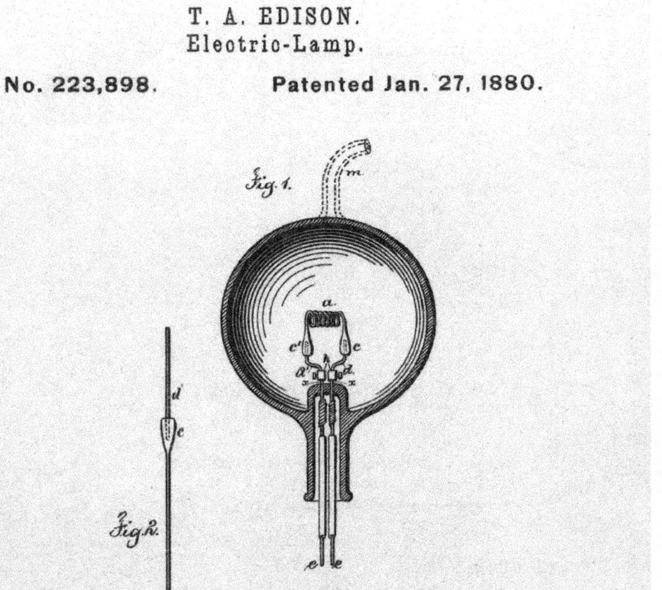

Figure 1.6 Edison's patent of incandescent lightbulb. Source: T. A. Edison, Electric Lamp. 1880. US Patent No. 223,898.

filed his patent (Figure 1.6) of an incandescent lightbulb with a service life of 40 hours (Figure 1.7). In the ensuing year he improved the filament materials and increased the service life to more than 1000 hours. Together with the distribution of electrical power lines, the incandescent lightbulb became emblematic of the second generation of lighting

Figure 1.7 Edison's first successful lightbulb demonstrated to the public. Source: https://commons.wikimedia.org/wiki/File:Edison_Carbon_Bulb.jpg#/media/File:Edison_Carbon_Bulb.jpg.

systems. In the middle of the twentieth century, fluorescent light tubes appeared and were considered the third generation of lighting systems [7]. Starting from the late 1990s, following the invention of the blue light LED and the implementation of white light illumination with a phosphor-converted LED (pc-LED), it became possible to use LEDs for general lighting applications. This marked the onset of the fourth generation of lighting systems. After two decades of development and improvement, LEDs have been widely used in all kinds of applications (Figure 1.8). Nowadays, people also term the fourth generation of lighting systems with LEDs as solid-state lighting (SSL) or semiconductor lighting [8]. It is interesting to note that the gap between generations of lighting systems appears to be approximately 60 years.

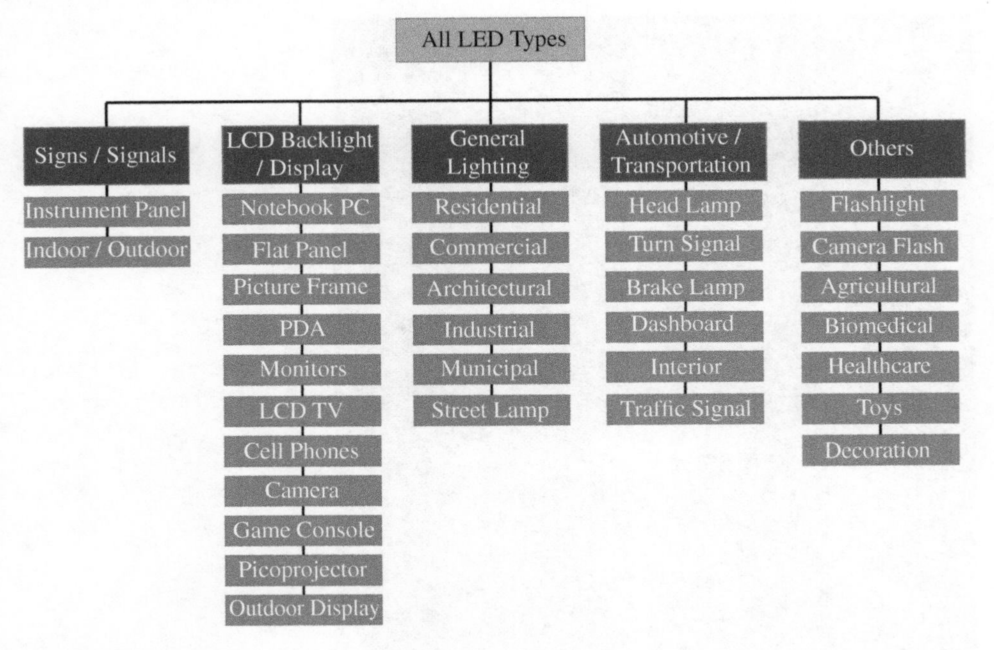

Figure 1.8 Examples of LED applications.

1.3 Historical Development of LEDs

The literature traces back the LED to the beginning of the twentieth century. The British scientist Henry Joseph Round of Marconi Labs discovered the electroluminescence phenomenon in 1907 using a crystal of silicon carbide (SiC) and a cat's whisker detector (Figure 1.9) [9]. In the mid-1920s, the Russian engineer Oleg Losev independently invented a diode that could emit light and filed the first patent for an LED [10]. For decades afterwards, there was little development of the LED.

The next milestone for the development of the LED occurred in 1962 when Dr. Nick Holonyak Jr. of General Electric invented the red light LED [11]. Subsequently, Dr. M. George Craford of Monsanto introduced an LED that could emit yellow light in 1972 [12]. The last piece of the jigsaw appeared 20 years later when Shuji Nakamura of Nichia demonstrated the first high brightness blue light LED based on indium gallium nitride (InGaN) in 1993 [13]. The presence of the blue light LED implied the completion of the visible light spectrum and the feasibility of implementing white light illumination using LEDs with color mixing schemes [14]. The holy grail of LED development was bestowed on the "three musketeers" who contributed the most to the blue light LED: Professors Isamu Akasaki, Hiroshi Amano, and Shuji Nakamura became the 2014 Nobel Prize Laureates in Physics (Figure 1.10). The award citation was "for the invention of efficient blue light-emitting diodes which has enabled bright and energy-saving white light sources" [15].

Figure 1.9 The first demonstration of electroluminescence with SiC crystal.Source: https://es.wikipedia.org/wiki/Led#/media/Archivo:SiC_LED_historic.jpg.

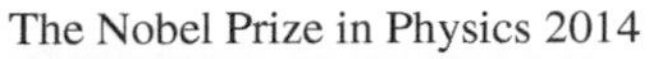

The Nobel Prize in Physics 2014 was awarded jointly to Isamu Akasaki, Hiroshi Amano and Shuji Nakamura "for the invention of efficient blue light-emitting diodes which has enabled bright and energy-saving white light sources".

Figure 1.10 Nobel Prize in Physics 2014 honored the blue light LED inventors [15]. Source: © Nobel Media AB, Photo Alexander Mahmoud.

Nowadays, LEDs can cover the whole visible light spectrum with various kinds of semiconductor compounds, as shown in Table 1.1. Although II–VI compounds may be used, most of them belong to the III–V compounds family. There are two main categories. The aluminum gallium indium phosphide (AlGaInP) quaternary compound may cover from red to green light and the InGaN ternary compound may cover from green to violet light [16].

Table 1.1 Recipes of semiconductor compounds for LEDs.

Color	Wavelength λ (nm)	Forward ΔV (V)	Semiconductor materials
Infrared	$\lambda > 760$	$\Delta V < 1.63$	GaAs, AlGaAs
Red	$610 < \lambda < 760$	$1.63 < \Delta V < 2.03$	AlGaAs, GaAsP, AlGaInP,GaP
Orange	$590 < \lambda < 610$	$2.03 < \Delta V < 2.10$	GaAsP, AlGaInP,GaP
Yellow	$570 < \lambda < 590$	$2.10 < \Delta V < 2.18$	GaAsP, AlGaInP,GaP
Green	$500 < \lambda < 570$	$1.9 < \Delta V < 4.0$	Traditional green: GaP AlGaInP, AlGaP Pure green: InGaN/GaN
Blue	$450 < \lambda < 500$	$2.48 < \Delta V < 3.7$	ZnSe, InGaN, SiC
Violet	$400 < \lambda < 450$	$2.76 < \Delta V < 4.0$	InGaN
Ultraviolet	$\lambda < 400$	$3 < \Delta V < 4.1$	InGaN, Diamond, BN, AlN, AlGaN, AlGaInN

Although infrared (IR) and ultraviolet (UV) emissions are possible as well, they are for nonlighting applications.

1.4 Implementation of White Light Illumination with an LED

The basic principle for generating white light is to mix red, green, and blue (RGB) lights. With various mixing ratios, different degrees of white light may be achieved. Such an RGB color mixing scheme will involve three kinds of LEDs that can emit red, green, and blue lights, respectively, as shown in Figure 1.11. Since different kinds of LEDs have different forward bias voltages, different chip sizes, different compound materials, and different service lives, the RGB color mixing scheme usually leads to a more complicated and expensive system which is more suitable for display applications [17]. For general lighting, it is often more cost sensitive. Therefore, a cost-effective approach for white light illumination is needed.

Another way of generating white light is by fluorescence with phosphor powders. The conventional fluorescent lamp is an example of this. The phosphors may be excited by blue light or UV emitted from an LED, as illustrated in Figure 1.12. Such an approach is termed phosphor-converted LED (pc-LED) [18]. If the light source is UV, the phosphors must be a mixture of RGB powders. Although this may offer more flexibility for various color tuning, it may also lead to higher cost and less uniformity.

The most cost-effective method for generating white light is to use a blue light LED to excite yellow phosphor, as shown in Figure 1.13. The relevant mechanism is illustrated in Figure 1.14 with the CIE 1931 chromaticity diagram. The blue light emitted from the LED has a color coordinate at the "blue corner" of the diagram. When the phosphor is added on the top of the LED, the excited yellow light will make the mixed light move toward the "yellow edge". This is a one-dimensional path and the degree of progress from blue to yellow depends on how much yellow light is excited. Therefore, the amount of yellow phosphor may be adjusted so that the resulting color coordinate lands at a precise location in the

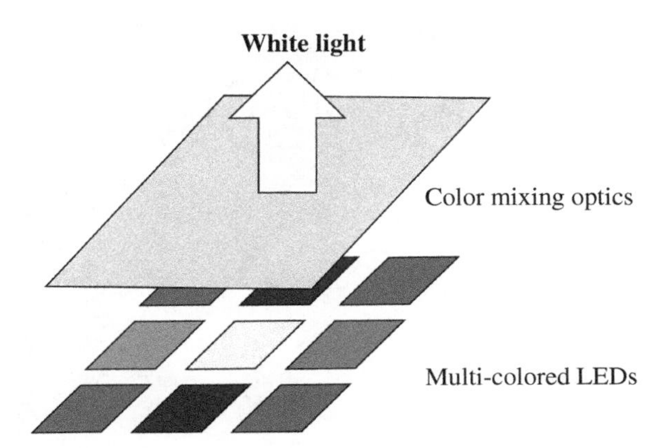

Figure 1.11 White light illumination with RGB LEDs.

Figure 1.12 White light illumination with phosphor-converted LED.

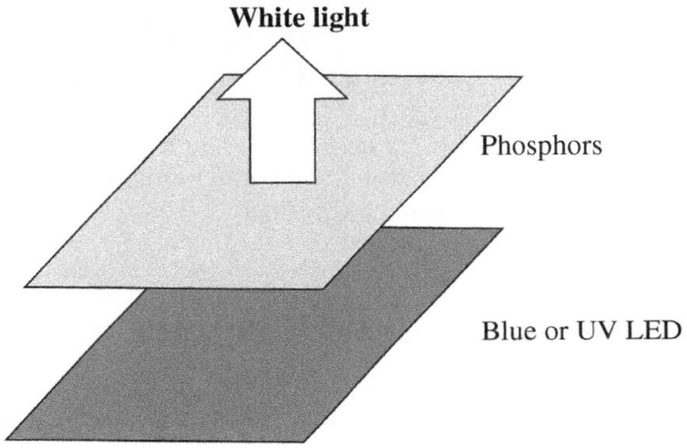

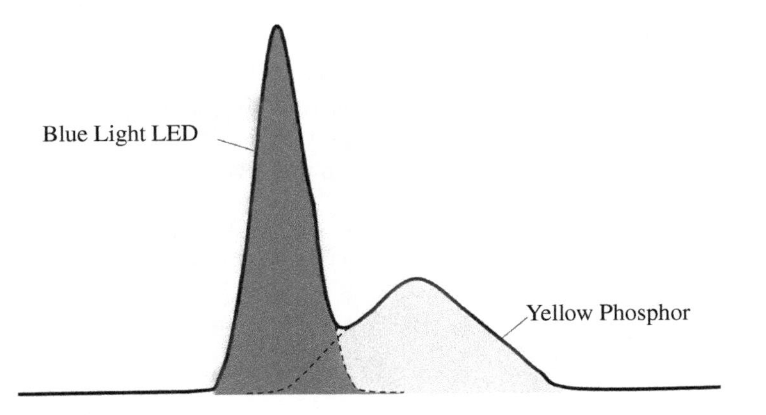

Figure 1.13 Blue light LED with yellow phosphor for white light illumination.

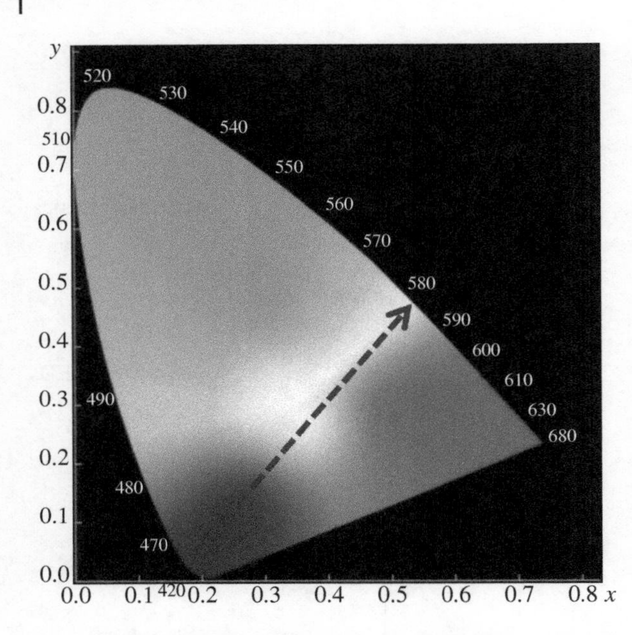

Figure 1.14 Mechanism of white light generation with blue–yellow color mixing.

"white zone". This is the most convenient and cost-effective way to generate white light for general lighting applications [19].

1.5 LEDs for General Lighting

The benefits of the LED – its high brightness, low power consumption, vivid color spectrum, compact size, and long service life – are well known. These make the LED suitable for many industrial and consumer applications. In particular, people have very much emphasized the general lighting applications of the LED in the 2010s [20]. According to the International Energy Agency, lighting accounts for nearly a fifth of electrical energy consumption worldwide (Figure 1.15). The efficiency of conventional incandescent lightbulbs is typically less than 5%. Therefore, tremendous amounts of energy have been wasted on lighting in the past. Although the efficiency of fluorescent lamps has been improved to some extent, the mercury contents in the light tube impose substantial threats to the environment. Therefore,

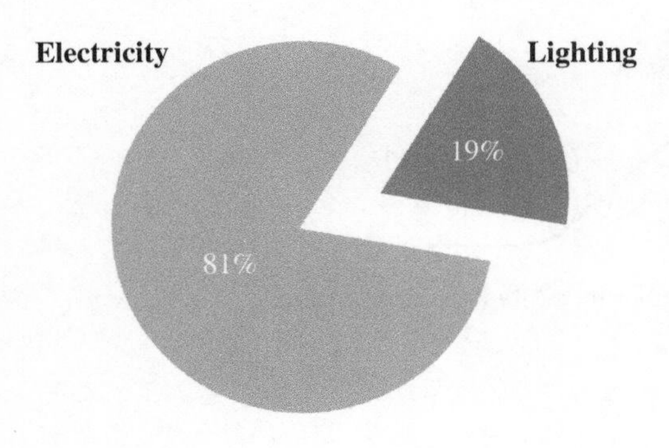

Figure 1.15 Percentage of lighting energy consumption in overall electricity supply.

Figure 1.16 Application of solid-state lighting in a Hong Kong subway train.

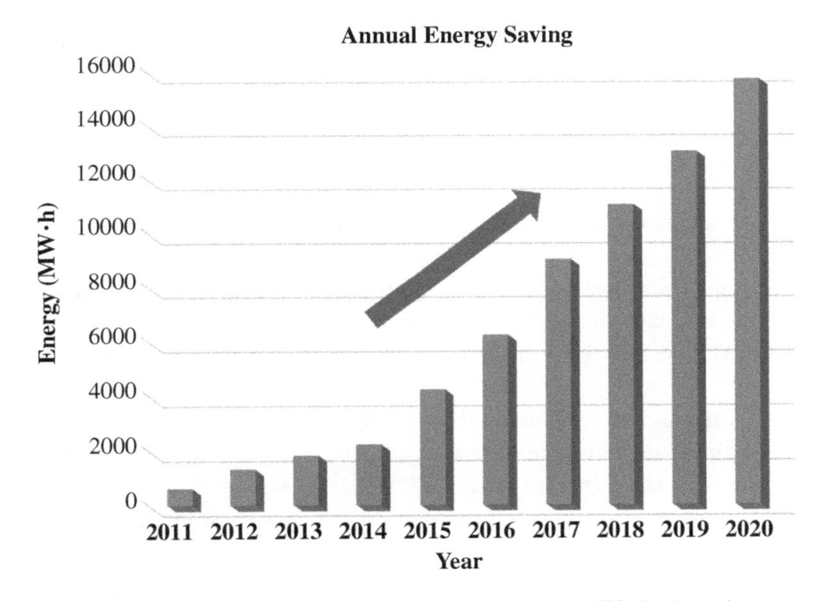

Figure 1.17 Energy saving due to the implementation of SSL in the subway systems.

both incandescent and fluorescent lamps must be banned. Many countries have enforced government policies to phase out the second and third generation lighting systems with very aggressive schedules. Eventually, SSLs with LEDs will be the only allowable and affordable lighting system [21].

In addition to consumer and household applications, there exist many examples of the use of LEDs for general lighting in the public domain. Figure 1.16 shows the first application of an SSL in a subway system, which was a collaborative project between the Mass

Transit Railway Corporation (MTRC) of Hong Kong and the Hong University of Science and Technology (HKUST). LED lighting modules were developed and used to replace the fluorescent light tubes on the trains for saloon interior lighting [22]. The first trial batch of prototypes was installed in 2007. After three years of pilot runs, the MTRC decided to launch massive scale deployment. Besides saloon interior lighting on trains, SSLs are now used for interior ceiling lighting and for advertisement box back lighting in all subway stations. The energy saving due to the implementation of SSL in the subway systems of Hong Kong over the years is given in Figure 1.17. The merit of SSL in terms of energy saving has been evidenced by numerous general lighting applications. This is also the best testimony to the award citation of 2014 Nobel Prize in Physics!

References

1. Kuphaldt, T.R. (2009). Lessons in electric circuits. In: *Semiconductors*, 5e, vol. III. https://www.ibiblio.org/kuphaldt/electricCircuits/Semi/SEMI.pdf (accessed 22 February 2020).
2. Schubert, E.F. (2003). *Light-Emitting Diodes*. Cambridge University Press.
3. Department of Energy. (2020). LED basics. https://www.energy.gov/eere/ssl/led-basics (accessed 3 September 2020).
4. Wikipedia (2020). List of light sources. https://en.wikipedia.org/wiki/List_of_light_sources (accessed 3 September 2020).
5. Pankove, J.I. (ed.) (1997). Electroluminescence. In: *Topics in Applied Physics*, vol. 17, 1–29. Springer-Verlag.
6. Braunstein, R. (1955). Radiative transitions in semiconductors. *Physical Review* 99 (6): 1892–1893.
7. Bellis M. (2019). The history of lighting and lamps. https://www.thoughtco.com/history-of-lighting-and-lamps-1992089 (accessed 3 September 2020).
8. Zheludev, N. (2007). The life and times of the LED: a 100-year history. *Nature Photonics* 1 (4): 189–192.
9. Round, H.J. (1907). A note on carborundum. *Electrical World* 19: 309.
10. Losev, O. (1928). Luminous carborundum detector and detection effect and oscillations with crystals. *Philosophical Magazine*, 7th series 5 (39): 1024–1044.
11. Holonyak, N. Jr., and Bevaqua, S.F. (1962). Coherent (visible) light emission from $Ga[As_{(1-x)}P_x]$ junctions. *Applied Physics Letters* 1 (4): 82–83.
12. Craford, M.G., Shaw, R.W., Groves, W.O. et al. (1972). Radiative recombination mechanisms in GaAsP diodes with and without nitrogen doping. *Journal of Applied Physics* 43: 4075–4083.
13. Nakamura, S. (1994). Growth of InxGa(1-x)N compound semiconductors and high-power InGaN/AlGaN double heterostructure violet-light-emitting diodes. *Microelectronics Journal* 25: 651–659.
14. Wierer, J.J., Steigerwald, D.A., Krames, M.R. et al. (2001). High-power AlGaInN flip-chip light-emitting diodes. *Applied Physics Letters* 78 (22): 3379–3381.
15. Nobel Media AB. (2014). The Nobel Prize in Physics 2014. www.nobelprize.org (accessed 3 September 2020).

16 Wikipedia. (2020). List of semiconductor materials. https://en.wikipedia.org/wiki/List_of_semiconductor_materials (accessed 3 September 2020).

17 Shubert, E.F. (2006). *Light-Emitting Diodes*, 2e. Cambridge University Press.

18 Nair, G.B., Swart, H.C., and Dhoble, S.J. (2020). A review on the advancements in phosphor-converted light emitting diodes (pc-LEDs): phosphor synthesis, device fabrication and characterization. *Progress in Materials Science* 109: 100622.

19 Lee, K.H. and Lee, S.W.R. (2006). Process development for yellow phosphor coating on blue light emitting diodes (LEDs) for white light illumination. In: *Proceedings of the 8th Electronics Packaging Technology Conference (EPTc)*, Singapore, 379–384. IEEE.

20 Liu, S. and Luo, X. (2010). *LED Packaging for Lighting Applications: Design, Manufacturing and Testing*. Wiley.

21 Department of Energy. (2020). LED lighting. https://www.energy.gov/energysaver/save-electricity-and-fuel/lighting-choices-save-you-money/led-lighting (accessed 3 September 2020).

22 Lee, S.W.R., Lau, C.H., Chan, S.P. et al. (2006). Development and prototyping of a HB-LED array module for indoor solid state lighting. In: *8th IEEE International Conference on High Density Microsystem Design, Packaging & Failure Analysis (HDP)*, Shanghai, China, 192–196. IEEE.

2

Packaging of LED Chips

2.1 Introduction

Packaging is a fundamental step in the manufacture of microelectronics and LED components. Devices which are fabricated on a wafer will not work without being packaged into a component and assembled as a system. The overall packaging process is categorized into different levels. The first level refers to the chip level packaging process which packages chips to components. Burn-in testing or sorting can be performed on the packaged components. In the second level, which refers to the board level assembly process, different packaged components are assembled onto a board, normally a printed circuit board (PCB). Finally, the board assemblies and other relevant accessories are integrated into a system as a final product for the end user.

Traditional integrated circuit (IC) packaging is commonly used for providing power, electrical connections with other components, mechanical support, protection from the environment, and heat dissipation paths. The performance and reliability of the components depend heavily on the package type, packaging process, and packaging materials. In LED chip packaging, extracting light emitted from the chip to the outside world and color tuning are other important considerations. These directly determine the performance of the LED component. Upon packaging of the LED components, burn-in testing, sorting, screening, binning, and other necessary processes can be performed thereon. Components with similar optical performance (e.g. lumen output, color temperature, color rendering index, etc.) are selected for the subsequence board level assembly. The overall functions of the package are illustrated in Figure 2.1.

This chapter discusses the LED chip packaging process in detail. Different types of LED packages currently available in the market are introduced. This chapter also covers various types of chip mounting and interconnection methods, as well as several phosphor deposition methods which are commonly adopted. The phosphor coating configurations of the latter will have a big influence on the optical performance of LED components.

In a traditional IC package, the IC is protected by a molding compound which is usually flat and not transparent. Such a molding compound is not suitable for LED packaging. Rather, transparent epoxy or silicone is generally used. The encapsulant material protects the chip and interconnects in the package and serves as a light-transmitting medium. In some applications, the encapsulation also serves as an optical lens to achieve the designed

From LED to Solid State Lighting: Principles, Materials, Packaging, Characterization, and Applications, First Edition.
Shi-Wei Ricky Lee, Jeffery C. C. Lo, Mian Tao, and Huaiyu Ye.

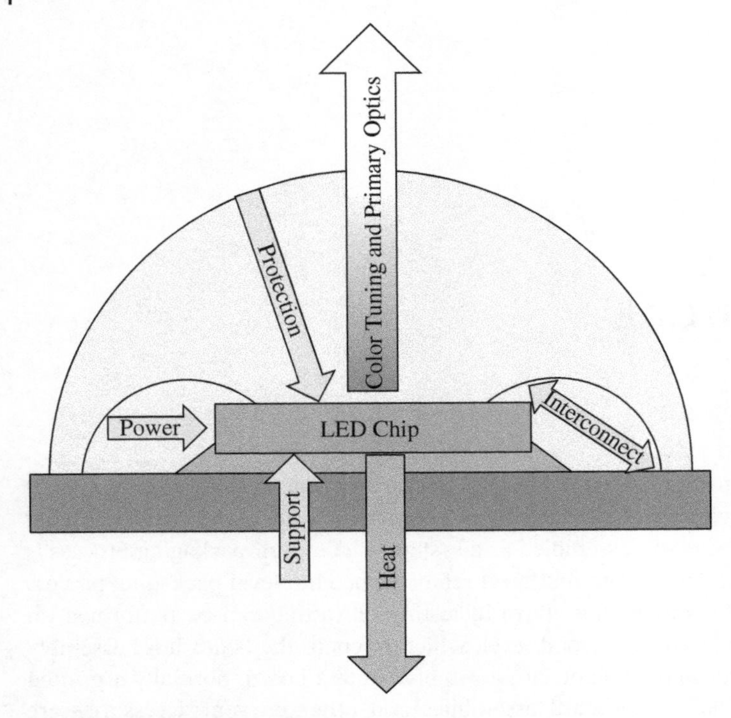

Figure 2.1 Functions of the package.

light output pattern. This chapter briefly discusses various types of encapsulation and molding methods.

For packaging LED chips, some major considerations include extracting more light emitted from the chip for increasing the lumen efficacy and achieving a uniform color distribution, etc. Nevertheless, different applications may require different light output patterns. For example, a rectangular light pattern is preferred to a circular one for street lighting. In such cases, the regular LED packaging process may not be able to cater for the specific requirements of each case. Secondary optics components (e.g. lenses) may be required to adjust the light pattern accordingly. This chapter covers the basic concepts of optical simulation, which is a powerful tool in optimizing the design of secondary optics components.

2.2 Overall Packaging Process and LED Package Types

Similar to traditional IC packaging, there are basically three types of packages in LED packaging, namely plated through hole (PTH), surface mount devices (SMDs), and chip-on-board (COB). This section focuses on the packing process involved in the PTH and SMD LED components. The fabrication and assembly processes of COB LED components are covered in Chapter 4.

The packaging of the LED involves various processes. The chips are first diced from a wafer and then mounted onto a substrate (metal lead frame, ceramic substrate, silicon wafer

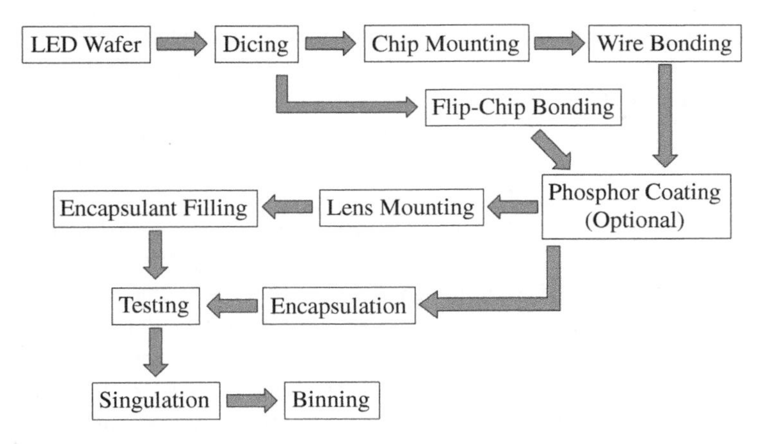

Figure 2.2 Overall LED packaging process.

submount, etc.) with proper die attach materials. The substrate and die attach materials provide mechanical support and a heat dissipation path. After that, wire bonds are made to electrically connect the chip to the circuit on the substrate. If flip-chip LEDs are used, the bumps will provide both mechanical support and electrical interconnects. A phosphor coating is applied for color tuning and is followed by encapsulation. Transparent epoxy or silicone is applied to protect the chip and interconnects. In some high-power LED packages, optical lens may be used to extract more light. Encapsulant is used to fill the gap between the chip and the lens. Individual components are then singulated from the panel after testing. The final step, i.e. binning, will divide the components into groups with similar optical performance. The overall packaging process is summarized in Figure 2.2.

2.2.1 PTH LED Component

PTH components were widely used in IC packaging decades ago. The chips are mounted on a lead frame. The lead frame has long and thick leads. Figure 2.3 shows some typical PTH components, such as transistor outline (TO) and dual in-line (DIP). During the assembly process, those leads are inserted into the copper PTHs of the PCB. The components are then soldered to the PCB by hand soldering or wave soldering.

Figure 2.3 Typical PTH components. Source: Texas Instruments.

Low-power LEDs are generally used as signal indicators or the light source of declarations. They cannot be used as the light source for general lighting. In some electronic devices (e.g. remote controller), infrared LEDs are used as sources of signal transmission. Since the power consumption of these LEDs is not high, heat dissipation and therefore thermal management are not major concerns. These LEDs are packaged in PTH form and assembled with other electronic components on a PCB.

As shown in Figure 2.4, the overall outlook of the PTH LED components is very similar to that of the PTH IC components. Since an LED has only two input/outputs (I/Os), cathode and anode, the lead frame of the PTH LED component has only two leads. The lead frame has a small reflective cup for collimating the light emitted from the side wall of the chip to the front of the package. One lead frame contains several units. Transparent epoxy is normally used as an encapsulant in PTH LED packaging. This type of package also has a special name (bullet head), which is due to the elliptical shape of the encapsulation.

2.2.2 SMD LED Component

PTH components have a number of weaknesses. For example, they require through holes on the PCB. They cannot be assembled on both sides of the PCB. They are large and each component occupies a large area on the board. Moreover, the PTH lead frames cannot accommodate devices with a large number of I/Os.

SMDs have a lot of advantages over PTH devices. The SMD package has a smaller package size. The components are soldered onto the surface of the PCB. It is possible to place components on both sides of the PCB. As a result, more components can be assembled on the PCB.

A lead frame is used in packaging SMD components if the number of I/Os is not large. The die is mounted on the die paddle of the lead frame. In some lead frame designs, heat is generated from the die and transferred (dissipated) to the environment through die attach materials, lead frame die paddle, and solder. Wire bonds are made to connect the bond pads on the chip to the leads. Upon encapsulation, individual components are singulated from the panel.

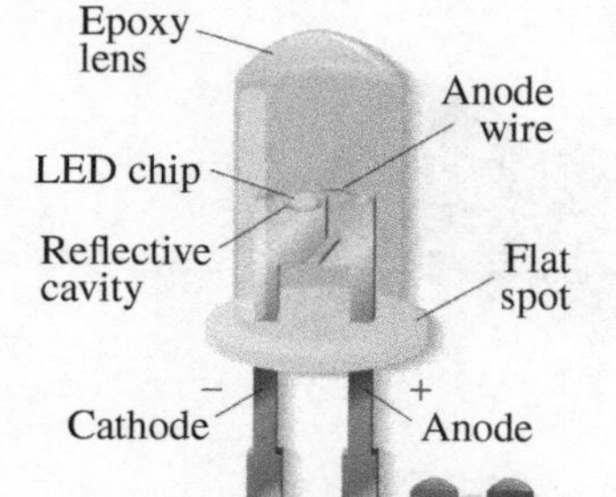

Figure 2.4 Typical PTH LED component. Source: Marubeni Opto.

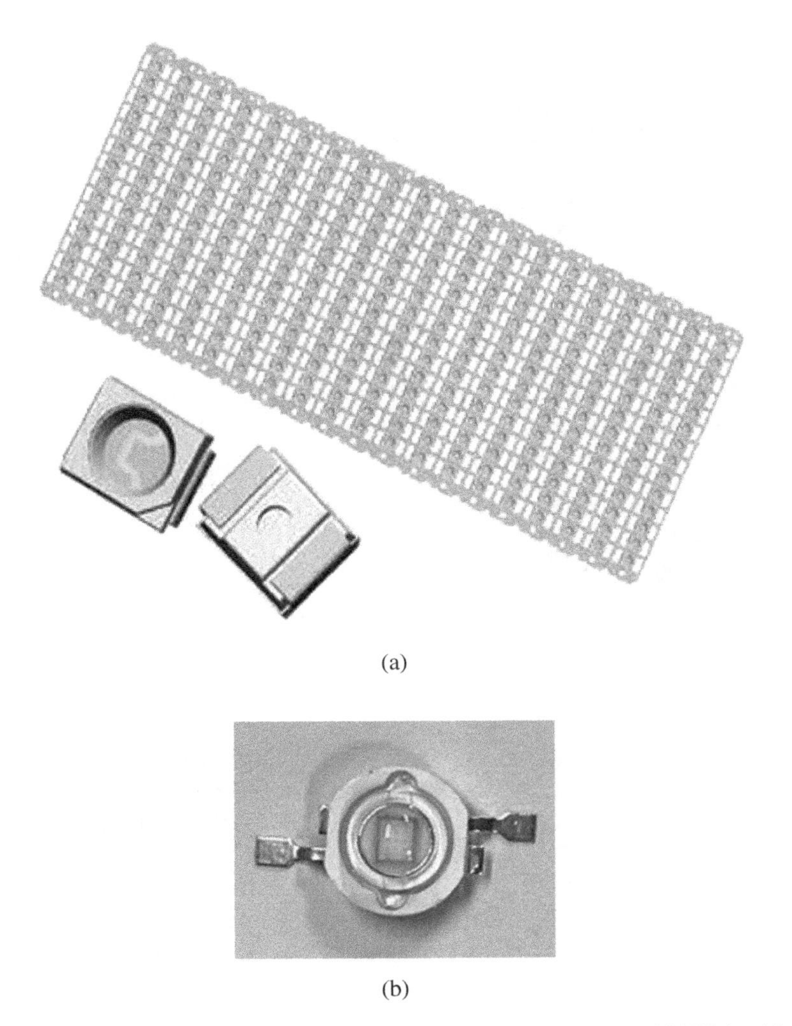

(a)

(b)

Figure 2.5 LED lead frame and typical SMD LED components: (a) LED lead frame; (b) K1 SMD LED.

LED chips can also be packaged in the SMD format. The lead frame design is different from the one used in a traditional IC component. A reflective cup is molded on top of the metal lead frame for reflecting light to the top. Also, it provides a cavity for phosphor and encapsulant deposition. Some lead frames are designed for high-power LEDs (K1 emitter, K2, Golden Dragon, etc.) with heat slugs under the die, allowing heat to transfer directly to the board or heat sink. Those without heat slugs are designed for low- to medium-power LEDs (3030, 3528, 5050). Figure 2.5 shows a typical lead frame for LED packaging and common SMD LED components. In some applications, multiple LEDs (red, green, blue) and controlling ICs are packaged into one component.

In lead frame type SMD components, the leads are allocated at the peripherals of the package, which is insufficient for devices with a large number of I/Os. An interposer, which can be an organic substrate (normally BT or FR4) or a ceramic substrate, is normally used to

Figure 2.6 Ball grid array (BGA) and land grid array (LGA). Source: ASE Group.

Figure 2.7 High-power LED package.

replace the lead frame. This multilayer substrate provides a platform for circuit routing and I/O redistribution. It also allows the bond pads to be arranged in an area array format for board level assembly. Ball grid array (BGA) and land grid array (LGA) are the most common formats for this type of package (Figure 2.6).

High-power LEDs are packaged in a similar manner (Figure 2.7). Normally, organic substrates are not suitable, owing to their low thermal conductivity. Ceramic substrates, on the other hand, have a higher thermal conductivity, which enables the efficient dissipation of heat from the chip. Thermal vias on the ceramic substrates are fabricated under the chip location to further reduce thermal resistance.

2.3 Chip Mounting and Interconnection

After wafer dicing, the chip is mounted onto a lead frame or a substrate. This chip mounting process involves die attach materials with the following functions [1].

a) Provide Mechanical Support
   Die attach materials will bond the chip on to the lead frame or substrate at the designed location. They are required to sustain the adhesion in extreme conditions and high working temperatures to ensure the reliability of the package.
b) Provide Heat Dissipation Path
   In an LED package, the heat generated by the chip is mainly transferred by conduction. The encapsulant (epoxy or silicone) has a low thermal conductivity. Die attach materials

provide a main heat conduction path underneath the die. Thermal resistance of the die attach materials depends on the materials used and their bonding quality. This is important in controlling the junction temperature of the chip [2]. The thermal conductivity of common die attach materials and packing materials are shown in Table 2.1.

c) Provide Electrical Connection
One of the electrodes locates at the bottom of the vertical LED. Die attach materials, which are electrically conductive, can provide an electrical path between the back side electrode and the bond pad on the substrate.

There are various types of die attach materials, of which die attach adhesives and eutectic solders are commonly used in LED packaging. This section briefly discusses the chip mounting process of these two types of die attach materials.

Interconnects between the chip and the substrate are made upon completion of the die mounting process. Since LEDs have a limited number of I/Os, wire bonding is still the most commonly used interconnection method in LED packaging. Some high-power LEDs are packaged in a flip-chip. The flip-chip bumps provide mechanical support and electrical connections. They also serve as a good heat dissipation path. Despite the fabrication and packaging costs of flip-chip LEDs being high, they do deliver better performance.

2.3.1 Die Attach Adhesive

Die attach adhesive is the most commonly used die attach material in both IC and LED packaging. It is normally made by epoxy materials, which provide mechanical strength and help bond the die to the substrate or lead frame. Epoxy is not electrically conductive and has poor thermal conductivity. Metal particles or flakes (usually silver) are suspended in the epoxy carrier to provide an electrical path and improve thermal conductivity. Nevertheless, the overall thermal conductivity of the die attach adhesive is still low as compared with other bonding materials (Table 2.1).

The thermal conductivity of the die attach adhesive depends on the amount of metal suspended in the epoxy. A higher thermal conductivity can be achieved by suspending more

Table 2.1 Thermal conductivity of common packaging materials.

Materials	Thermal conductivity (W/(m·°C))
Sapphire	25
Silicon	124
AlN	170
Silver epoxy	3–40
Au/Sn eutectic solder	57
SAC solder	70
Copper	390
Aluminum	210
Silicone	0.3–0.5

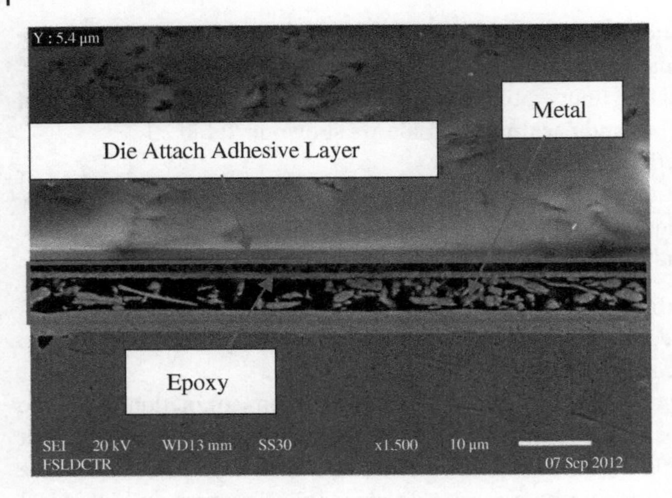

Figure 2.8 Cross-section inspection of die attach adhesive with less metal content.

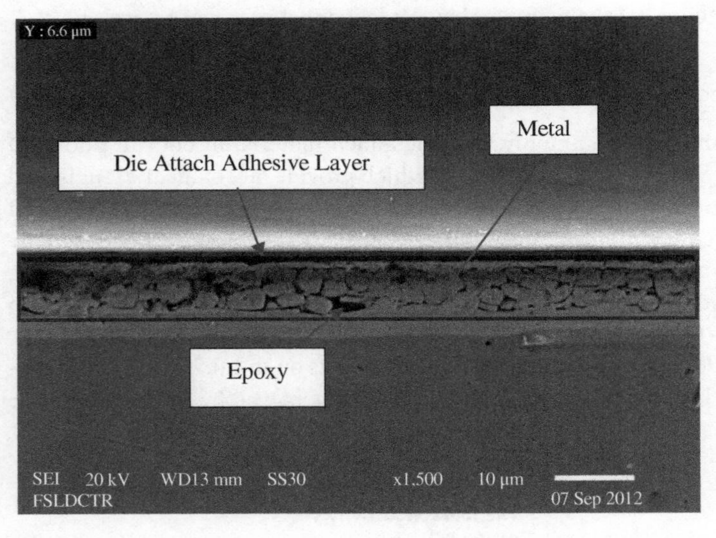

Figure 2.9 Cross-section inspection of die attach adhesive with more metal content.

metal in the epoxy. This will, however, correspondingly increase the overall cost of the material. Figure 2.8 shows a cross-section inspection of a die attach adhesive with a relatively low thermal conductivity (2.5 W/(m·°C)). In the cross-section inspection, only a small amount of metal flakes is found. Figure 2.9 is the cross-section inspection of another die attach adhesive with a higher thermal conductivity (4.8 W/(m·°C)). More metal particles are found in this die attach adhesive layer. Other material properties, such as viscosity and adhesion strength, vary with the amount of metal suspended, which also affects the bonding process and its quality.

Die attach adhesive is generally available in a paste form in which the epoxy is not cured yet. In the LED packaging process, the first step is to transfer the die attach adhesive to the

Figure 2.10 Die attach adhesive dispensing pattern. Source: Henkel.

substrate or lead frame. There are two material transfer methods, namely dispensing and dipping. The former is adopted if a large chip is used or a tight control in adhesive volume is required. In such cases, the adhesive is dispensed in a certain pattern (Figure 2.10) so that the adhesive will spread more uniformly during the die placement process and hence reduce the chance of void formation.

If a small chip is used (less than 2 mm), the adhesive is transferred to the substrate by dipping. This process is much faster than the dispensing process and hence can increase the throughput. A dipping head is first dipped into a pot with die attach adhesive. The dipping head, with adhesive attached thereon, is then stamped onto the substrate, transferring the adhesive material to the substrate. The volume of the adhesive transferred is determined by the viscosity of the adhesive, the dipping depth, and the shape of the dipping head. However, it is difficult to have a tight control on the adhesive volume transferred to the substrate during each dipping process. In some cases, dipping is required for several times with the aim of transferring a sufficient amount of adhesive materials to the substrate.

Adhesive volume is one important process parameter. Insufficient adhesive materials will not be able to cover the entire LED chip bottom surface, adversely affecting the mechanical strength and the heat dissipation efficiency. Figure 2.11 shows a sample with insufficient die attach materials. The X-ray inspection reveals that adhesive material does not reach the

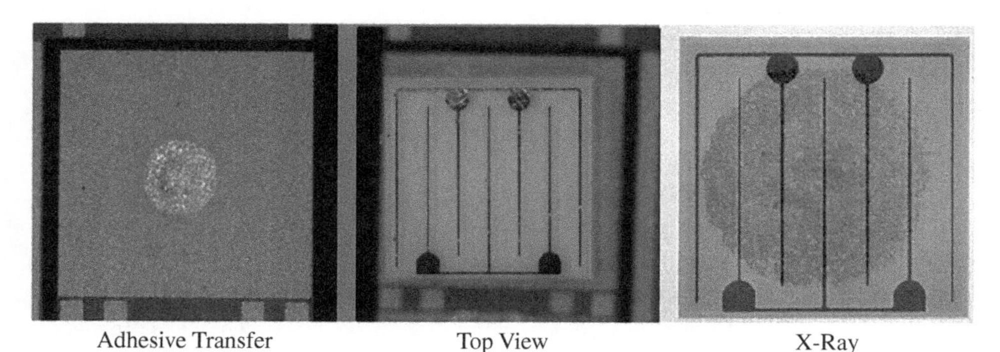

Figure 2.11 Insufficient die attach adhesive material.

corner of the chip. On the contrary, the adhesive will spread further away on the substrate if too much adhesive is applied. This occupies more space and the adhesive may cover the wire bond pads on the substrate, making the fine pitch die placement impossible. If a thin die is used, the excessive adhesive may even reach the die surface.

After the adhesive transfer, the LED chip is aligned with and placed onto the substrate. The accuracy of the aligning process is another important parameter. The die placement process controls the position of the LED chip on the substrate. If the chip is misaligned with the designed location, the optical performance will be affected. Figure 2.12 shows an example of the effect of chip misalignment on the light pattern by optical simulation. In this study, a package with a hemispheric dome lens is used. The chip center is supposed to be aligned with the center of the dome lens. Figure 2.12a shows the light pattern (i.e. a symmetric pattern) when the chip is well aligned with the lens. However, if the chip is offset from the center, the light pattern is no longer symmetric. Figure 2.12b shows the distorted pattern. The amount of light extraction is also related to the chip location. Figure 2.12c shows the effect of chip misalignment on the amount of light extraction. The amount of light extracted is reduced if the chip is offset from the center of the lens.

During the die placement, a small force is applied by the bond head. This controls the final thickness of the die attach adhesive. Finally, the die attach adhesive is cured at high temperature. The curing temperature and time depend on the material. Normally, it is less than 180 °C for 30 minutes. The overall die mounting process is shown in Figure 2.13.

2.3.2 Soldering and Eutectic Bonding

For an LED package structure, die attach materials have the highest thermal resistance and often adversely affect the thermal management of LEDs. They may not be able to provide a good thermal dissipation path in high-power LEDs. The large amount of heat generated by the LED chip cannot be dissipated to the environment efficiently and hence increases the junction temperature. In addition, epoxy in the die attach adhesive degrades at high temperature and high humidity. The adhesion strength decreases and the thermal resistance increases after aging. This creates reliability issues and lumen drop.

Metals or solder alloys have a much higher thermal conductivity as compared with regular die attach adhesives. They can be used as die attach materials to lower the thermal resistance. The most commonly used materials are tin-based lead-free solders and gold–tin eutectic solders. Yin et al. conducted a study to compare the thermal performance of different die attach materials [3]. In their study, silver epoxy (25 W/(m ⋅°C)), SAC 305 solder (67 W/(m ⋅°C)), and AuSn eutectic solder (57 W/(m ⋅°C)) were used as the die attach materials, and high-power LED chips were bonded onto a ceramic substrate. Their study concluded that the thermal resistances of AuSn eutectic solder and SAC solder were much lower than that of silver paste.

Sn–Ag–Cu (SAC) solder alloy is the common lead-free solder used in die bonding applications. There are various types of SAC solders with different weight percentage of the metals. SAC 305 (96.5% Sn, 3.0% Ag, 0.5% Cu), SAC 387 (95.5% Sn, 3.8% Ag, 0.7% Cu), SA (SnAg alloy), etc., are widely available in the market. The soldering process is completely different

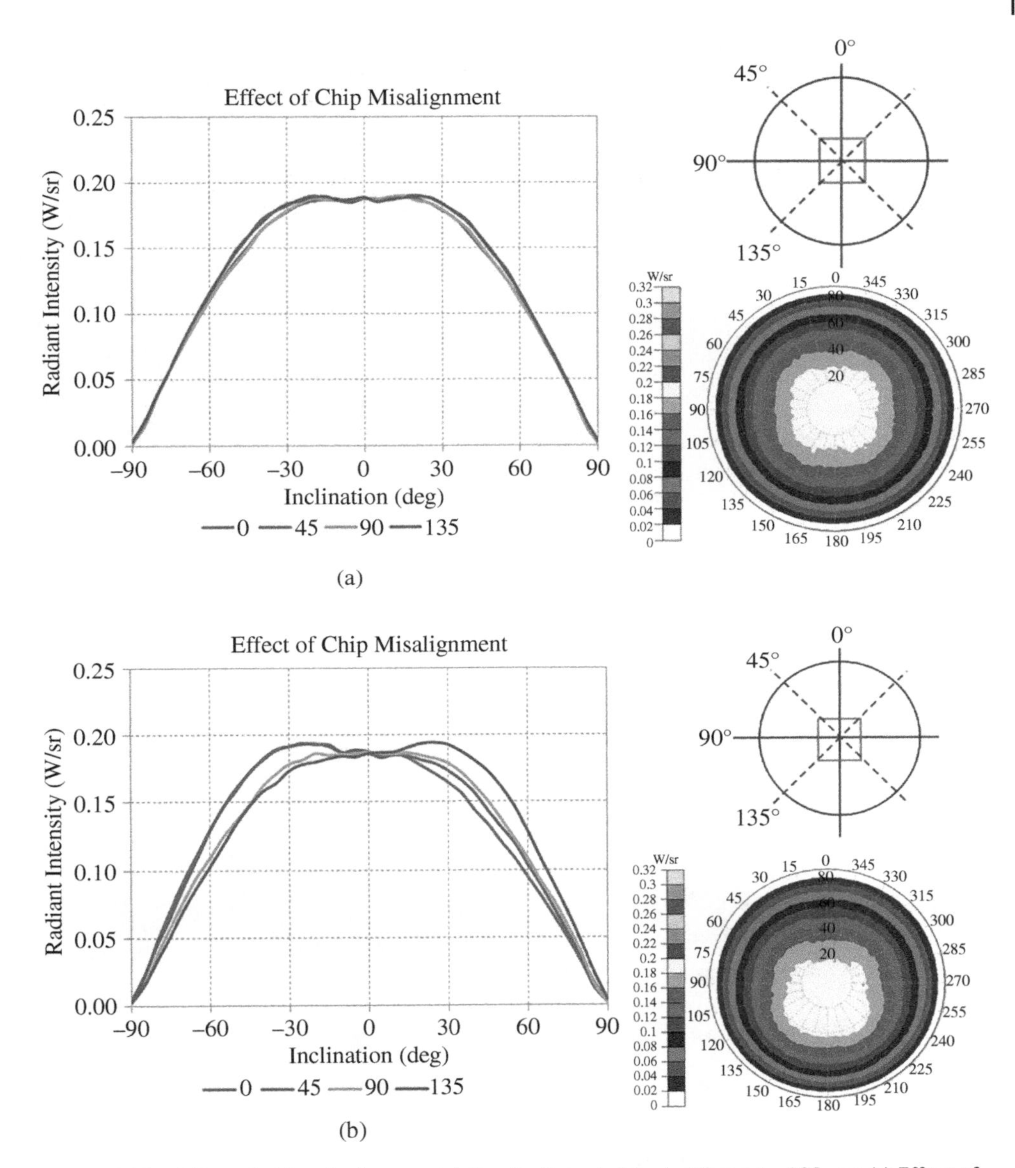

Figure 2.12 Effect of chip misalignment: (a) well aligned; (b) misaligned by 100 μm; (c) Effect of chip misalignment of light output.

from the die mounting process of die attach adhesive. Also, there are other requirements on the bonding surfaces of substrates and chips.

In the bonding process, the solder is first heated up to a reflow temperature which is higher than its melting point. The typical reflow temperature of a SAC solder is higher than 240 °C. The molten solder then reacts with the metal on the bond pad. An intermetallic compound (IMC) layer is formed between the solder material and the bond pad upon solidification. This IMC layer joins the solder and bond pad together and provides mechanical

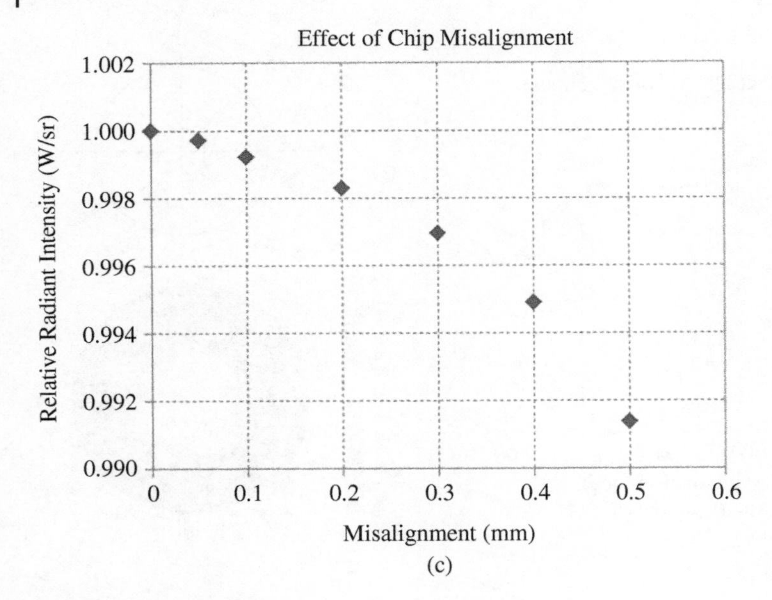

Figure 2.12 (*Continued*)

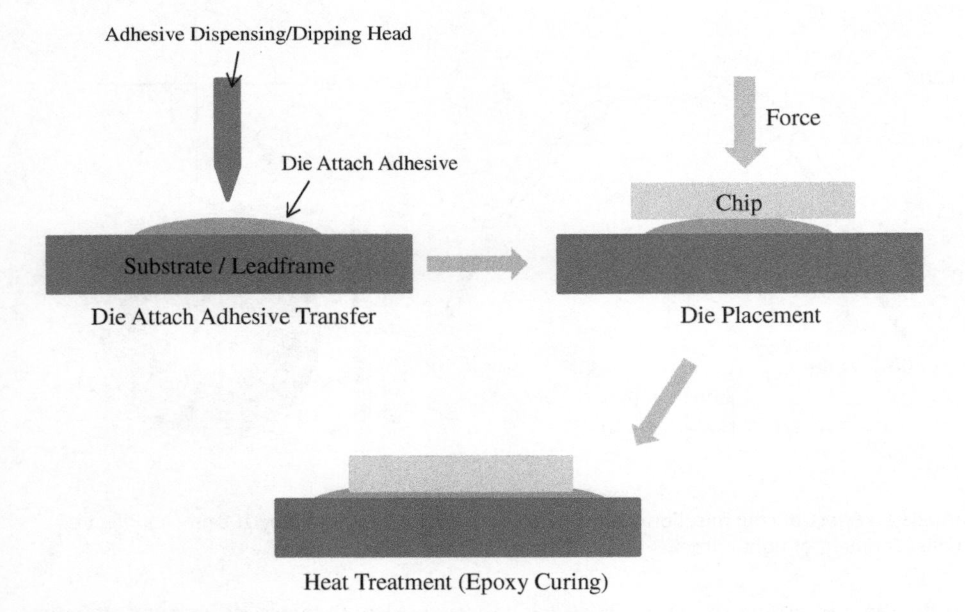

Figure 2.13 Die mounting process (die attach adhesive).

strength. In some applications, LED chips are normally fabricated on sapphire, SiC wafers, and silicon. These materials are not compatible with solder, and therefore an IMC layer cannot be formed unless a proper metallization is applied on the bonding surface. Copper and nickel are most commonly used as the bond pad materials. On top of that, a very thin layer of gold is deposited as a passivation layer to prevent oxidation. A similar metallurgy is required on the lead frame or substrate side.

There are different methods to apply solder materials [4]. One of them is solder wire dispensing. In this method, the substrate is first heated up to a temperature above the melting point of the solder. A solder wire is then dipped onto the heated substrate, in which the solder melts and wets on the bond pad. When enough solder materials is transferred to the substrate, the solder wire is lifted up, and the LED chip is aligned and placed on the molten solder by applying a proper bonding force. The substrate is then cooled down to allow the solder to solidify. The whole process is carried out in an inert environment, such as a nitrogen purged chamber, to prevent oxidation. The solder wire dispensing method, however, has a major drawback: the entire substrate is exposed to a high temperature throughout the bonding process. The overall process is shown in Figure 2.14.

The solder paste method is another way of applying solder onto a substrate. Solder paste is a mixture of tiny solder particles and flux. The latter is a reducing agent which can remove oxide on the solder and bond pad at high temperatures. This provides a clean and fresh surface for soldering. The solder paste material is screen printed on the substrate with a stainless steel stencil. The opening and thickness of the stencil define the solder volume applied to the substrate. This allows solder materials to be applied to multiple locations at the same time, and hence increases the throughput. A flat substrate is required to ensure a good contact with the stencil. If a lead frame with reflective cups is used, the solder paste will be applied by a dispensing method.

Once solder paste is applied, the chip is aligned and placed on the substrate. The whole substrate then undergoes a reflow process. The chip is bonded on the substrate upon solidification of the solder. Since solder paste contains flux, there are flux residues on the substrate after the bonding process. An extra cleaning step is therefore required. Different cleaning methods will be adopted for different types of flux used. For water-soluble flux, the flux residues are corrosive. It is therefore necessary to remove the residues with warm water.

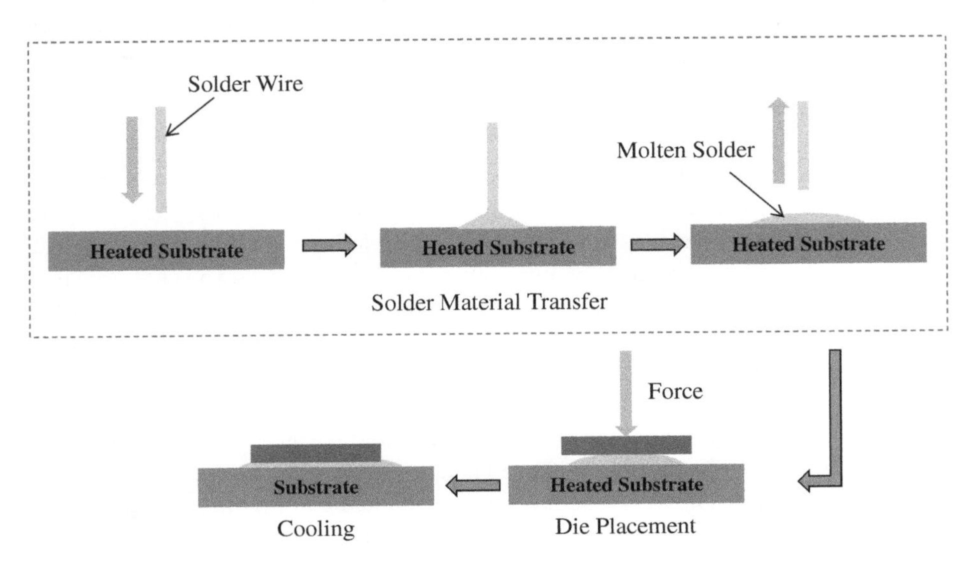

Figure 2.14 Solder wire dispensing method.

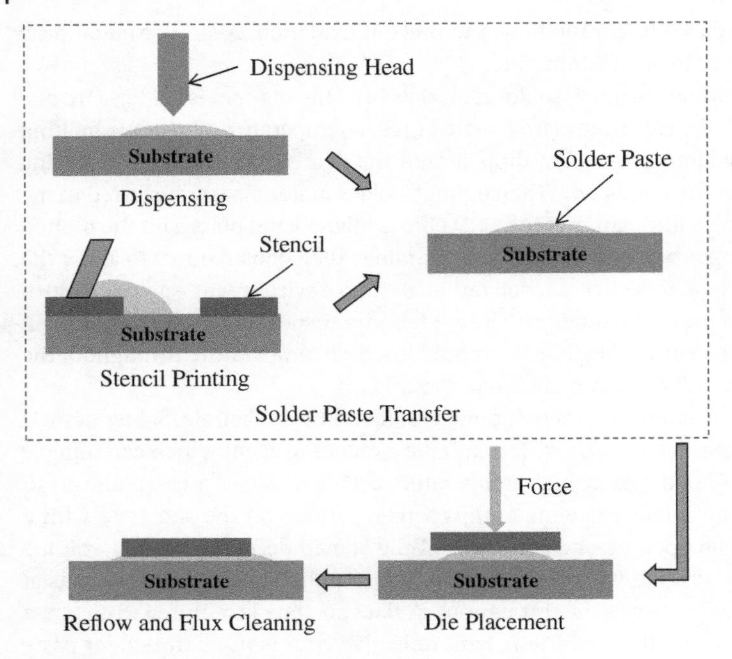

Figure 2.15 Solder paste method.

For no-clean flux, the flux residues are not corrosive. They may, however, contaminate the light-emitting surface or wire bond pad. This will affect the optical performance and lower the wire bonding yield. In such cases, it is necessary to clean the flux by using an organic solvent. The overall process flow is shown in Figure 2.15.

In addition to lead-free solder alloy, gold–tin (AuSn) solder is widely used as a die attach material in microelectromechanical systems (MEMS), optoelectronics systems, and high-power LED packaging [5–7]. AuSn solder is usually used in a eutectic composition [8].

AuSn solder has two eutectic points:

a) Au90Sn
Au90Sn contains 10% gold and 90% tin (weight percentage). It has the lowest melting point. The alloy melts at 217 °C, which is even lower than the melting point of pure tin (232 °C). It has a lower cost as its gold content is relatively low (10% only). The low melting point and low cost are attractive properties. However, Au90Sn is not used as a die attach material, owing to its poor reliability. It contains $AuSn_4$ which is a very brittle IMC. The die shear strength of the samples fabricated by Au90Sn degrades quickly after high temperature aging test [8].

b) Au20Sn
Au20Sn contains 80% gold and 20% tin. It is more suitable as a die attach material despite its higher melting temperature (280 °C) and greater cost. Two IMCs coexist in Au20Sn, namely AuSn and Au_5Sn. The weight percentage of gold and tin must be tightly controlled at the eutectic ratio. The liquidus lines are very steep on both sides of the eutectic melting point of Au20Sn. A slight alternation in weight composition of gold and tin will

change the melting point significantly. An increase in 1% of gold will lead to an approximately 30 °C increase in the melting temperature.

The Au20Sn eutectic solder has several advantages [9]:

a) High thermal conductivity
The thermal conductivity of AuSn eutectic solder is 57 W/(m·°C). This high thermal conductive material allows heat generated from the chip to dissipate in a more efficient way.
b) Fluxless process
Owing to the high gold content, the bonding process can be fluxless. There will not be any flux residues after the bonding process. The light-emitting surface will not be contaminated.
c) High creep resistance
AuSn eutectic solder is more resistant to creep. This is an advantage to those devices which work at high temperatures, such as high-power LEDs.
d) Low process temperature
The melting temperature of an Au20Sn eutectic solder is 280 °C. The process temperature is about 20–30 °C higher than the melting temperature, i.e. 300–310 °C. This process temperature is relatively high. In the subsequent board level assembly process, SAC solder is commonly used and the reflow temperature is around 240–260 °C where Au20Sn will not melt again.

Au20Sn eutectic solder also requires a proper metallization on the bond pad and backside of the chip. Common metallization materials are nickel, copper, and gold. The study conducted by Tsai et al. revealed that the Au_5Sn phase in the Au20Sn eutectic solder has a much higher solubility at high temperatures in copper than in nickel [10]. In their study, the consumed thickness of nickel bond pad in the high temperature aging test was much smaller than that of the copper bond pad. A thick gold bond pad on the substrate is also recommended by some chip manufacturers given that Au20Sn solder has a good wettability on gold surface. In the bonding process, the gold on the pad diffuses into the Au20Sn solder, which increases the gold content of the solder, making the final solder joint no longer maintained in a eutectic composition. As a result, the melting point of the solder joint is much higher than 280 °C after the bonding process.

Au20Sn solder is available in different forms, such as solder paste and solder preform. If the former is used, the die attach process will be the same as the solder paste method as mentioned in earlier paragraphs. Although Au20Sn can be used without flux, Au20Sn solder paste contains flux, making it necessary to clean the flux residues after the bonding process to avoid contamination.

Solder preform is a thin solder plate as shown in Figure 2.16. It can be trimmed into a specific size to match with the chip. It is first pick-n-placed onto the substrate. The chip is then aligned and placed onto the substrate. Heat and a suitable bonding force are applied to melt the solder preform. This process is normally taken in an inert environment to prevent oxidation. Upon solidification of the solder, the chip is bonded onto the substrate. The overall process is shown in Figure 2.17 [11].

For Au20Sn eutectic die bonding, LED dies usually have a layer of Au20Sn eutectic solder on the backside. This solder layer is normally around 3 μm thick and is deposited by

Figure 2.16 Solder preforms. Source: Indium Corporation.

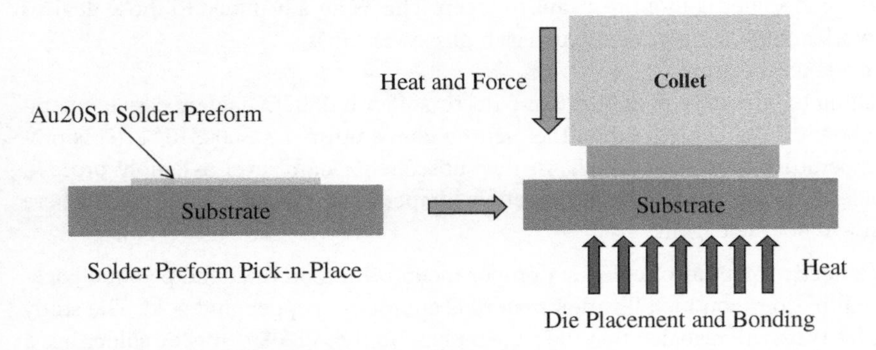

Figure 2.17 Solder preform method.

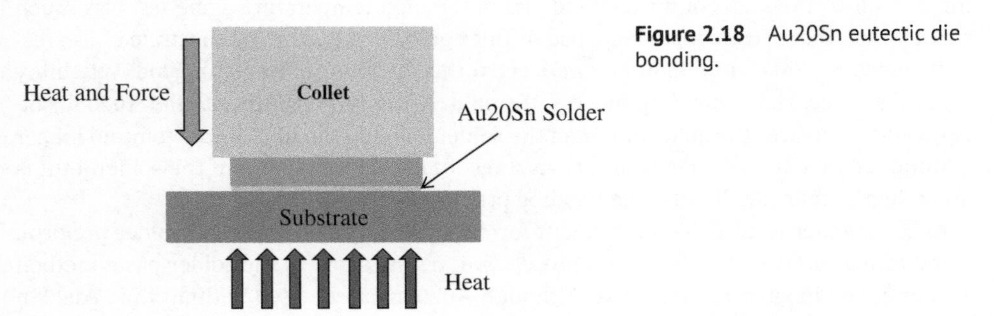

Figure 2.18 Au20Sn eutectic die bonding.

evaporation or the plating method. The gold/tin ratio is tightly controlled in the eutectic composition. The chip can be bonded onto the substrate directly by applying force and heat. The substrate is preheated and the chip is heated up by the collet. The whole process is flux-less and carried out in an inert environment. The temperature is increased to above 300 °C for between 3 and 20 seconds [12]. The time required for the bonding cycle is longer than that for the solder paste method, and hence the former has a lower throughput. Figure 2.18 shows the process flow. Since the solder material is thin and the volume is well controlled, the solder will not spread out after the bonding process. The footprint is the same as the die size. This allows placing multiple LED dies in a finer pitch.

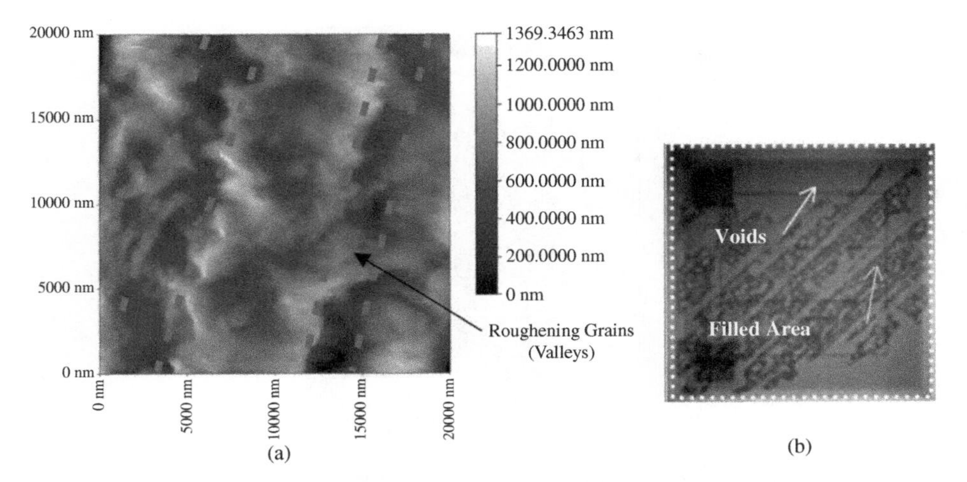

Figure 2.19 Effect of bonding surface roughness [13]: (a) AFM inspection of lead frame surface; (b) X-ray inspection. Source: Li, Z.T., Tang, Y., Ding, X., et al. Reconstruction and Thermal Performance Analysis of Die-bonding Filling States for High-power Light-emitting Diode Devices. Applied Thermal Engineering, 2014, 65: 236-245.

The Au20Sn solder layer on the chip is very thin. The bond pads on the substrate are required to be very flat. The co-planarity of the substrate is the key factor which affects the quality of the bonding layer. If the substrate is not flat, large voids will be observed. The study conducted by Li et al. showed the correlation of substrate roughness on void formation in the bonding layer (Figure 2.19) [13]. Large voids reduce bonding strength and heat dissipation. In general, regular lead frames cannot meet the co-planarity requirement. In this application, ceramic or silicon substrates are normally used.

Basically, soldering and eutectic die bonding provide a better thermal dissipation path. Nevertheless, the bonding process is more complicated and requires proper metallization on both the substrate and the backside of the chip. The process temperature is usually higher, and the temperature profile must be optimized and well controlled. Otherwise, large voids will be formed in the bonding layer, which will affect the performance.

2.3.3 Wire Bonding

Wire bonding is the most common interconnection method in both traditional IC and LED packaging. It is considered the most cost-effective and flexible solution to electrically connect the chip to the substrate. A modern automatic wire bonder can complete more than 20 wire bonds in one second. The wires are normally made of aluminum, gold, or copper. There are two types of wire bonding methods, namely wedge bonding and ball bonding.

Wedge bonding can be used to bond both aluminum and gold wires. In this process, the wire is first fed into a wedge tool. The tool is then lowered down to the chip bond pad. Ultrasonic power and force are applied to form the first bond. The process can be done at room temperature if aluminum wire is used. If gold wire is used, the sample must be heated to above 120 °C. Upon formation of the first bond, the wedge tool is lifted up and moved to the bond pad on the substrate. This move forms the wire loop and defines the loop height.

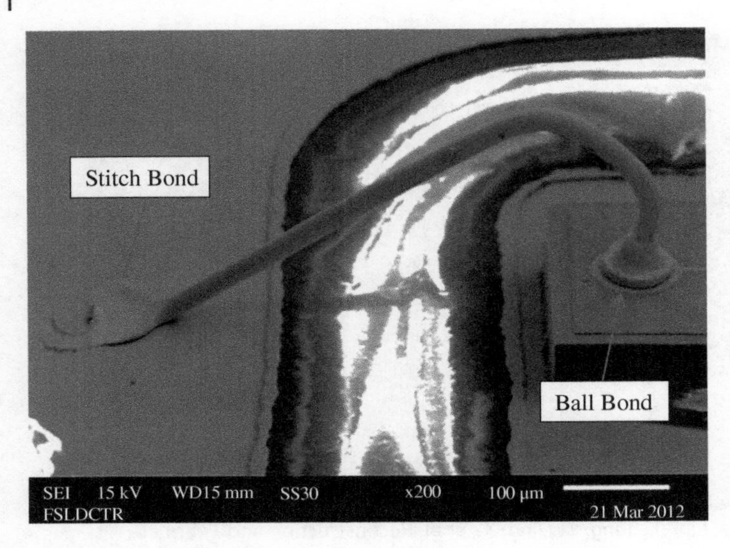

Figure 2.20 Typical gold wire bonds.

The second bond is formed on the substrate and the wire is broken off. The bonding cycle is then completed and ready for the next bonding [14] The footprint of the wedge bond is smaller, which enables a fine pitch wire bonding. However, the wedge bond is directional. Upon formation of the first bond, the wedge tool must be moved parallel to the wedge bond direction. The sample may have to be rotated accordingly after completing each bond. This slows down the whole process.

Ball bonding is extensively used in IC and LED packaging. Gold wire (0.8–1.2 mil diameter) is first fed in a capillary bond head. Electronic flame-off (EFO), which applies a high electrical voltage spark, is used to melt the wire to form a tiny gold ball (free-air ball). The diameter of the free-air ball is larger than that of the wire. The free-air ball occupies a larger footprint as compared with the wedge bond. It is then brought to the bond pad of the chip. The sample is preheated to above 120 °C. Pressure and ultrasonic power are applied to form a first bond, which is also called the ball bond because of its shape. Upon formation of the first bond, the capillary bond head is moved to the bond pad of the substrate, forming a loop. Pressure and ultrasonic power are then applied again to form a second bond [14]. This is called a stitch bond. Figure 2.20 shows a typical ball bond. Ball bonding is nondirectional, which is much faster and more flexible than wedge bonding.

Some applications require a low loop height. In such cases, a ball bond is first formed on the chip bond pad. Instead of moving the capillary to the bond pad of the substrate, the wire is broken off to form a gold stud on the chip bond pad. Another ball bond is then formed on the substrate and the corresponding stitch bond is formed on top of the gold stud. The final wire bond will then have a very low loop height [15, 16]. This method is usually called reverse wire bond or ball stitch on ball (BSOB). Figure 2.21 shows a stitch bond on a gold stud and Figure 2.22 shows a typical reverse wire bond.

Copper wire is sometimes used to replace gold wire in order to reduce the material cost. It is easily oxidized during the bonding process. A nitrogen purged environment is required to prevent oxidization. Formic gas may be required in forming the free-air ball. Copper is a

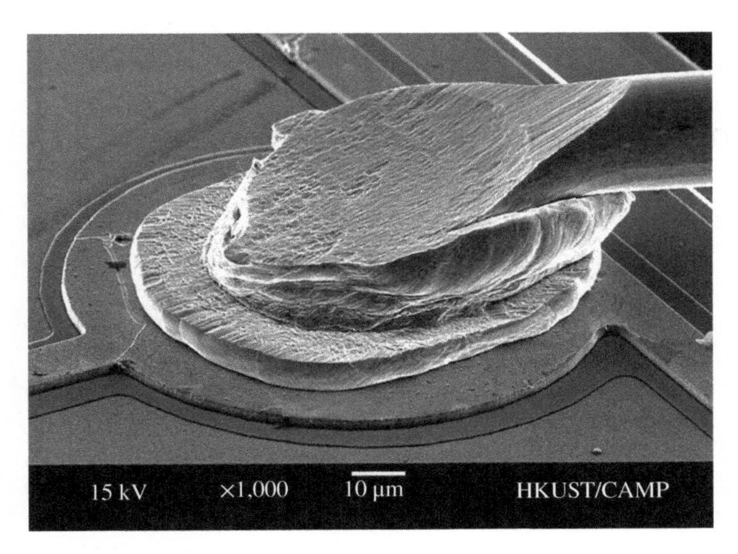

Figure 2.21 Stitch bond on gold stud.

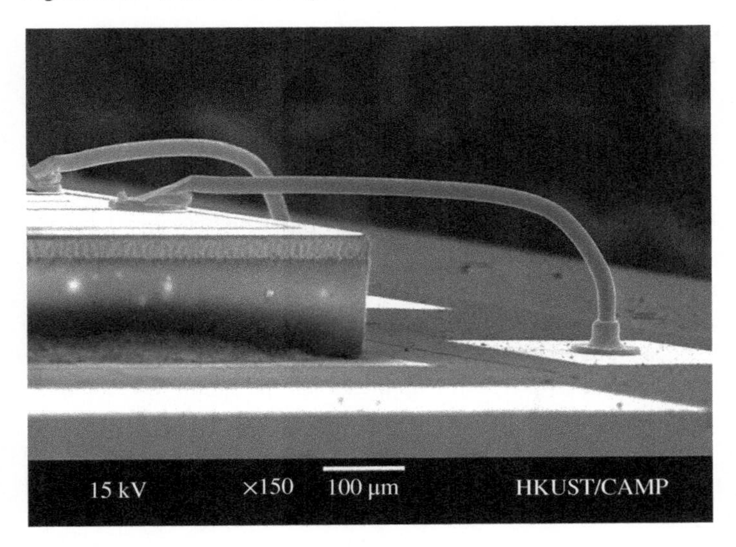

Figure 2.22 Reverse wire bond.

harder material than gold, which is easier to damage the bond pad on the chip and cause pad cratering when a high bonding force is applied.

In LED packaging, gold ball bonding, which is fast and flexible, is commonly used as the interconnect. Given that the number of wires in an LED package is low, the material cost of gold wire only contributes a small portion of the whole packaging cost. It is therefore not cost effective to change to copper wire.

Gold wires can be bonded on aluminum or gold bond pads. If aluminum bond pads are used, Au–Al IMCs and the associated Kirkendall voids will be formed. The formation of Au–Al IMCs and Kirkendall voids are accelerated at high temperature during the operation.

This reduces the bond strength and creates reliability issues. Gold wires on gold bond pads are more reliable as the bonds are not subject to interface corrosion and IMC formation. In order to achieve a good bondability and high bond strength, the gold layer on the bond pad must be thick enough. Oda et al. studied the effect of the gold layer thickness on the pull strength [17]. In their study, gold wire bond on a bond pad with a 0.3 μm thick gold layer had a good wire pull strength.

2.3.4 Flip-Chips

Flip-chip is one of the advanced interconnection technologies. In wire bond devices, the active surface of the chip and the bond pad faces upward. In flip-chip devices, the chip is "flipped over" and the active surface faces toward the substrate. The I/Os and interconnects are under the chip. This type of packaging technology is therefore known as a flip-chip.

A flip-chip has a lot of advantages over a wire bond. For example, if wire bonds are used, the bond pads have to be arranged in a peripheral array. This limits the number of I/Os. In a flip-chip, however, the interconnects are under the chip, which can be arranged in an area array. This offers a platform for high-density interconnects. In addition, the second bonds on the substrate of the wire bond devices lead to larger footprints. On the other hand, the form factor of a flip-chip is much smaller.

The flip-chip interconnects provide an electrical connection path, heat dissipation path, and mechanical support. The electrical connection of a flip-chip is much shorter than that of the wire bond and hence it has an excellent electrical performance. The flip-chip joints are normally metals which have a higher thermal conductivity than regular die attach adhesives. Heat can be dissipated in a more efficient way.

The configuration of flip-chip LEDs is different from regular LED chips (Figure 2.23). For flip-chip LEDs, a reflective layer is deposited on the p-GaN to reflect downward light to the top surface. The light emitted from the multi-quantum well (MQW) is extracted through the sapphire to the encapsulant. The refractive index of sapphire (1.78) is in between the n-GaN (2.54) and silicone encapsulant (1.40–1.60). This graded refractive index structure enhances the overall light extraction [18]. In some applications, the sapphire is removed by laser lift-off. This type of structure is called a thin film flip-chip (TFFC) LED [19]. The surface of n-GaN is roughened or patterned with different microstructures that help reduce the incident angle of light from the n-GaN. As a result, less light is trapped by total internal reflection and hence the light extraction efficiency is enhanced [20].

Flip-chip LEDs have several advantages over regular LED chips. The flip-chip configuration allows more bond pads to be put under the chip. This can spread the current more

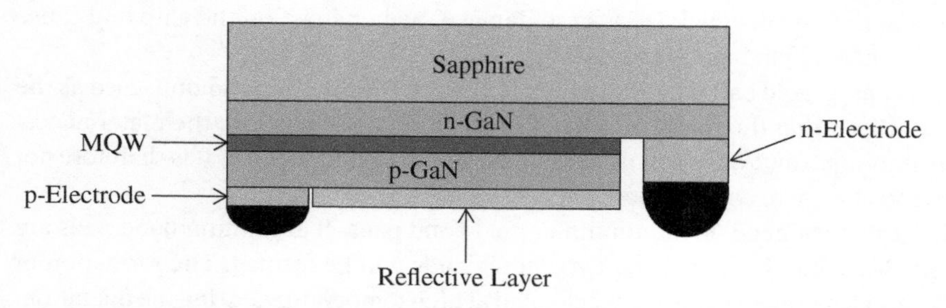

Figure 2.23 Schematic layout of flip-chip LEDs.

uniformly and has a lower contact resistance. It also has a lower forward voltage and lower diode resistance [18]. This largely enhances the electrical performance. Flip-chips also provide a better thermal management solution for high-power LED packages [21]. Heat generated by the active layer is directly conducted to the metal bumps without passing through the sapphire, which has a low thermal conductivity. The thermal resistance is associated with the number of bumps under the chip. If more bumps are used, better temperature uniformity on the chip and lower thermal resistance can be achieved [22–24].

There are different types of flip-chip, which can be divided into two categories based on their bonding mechanism. The first one is adhesive based. Adhesive materials are used to bond the chip to the substrate. The metal particles suspended in the adhesive materials link up the electrical conduction paths between the bond pads on the chip and the substrate. There are two types of adhesive materials, namely conductive adhesive and anisotropic conductive adhesive [25]. Figure 2.24 shows the corresponding schematic diagrams.

Conductive adhesive is used when the number of I/Os is low. The chip with gold stud bumps on the bond pad is first dipped with conductive adhesive. It is then aligned with and placed on the substrate. The process is completed after curing the adhesive. Anisotropic conductive adhesive is extensively used in chip-on-glass (COG) applications. It supports fine pitch and high-density interconnects. Metal particles coated with a thin layer of polymer are suspended in the adhesive. The polymer layer prevents the particles from being electrically shorted together. In the bonding process, a high pressure is applied. Only the particles under the bond pads deform. Subsequently, the polymer layers of the deformed particles break and the metals are exposed. This connects the electrical path between the bond pads on the chip and the substrate without shorting the adjacent I/Os. The contact resistance is normally high and depends on the concentration of the particles, bonding pressure, and temperature. The application of both conductive adhesive and anisotropic conductive adhesive has a low processing temperature. The thermal resistance is relatively high, which is not suitable for high-power LED packaging.

The bonding mechanism of another category of flip-chip is by formation of the IMC between the flip-chip bumps and bond pads. This involves metal bumps and hence has a better thermal conductivity. There are three types of bonding methods, namely solder reflow, thermo-compression, and thermo-sonic. The bonding process of the solder reflow method is the same as the soldering and eutectic bonding process mentioned in the previous session. Au20Sn solder is extensively used in flip-chip LEDs, owing to its excellent thermal and mechanical properties. It is deposited on the chip bond pads either by plating or evaporation. Since the process requires melting of the solder, the process temperature is the highest among different types of flip-chip. The bonding surface can be activated by plasma

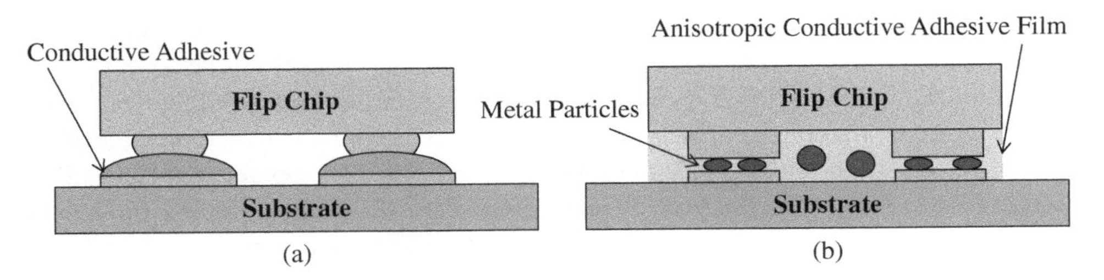

Figure 2.24 Adhesive type flip-chips: (a) conductive adhesive; (b) anisotropic conductive adhesive.

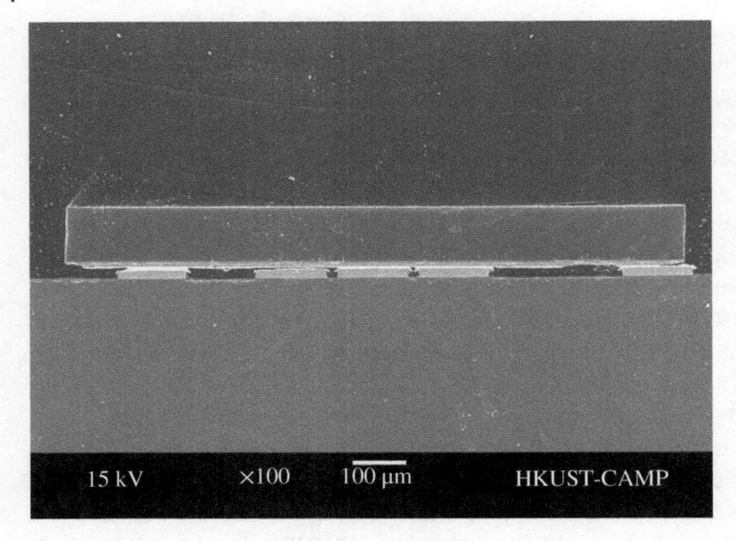

Figure 2.25 Flip-chip LED with Au20Sn eutectic solder bonded on silicon substrate.

Figure 2.26 Gold stud bump (before coining).

pretreatment to lower the process temperature [26]. Figure 2.25 shows a cross-section view of the flip-chip LED with Au20Sn eutectic solder bonded on a silicon substrate.

Gold stud bumps are used in thermo-compression and thermo-sonic bonding. Gold studs can be formed on the chip bond pads by electroplating or ball bonding. A regular wire bonder can be used to form the gold stud on the chip bond pad. The bond pad metallization requirement is similar to gold wire bonding. After the ball bonding process, the height of each bump on the chip is not uniform (Figure 2.26). The poor co-planarity will lower the bonding yield. The variation of the bump height can be reduced by a coining process.

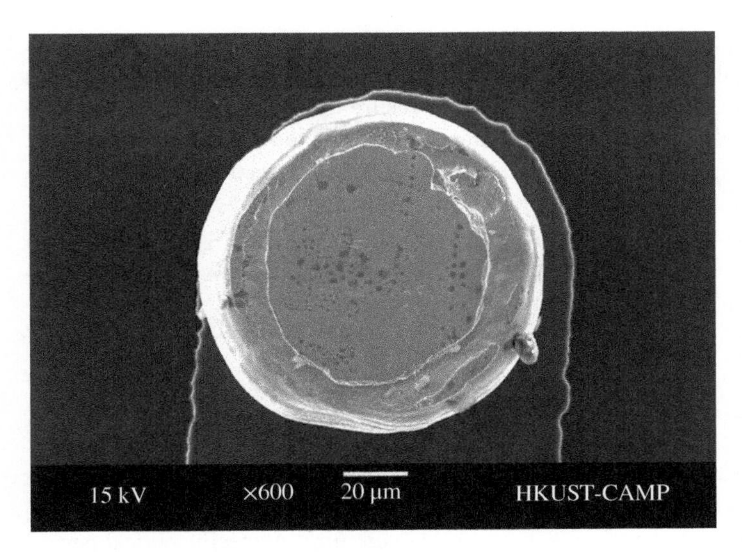

Figure 2.27 Gold stud bump (after coining).

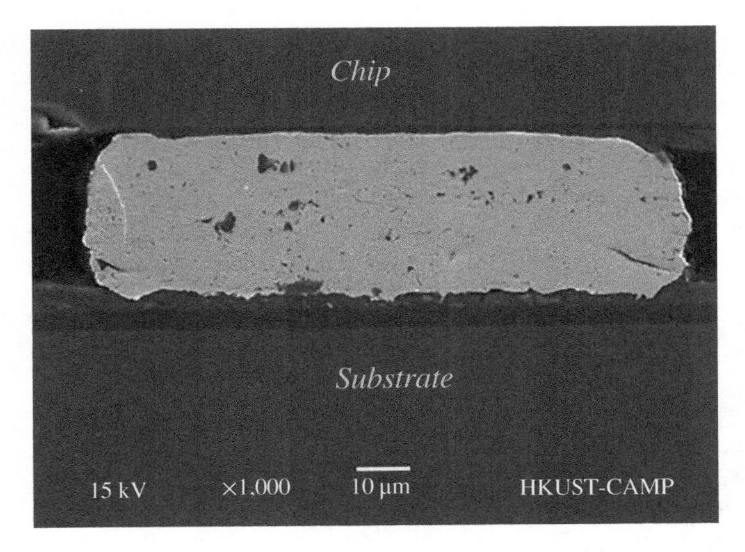

Figure 2.28 Flip-chip with gold stud bump (thermo-sonic).

The bump is pressed against a flat surface. Gold, being a very soft material, can be easily deformed by a mechanical force. Figure 2.27 shows a gold stud bump after coining. After aligning the chip with the preheated substrate, pressure is applied to promote the formation of IMC between the bump and the bond pad. In a thermal-sonic process, an ultrasonic power is also applied. Figure 2.28 shows a cross-section view of the gold bump flip-chip after the thermo-sonic process.

Flip-chips offer a better solution to high-power LED packaging. On the other hand, they have higher requirements for the substrate and equipment. In order to achieve better thermal and electrical performance, chip manufacturers try to maximize bond pad sizes. This only leaves a small gap (<200 μm) between the p- and n-electrodes. Also, the thin solder layer on the chip requires a flat substrate to ensure the bonding yield and reduce chance of forming large voids. Generally, regular lead frames and MCPCBs cannot meet those requirements. Expensive substrates such as silicon and ceramic substrates are normally used. These substrates are flat and easy to define fine features using photolithographic processes. Flip-chip bonders with a high alignment precision are also required. Slight die misalignment during bonding process may short the electrodes. All of this increases material and processing costs.

2.4 Phosphor Coating and Dispensing Process

Phosphor materials are applied after the die mounting and interconnection processes. The major role of phosphor coating is color tuning. There are different types of phosphor materials, each with their unique excitation and emission spectrums. A specific optical performance can be achieved by matching proper LED chips with phosphor materials. Blue light LED with yellow phosphor is the mainstream in generating white light for illumination in the industry. Blue light emitted by the chip excites the yellow phosphor to emit yellow light. By mixing blue light from the chip with yellow light from the phosphor, white light is generated. Yttrium aluminum garnet with cerium ($Y_3Al_5O_{12}$:Ce) is a widely used yellow phosphor [27]. In some applications, more than one type of phosphor, such as red or green phosphor, is used to improve the color rendering index [28–30].

In LED packaging, phosphor particles are used. They are irregularly shaped (Figure 2.29) with sizes ranging from a few microns to less than 20 μm. When light reaches a phosphor particle, a portion excites the phosphor for color conversion. Another portion is either being

Figure 2.29 Phosphor particles.

reflected or scattered by the particle. These optical properties affect the final outcome of the package and are related to the particle size [31, 32].

Phosphor coating is the most time-consuming and expensive process among all LED packaging processes. There are various types of phosphor coating configurations, including dispersed dispensing, conformal coating, and remote phosphor coating. Each configuration is associated with different coating methods and phosphor material consumptions. The lumen output, color temperature, angular color, and light intensity distribution of the package are related to the type of phosphor coating configuration selected [33].

2.4.1 Dispersed Dispensing

Dispersed dispensing is the simplest and most flexible phosphor coating method. The phosphor particles are mixed with encapsulant (epoxy or silicone) and directly applied to the chip. The optical performance of the package heavily depends on the concentration of the phosphor particles in the encapsulant. A higher phosphor concentration will convert much blue light emitted from the chip to yellow light, giving a lower CCT (correlated color temperature) value [34]. Dispersed dispensing can be applied to both in-cup and lens types of LED package. Figure 2.30 shows the schematic diagram.

The needle dispensing method is extensively used in applying phosphor materials to in-cup LEDs. In this method, phosphor materials are first mixed with a binder material in a specific concentration to form slurry. The slurry is then dispensed into the reflective cup by a dispenser through a needle. There are several types of dispensing valves, including time-pressure, fixed volume, auger, etc. These valves have their pros and cons in cost, volume control, dispensing speed, etc.

Dispensing volume is a key parameter which controls the amount of phosphor materials applied. The curvature of the phosphor layer changes from concave to convex with the increase in dispensing volume. The geometry of the phosphor layer also serves as a lens of the package. According to the study conducted by Shuai et al., color uniformity of an in-cup LED is related to the curvature of the phosphor layer [35]. By adopting a fixed volume valve and auger valve, the dispensing volume can be controlled in a more precise way. Nevertheless, the final volume applied to the package varies because of stringing or dripping, which is related to the distance between the needle and the sample (Figure 2.31). If the needle is too close to the sample, the slurry may touch the side wall of the needle. On the other hand, the slurry will not be snapped off properly if the needle is too far away from the sample.

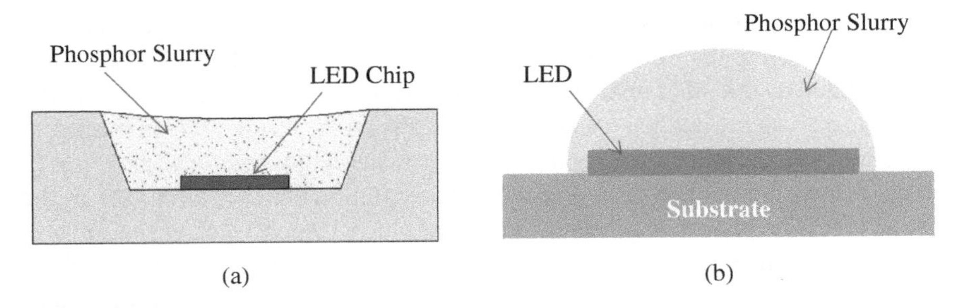

Figure 2.30 Dispersed dispensing method: (a) in-cup; (b) lens type.

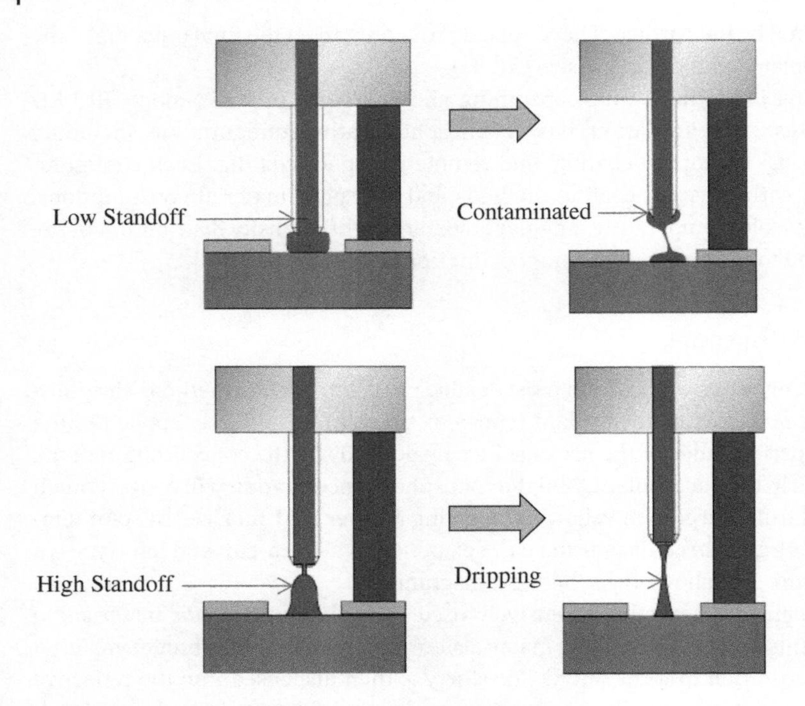

Figure 2.31 Dripping and stringing in needle dispensing.

Jet dispensing has a better control in volume. In each shot, a fixed volume of slurry is punched out by the jet head. The reflective up is filled with slurry by multiple shots. In this dispensing process, a needle is not required and there is no stringing or dripping during the jetting process. The dispensing head is not required to be placed close to the sample and hence the z-retraction time is eliminated. The jetting frequency can be synchronized with the head movement in x- and y-directions. This reduces the process time tremendously. Figure 2.32 compares the regular dispensing and jet dispensing processes. The punching force used in the jet dispensing process depends on the viscosity of the slurry. The viscosity of the binder material and the phosphor particles' concentration control the viscosity of the slurry. High phosphor concentration leads to high slurry viscosity. Normally, the jet dispensing process cannot handle phosphor slurry with a high viscosity.

Direct dispensing is not suitable for lens type LEDs. As there is no boundary, the slurry will spread and flatten out after dispensing. Compression molding is used to form the lens. Slurry is first dispensed in the mold cavity. The shape of the lens which affects the optical properties is defined by the mold cavity [36]. Substrate is then aligned and placed onto the mold. Pressure and heat are applied to cure the slurry. Figure 2.33 shows the procedure.

Precipitation of the phosphor particles in the binder material is a technical challenge in the dispersed dispensing method. The phosphor concentration changes with time in the dispensing syringe. The phosphor concentration of the slurry dispensed on each sample is slightly different. This increases the product variations. Phosphor particles also precipitate during the compression molding process. They may be clustered in certain parts of the lens and hence increase the angular color's nonuniformity.

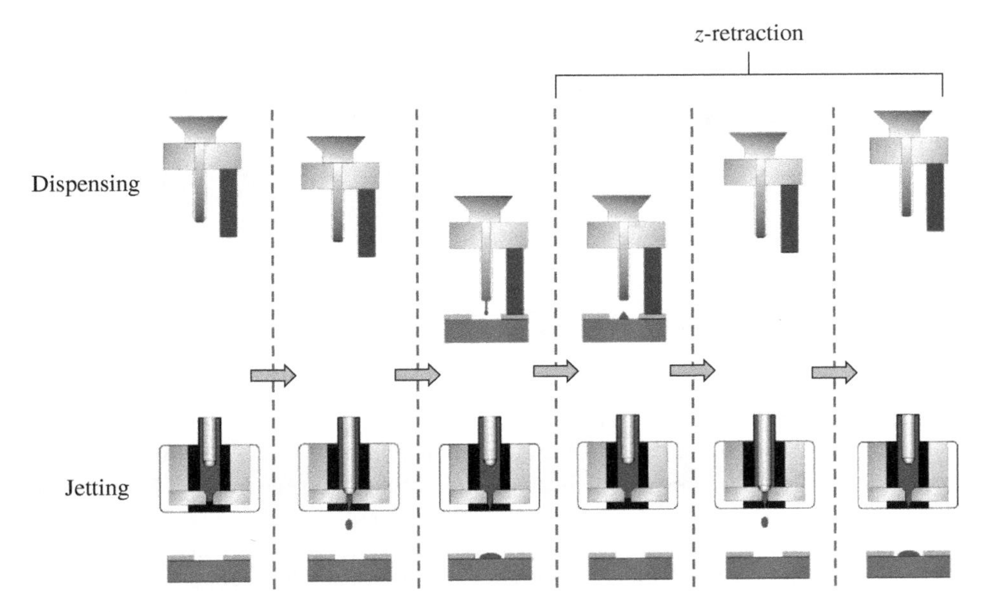

Figure 2.32 Dispensing vs. jetting.

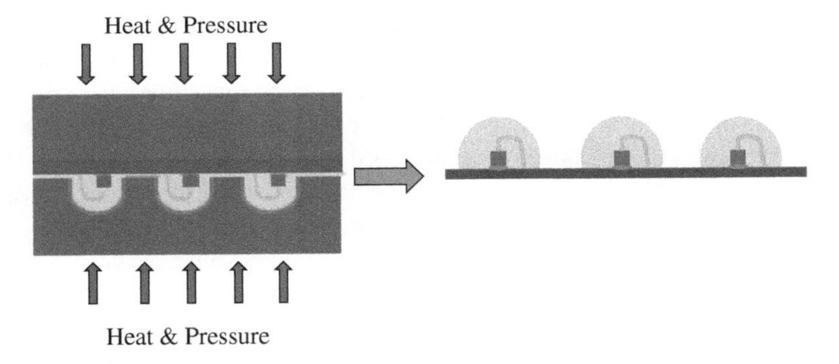

Figure 2.33 Compression molding.

The dispersed dispensing method has the poorest color uniformity. The phosphor concentration and thickness vary a lot at different directions. The amount of light transmitted through and being converted by the phosphors is not the same at different angles. This leads to different color temperatures at different viewing angles. Yellow rings or yellow spots are the most common phenomenon of the LED package using the dispersed dispensing method. This poor angular color uniformity is not acceptable in some applications.

2.4.2 Conformal Coating

With the conformal coating method, a thin layer of phosphor material is applied directly to the chip. The thickness of the layer is usually less than 50 μm. The layer consists of phosphor

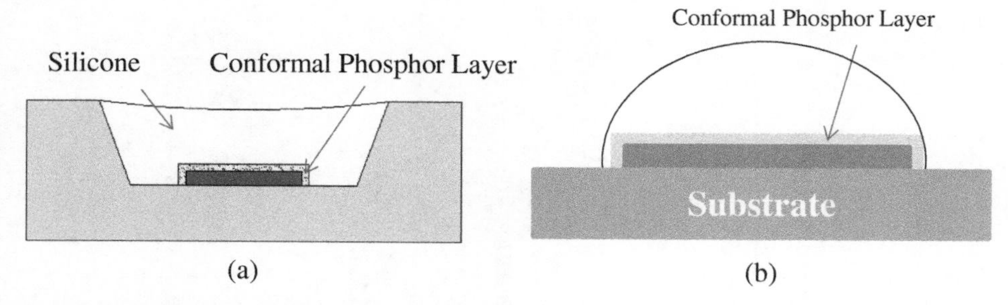

Figure 2.34 Conformal coating: (a) in-cup; (b) lens type.

particles and binder materials. The concentration of the phosphor particles in the conformal layer is more uniform. A uniform angular color temperature distribution can be achieved by optimizing the geometry and the concentration of the phosphor layer [37]. After the phosphor deposition process, transparent encapsulant is applied. For the in-cup LED package, clear silicone is dispensed directly to fill the cup. Compression molding is used to form a silicone lens for the lens type LED package. Figure 2.34 shows the schematic diagram of conformal phosphor coating being used in in-cup and lens type LED packages.

Conformal coating is usually applied to flip-chip LEDs. In flip-chip LEDs, all interconnects are under the die. There is no wire bond on top which provides a flat platform for the conformal coating layer. If a thin film flip-chip LED is used, conformal coating layer on the top surface is sufficient to achieve a uniform angular color distribution. Otherwise, light is emitted from four sides of the chip if the sapphire substrate is not removed. In this case, the side walls of the LED chips are also required to be covered by the phosphor layer [38]. Conformal phosphor can be used in wire bond type LEDs. Special arrangements, such as a reverse wire bond, are required to avoid breaking the wires during the coating process.

Conformal phosphor coating can be applied to the chip surface by the electrophoretic deposition (EPD) method [39, 40]. This method was patented by Lumileds in 2003 [41]. In the EPD process, the substrate is placed in an EPD bath. The phosphor particles are suspended in isopropyl alcohol (IPA). The bath contains nitrate salt, normally $Mg(NO_3)_2$, and a very small amount of water. The nitrate salt provides ions to positively charge the particles. An electric field, up to 700 V, is applied to transport those charged particles to the substrate. The particles are deposited and attached onto the substrate and chip surface. Figure 2.35 shows the setup of a regular EPD process.

EPD can deposit a uniform phosphor particle layer on the chip surface. The thickness is about 30 μm with a high packing density. The thickness, concentration, and packing density can be controlled easily. However, the adhesion strength of the EPD deposit is lower than the layer deposited by other methods [42]. The adhesion strength can be improved by post-deposit baking or spraying an additional adhesion layer on top.

Spray coating is another method for depositing a phosphor layer on the chip surface [43–45]. Phosphor particles are first mixed with silicone to form slurry. The slurry is sprayed on the sample directly with the aid of compressed air. The mechanism is the same as spray painting. This method is usually used to deposit a phosphor layer over a large area. Multiple layers with different phosphor materials can be sprayed on the substrate to achieve a better optical performance [46]. The phosphor consumption is high, causing a high material

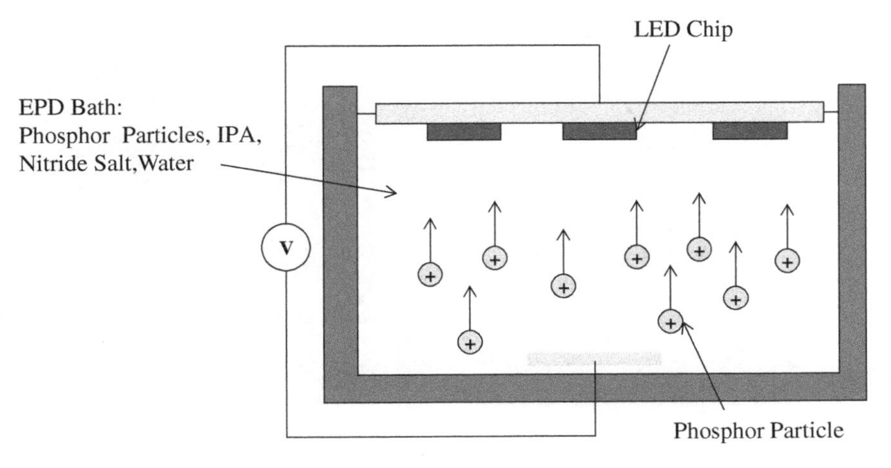

Figure 2.35 Electrophoretic deposition setup.

Figure 2.36 Spray coating setup.

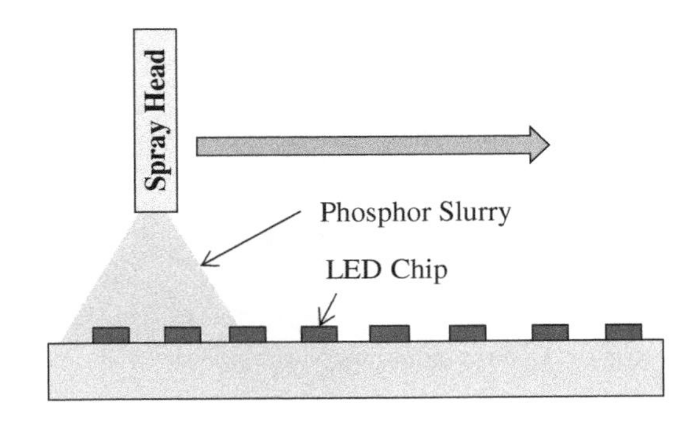

cost. Figure 2.36 shows the schematic diagram of spray coating and Figure 2.37 shows some examples.

A conformal phosphor layer can be applied to the chip by stencil printing [47]. This mechanism is the same as solder paste stencil printing. The stencil made by stainless steel is first aligned with the sample. A flat sample surface is required to ensure a good contact with the stencil to avoid bleeding. This method is normally inapplicable to an in-cup type LED, which has a reflective cup. Phosphor slurry is then printed on the chip. The aperture of the stencil defines the geometry and the volume of the phosphor layer. The side walls of the chip are also covered with phosphor. This process is shown in Figure 2.38.

Lee et al. proposed a modified stencil printing method for flip-chip LEDs [48]. The stencil was made by a silicon wafer. The aperture size was precisely defined by photolithography. Dry phosphor powders were printed onto the chip, followed by dispensing binder material. This allows the phosphor powders to be recycled. The binder material was UV curable and half-cured before separating the substrate with the stencil. This prevented the phosphor layer from collapsing after removing the stencil. Finally, the binder material was fully cured under a UV source. This method is also applicable to wire bond type LEDs [49]. A reverse wire bond was used to reduce the loop height. It has a very low profile, which allows a thin

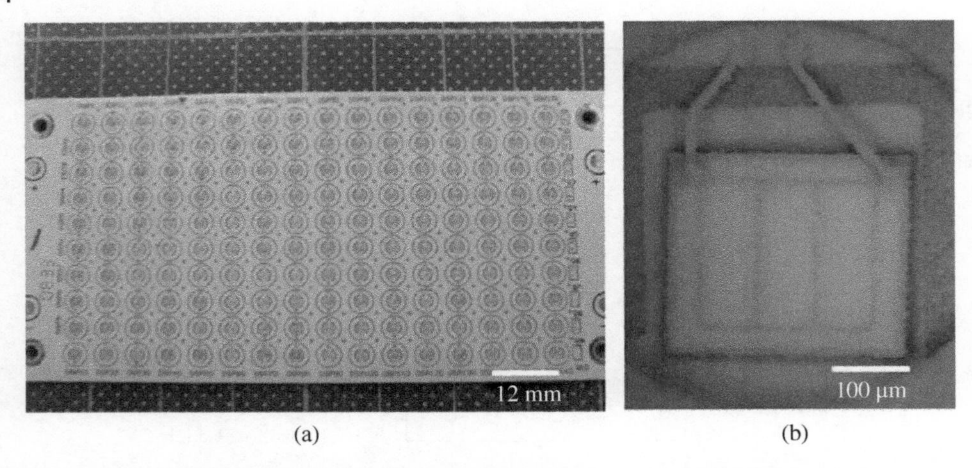

Figure 2.37 Spray coating [45]. Source: Huang, H.T., Tsai, C.C., Huang, Y.P. Conformal phosphor coating using pulsed spray to reduce color deviation of white LEDs. *Optics Express*, 2010, 18(S2): A201–A206.

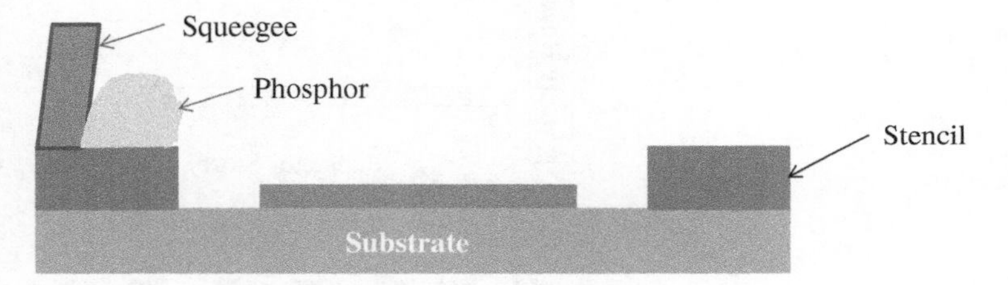

Figure 2.38 Stencil printing method.

layer of phosphor to be printed on the chip surface without damaging the wire. The stencil printing results are shown in Figure 2.39.

As mentioned before, it is not necessary to cover the side wall of the chip with phosphor particles if a thin film flip-chip LED is used. A ceramic phosphor preform [50, 51] can be attached directly on top of the LED chip. This method was first proposed by Lumileds and termed lumiramic technology [19]. The optical properties of the chip and the phosphor preform can be characterized separately. The phosphor preform is then matched with and attached to the chip. This method has better quality control. The process is shown in Figure 2.40.

Conformal phosphor coating can be applied to the chip by a settling method. Phosphor particles are mixed with a solvent and dispensed on the chip. The solvent is evaporated by heat. Phosphor particles are then left on the chip's surface. Self-exposure of the phosphor slurry is another method [52, 53]. Phosphor particles are mixed with photosensitive epoxy. The slurry is then dispensed on the chip. The epoxy, upon powering up the LED chip, is under the exposure of blue light emitted from the chip. It is cured subsequently and hot water is used to develop the photosensitive epoxy.

Not all of the aforementioned methods can be used to apply a conformal phosphor layer to in-cup type LEDs. EPD and spray coating methods can be used, but these methods

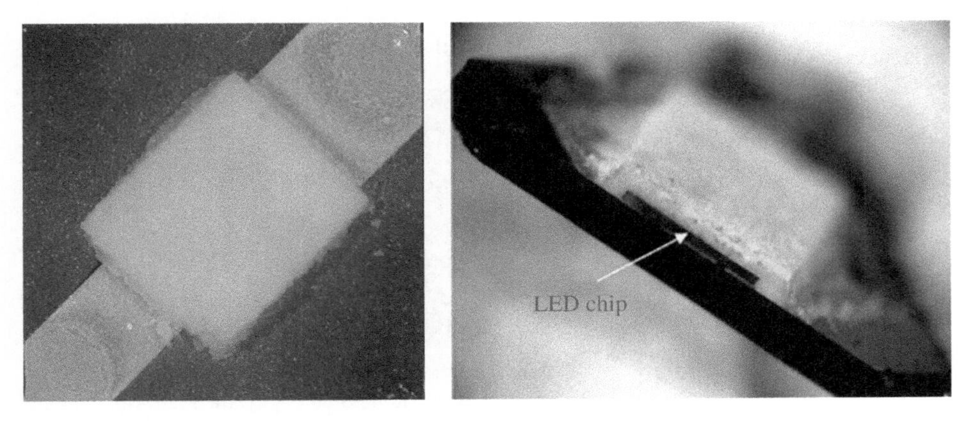

Figure 2.39 Conformal phosphor coating by stencil printing method.

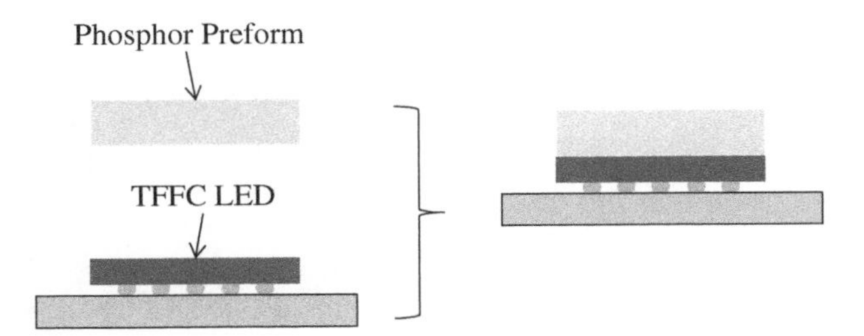

Figure 2.40 Conformal coating with phosphor preform.

have a high setup cost and material cost. Lee et al. proposed an innovative method called quasi-conformal phosphor coating [54]. In their study, a dummy platform was added underneath the LED chip. The size of the dummy platform must be larger than the LED chip to a certain extent. The phosphor slurry was dispensed directly on top of the chip. The step along the edges of such a platform stopped the dispensed phosphor slurry flow. Figure 2.41 shows the results of using this method on a 3528 and K1 packages. A quasi-conformal phosphor coating was formed on the chip. The equipment and setup costs of this method are low as it only needs a regular dispenser. The optical performance of the package fabricated by this method was found to be similar to that of the package with dispersed phosphor. The phosphor consumption of the quasi-conformal phosphor coating method was only half that of the dispersed dispensing method. A similar approach called dip-transfer phosphor coating was also proposed by Zheng et al. [55, 56].

2.4.3 Remote Phosphor

In both dispersed dispensing and conformal coating methods, phosphor materials are applied directly on top of the LED chip. When the LED is powered up, the phosphor materials on top are heated up by the high junction temperature of the chip. The emission efficiency and the emission spectrum of the phosphor materials depend heavily on the

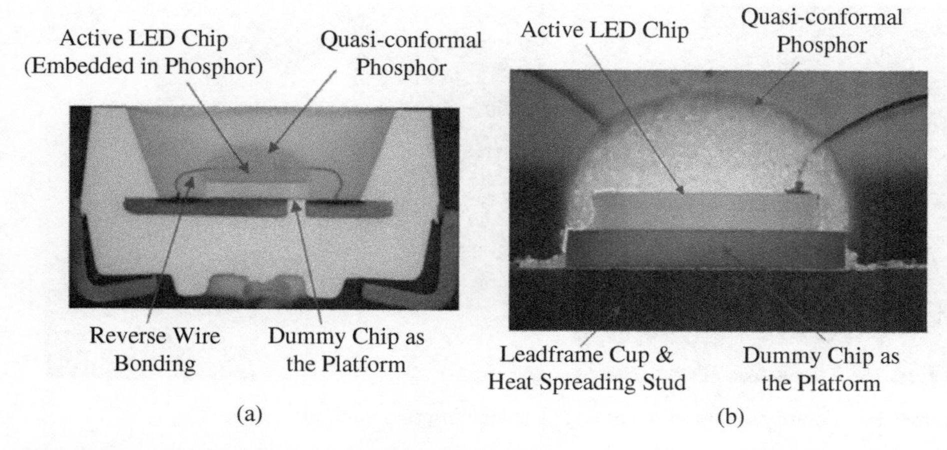

Figure 2.41 Quasi-conformal coating: (a) 3528; (b) K1.

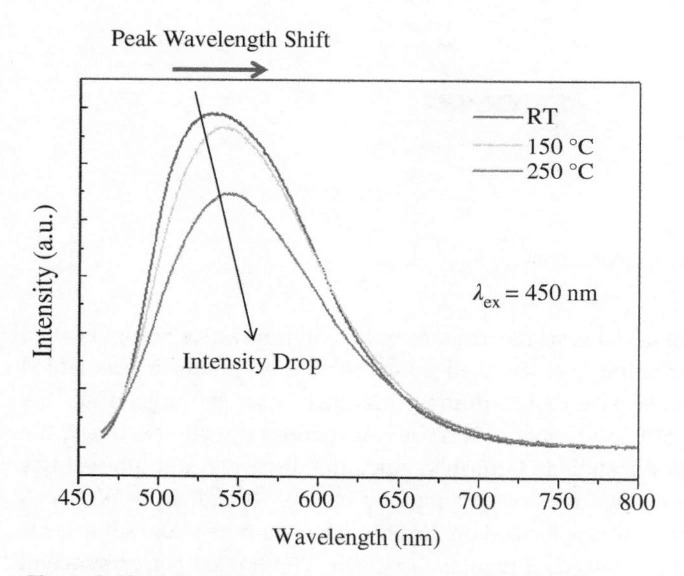

Figure 2.42 Effect of temperature on phosphor emission spectrum.

working temperature. Figure 2.42 shows the emission spectrum of a green phosphor measured at different temperatures. It is clear that with the increase in temperature the emission intensity decreases and the peak wavelength shifts. In addition, high working temperatures may cause degradation of the phosphor layer and induce reliability problems [57].

Packages with conformal phosphor coating have a better angular color uniformity. However, nearly 50% of the light is back-scattered to the chip by the phosphor layer, which is placed right on top of the LED [58]. The LED chip has a relatively low reflectivity. A large

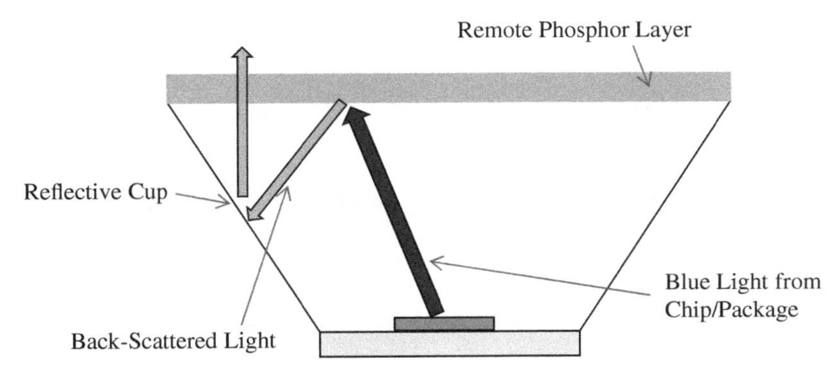

Figure 2.43 Remote phosphor configuration.

portion of the back-scattered light is absorbed by the chip instead of reflected back to the forward direction. This optical loss lowers the overall efficacy of the package.

A remote phosphor configuration is proposed to solve the heat and optical loss problems [59–61]. Phosphor materials are placed apart from the chip and there is a reflective cup between the phosphor materials and the blue light source (Figure 2.43). The back-scattered light is reflected forward by the reflective cup. Only a small amount of light reaches the chip and is absorbed. This reduces the optical loss tremendously and hence has a higher luminous flux [62]. The quality and the reflectivity of the reflective cup are very important as they affect the luminous output [63–65]. The phosphor materials are also not heated up by the chip directly.

Remote phosphor can be applied in a system level [66–68]. Blue LED packages without phosphor are used. Phosphor materials are deposited on a plate [69] or a diffuser, which is placed at a certain distance from the LED package. A large reflective cup is used to reflect the back-scattered light from the phosphor plate. The large surface area of the phosphor plate consumes a huge amount of phosphor materials, which increases the cost substantially.

Package level remote phosphor can reduce the consumption of phosphor materials. There are different types of package level remote phosphor configurations [70–73]. In a package level remote phosphor, the gap between the phosphor layer and the bare chip is filled by encapsulant. This reduces the refractive mismatch and hence increases the amount of light that can be extracted from the chip. For the in-cup type LED, transparent encapsulant, such as silicone, is first dispensed and cured. A phosphor layer is then dispensed or sprayed on top of the silicone. Multiple remote phosphor layers with different phosphor materials can be applied to improve the chromaticity performance [74]. For the lens type LEDs, a remote phosphor layer can be achieved by conducting compression molding twice. A transparent lens is first molded on the substrate. A phosphor layer is then molded on top of the transparent lens. Figure 2.44 shows the schematic diagram of the packages with a remote phosphor layer. Several innovative remote phosphor deposition methods for wafer level LED packaging are proposed and are discussed in Chapter 3.

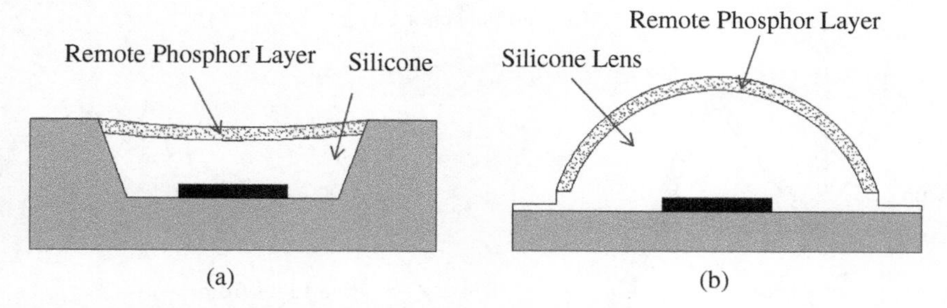

Figure 2.44 Package level remote phosphor configuration: (a) in-cup; (b) lens type.

2.5 Encapsulation and Molding Process

In IC packaging, one of the major roles of encapsulant is protection. It prevents the chip and interconnects from being exposed to the environment. It is usually flat, thin, and not transparent. In LED packaging, the encapsulant has other key roles in addition to protection. It serves as an optical medium for light transmission. Therefore, transparent epoxy or silicone is extensively used. The refractive index of the encapsulant is in between the GaN and the air. This reduces the refractive index mismatch, and hence less light is trapped by total internal reflection. The encapsulant also serves as the primary optics of the package. The light pattern of the LED package depends directly on the geometry of the encapsulation.

The encapsulation procedure of in-cup LEDs is simple and direct. Phosphor slurry (dispersed dispensing) or clear silicone (after conformal phosphor coating) is dispensed into the reflective cup. The process is completed after the curing of the encapsulant. For the lens type LEDs, there are two major encapsulation methods.

2.5.1 Encapsulant Filling with Lens

In some LED types, such as a K1 emitter, a plastic lens is attached to the lead frame after the die mounting, wire bonding, and phosphor coating processes. The lens is usually hemispherical in shape, as shown in Figure 2.45. After attaching the lens, encapsulant is used to fill the gap between the lens and the chip by needle dispensing (Figure 2.46). The lens is then bonded to the lead frame after the encapsulant is fully cured.

The plastic lens is normally made by a transparent epoxy. The thermal stability of the epoxy is the major concern. In the subsequent assembly process, the package is subject to high temperature treatment, such as reflow. The reflow peak temperature may reach up to 260 °C. The epoxy may not be able to sustain at such high temperatures. Also, it may degrade because of the heat generated by the chip during operation. Yellowing is one common failure phenomenon of epoxy. This leads to lumen output drop and color shift.

2.5.2 Lens Molding

A lens can be fabricated on the substrate directly by compression molding. The geometry of the lens is defined by the mold cavity. Silicone has a better thermal stability and is usually used as the encapsulant.

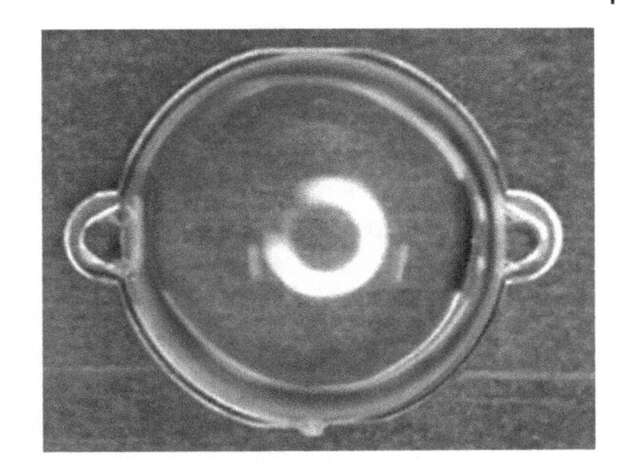

Figure 2.45 Plastic lens for K1 emitter.

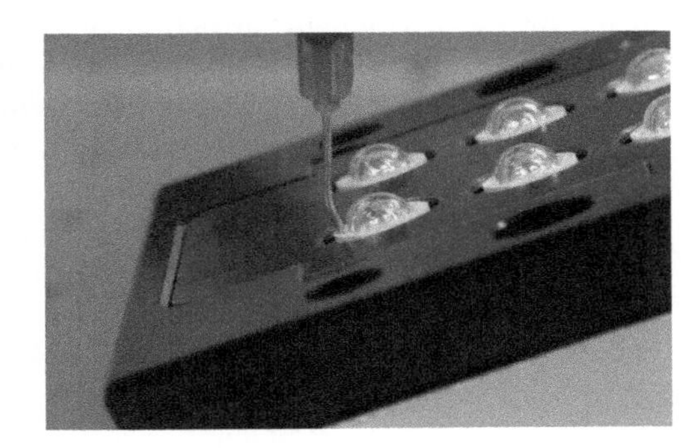

Figure 2.46 Encapsulant filling with lens.

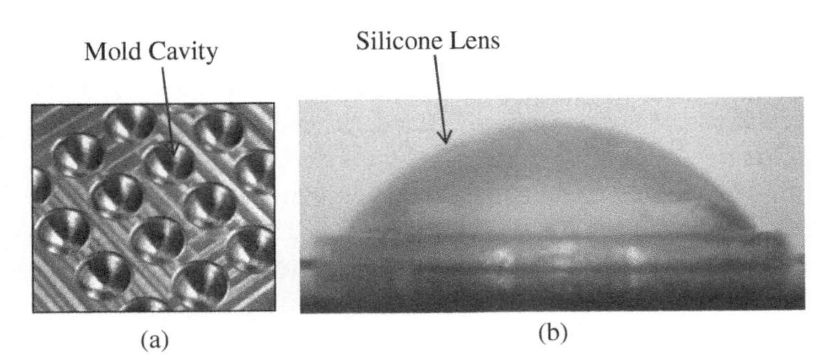

Figure 2.47 Lens formation by compression molding: (a) mold; (b) silicone lens formation after molding.

A mold releasing agent or film is first applied to the mold. This prevents the silicone lens from bonding to the mold and becoming damaged after de-molding. If phosphor coating is already applied to the substrate, clear silicone is dispensed into the mold cavity. Otherwise, phosphor slurry is used. The substrate is then aligned with the mold. Heat and pressure are applied to cure the silicone. Void formation can be suppressed if the process is carried out in a vacuum. After de-molding, the whole process is completed. The process is the same as the one described in Figure 2.33. Figure 2.47 shows the mold and the silicone lens fabricated on a silicon substrate.

Compression molding is an effective encapsulation method. All components on the substrate are molded at the same time. In order to achieve a high yield and a good quality, some parameters must be optimized (molding temperature, pressure, and time) [75]. The viscosity of the encapsulant is another important factor. It may be difficult to fill the mold cavity completely if a very high-viscosity encapsulant is used. On the other hand, bleeding may occur if the encapsulant viscosity is too low.

2.6 Secondary Optics and Lens Design

The phosphor layer, lens, lead frame reflective cup, etc. as described in the earlier sessions are the primary optics of the package. The major considerations regarding the optical performance include maximizing the optical output and improving color uniformity. Therefore, it is common to have a hemispherical lens type LED. A hemispherical lens reduces the amount of light trapped by total internal reflection, and hence increases the amount of light being extracted. The light pattern is normally circular in shape and close to Lambertian. It may not be suitable for all kinds of applications.

Secondary optics, also known as a total internal reflection (TIR) lens, is used to adjust the light pattern [76–79]. It collects the light emitted from the package and changes the direction by total internal reflection. Figure 2.48 demonstrates the working principle of secondary optics. The lens can be attached to each individual LED component or cover an LED array.

There are various types of lens design. A collimator lens is used to focus the light. The batwing lens improves the view angle. It has a more uniform light pattern on a large flat surface, which can be used to replace the diffuser. It is commonly used in traffic signal lamps. Side emitting lenses are commonly applied to the backlight module of LCD and reflection type head lamps. Figure 2.49 shows some TIR lenses. Free form lenses are used when the requirement of the light pattern on the target plane is known [80].

Optical simulation by ray tracing is a powerful tool for lens design. Light path traveled in a dedicated direction, refraction, and specular reflection are modeled by ray tracing. Scattering and diffraction are treated as random processes and are simulated by the Monte Carlo method. The general procedure in optical simulation is illustrated in Figure 2.50.

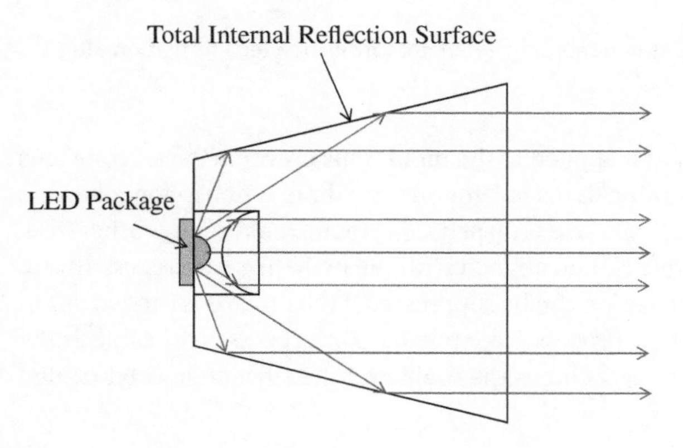

Figure 2.48 Secondary optics (total internal reflection lens).

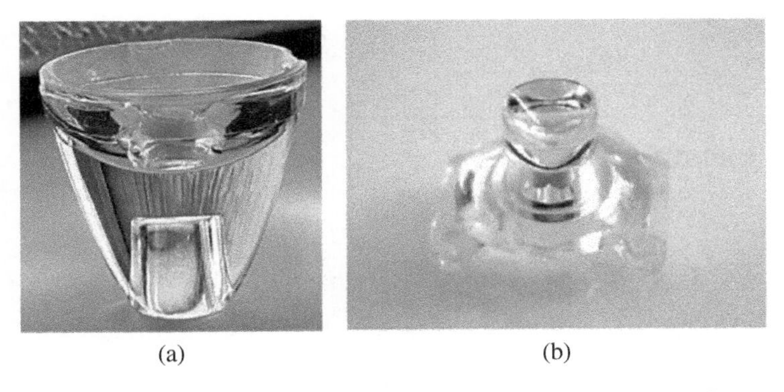

Figure 2.49 TIR lenses: (a) collimating lens; (b) side-emitting lens. Source: Ledil and LED-TECH.DE.

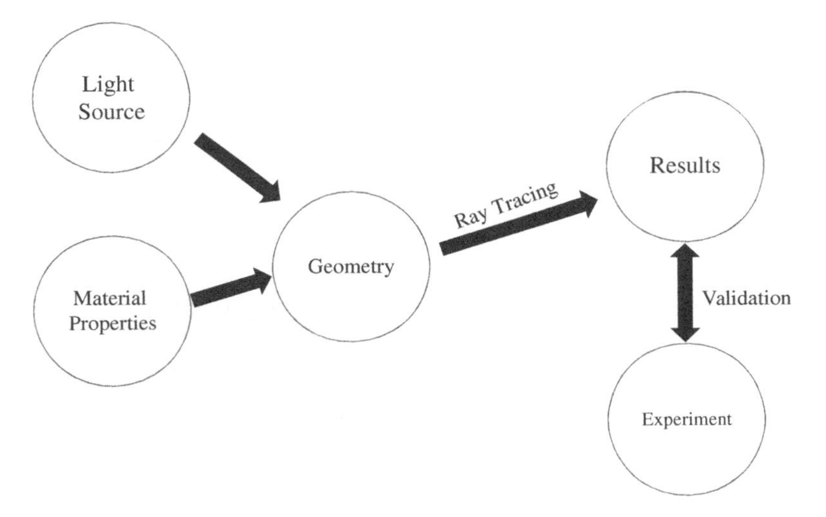

Figure 2.50 Optical simulation procedures.

The first step in optical simulation is to define the geometry. This involves the size and location of the chip/package, phosphor layer, reflective cup, diffuser, lens, etc. This step also defines the interface between each component. The next step is to describe the light source. The light-emitting surface and type of light-emitting pattern are required to be defined properly. The light pattern can be either assumed to have a standard pattern (Lambertian or uniform) or obtained by direct measurement. A source imaging goniometer is used to measure the actual light pattern of the source. The measurement result is then imported to a simulation tool for describing the light source. The light-emitting pattern of the source directly affects the results.

Material properties are required to be defined properly in order to have a meaningful and reliable result. Light incidents to a silicone layer suspended with phosphor particles is taken as an example here (Figure 2.51). It involves many optical processes, including reflection, refraction, absorption, scattering, and luminesce. The optical properties of the materials corresponding to the optical processes are required for the simulation.

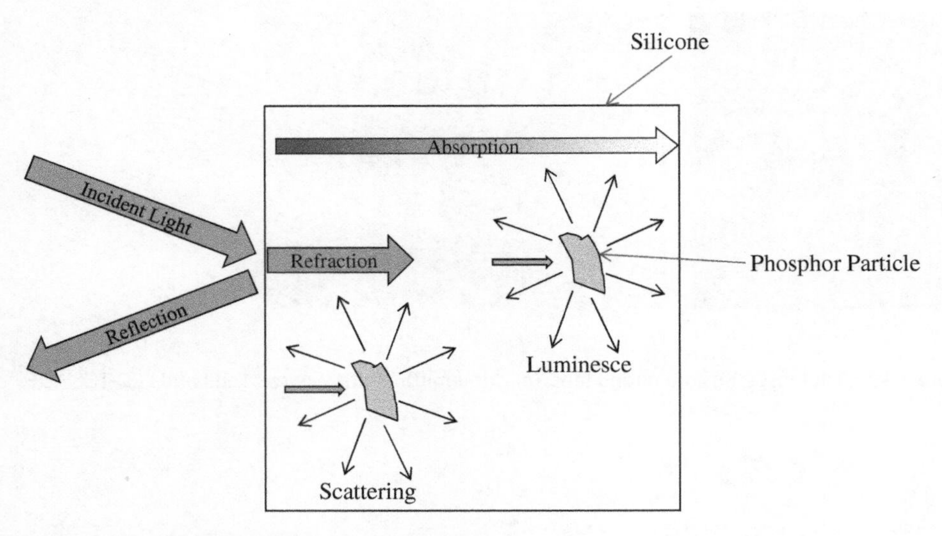

Figure 2.51 Optical processes involve in simulation.

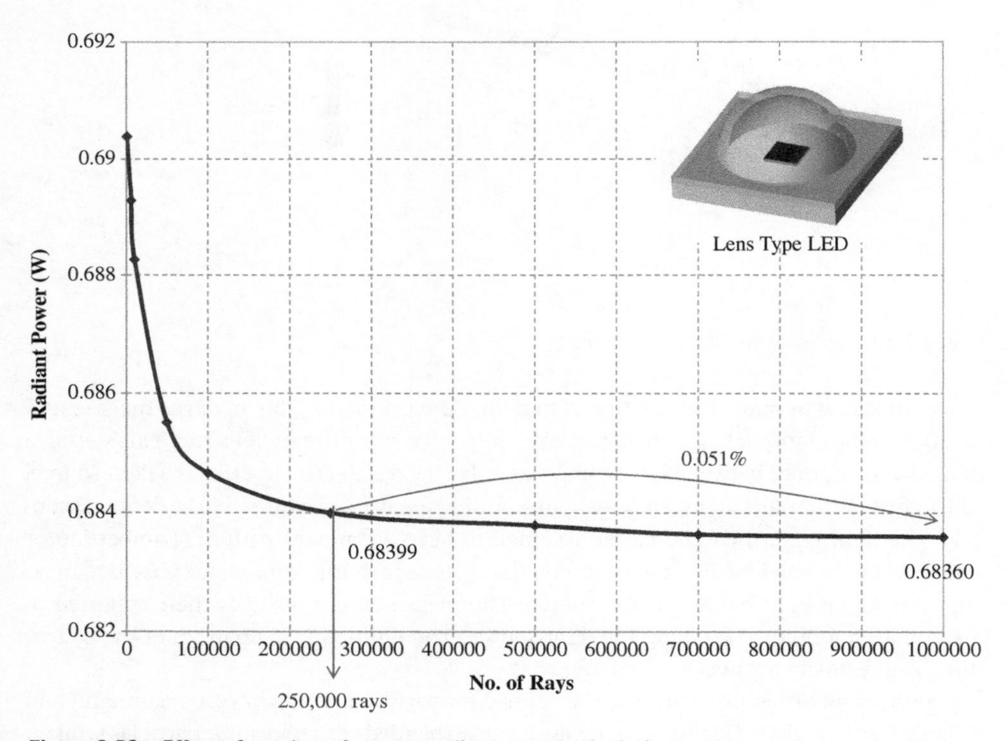

Figure 2.52 Effect of number of rays on radiant power calculation.

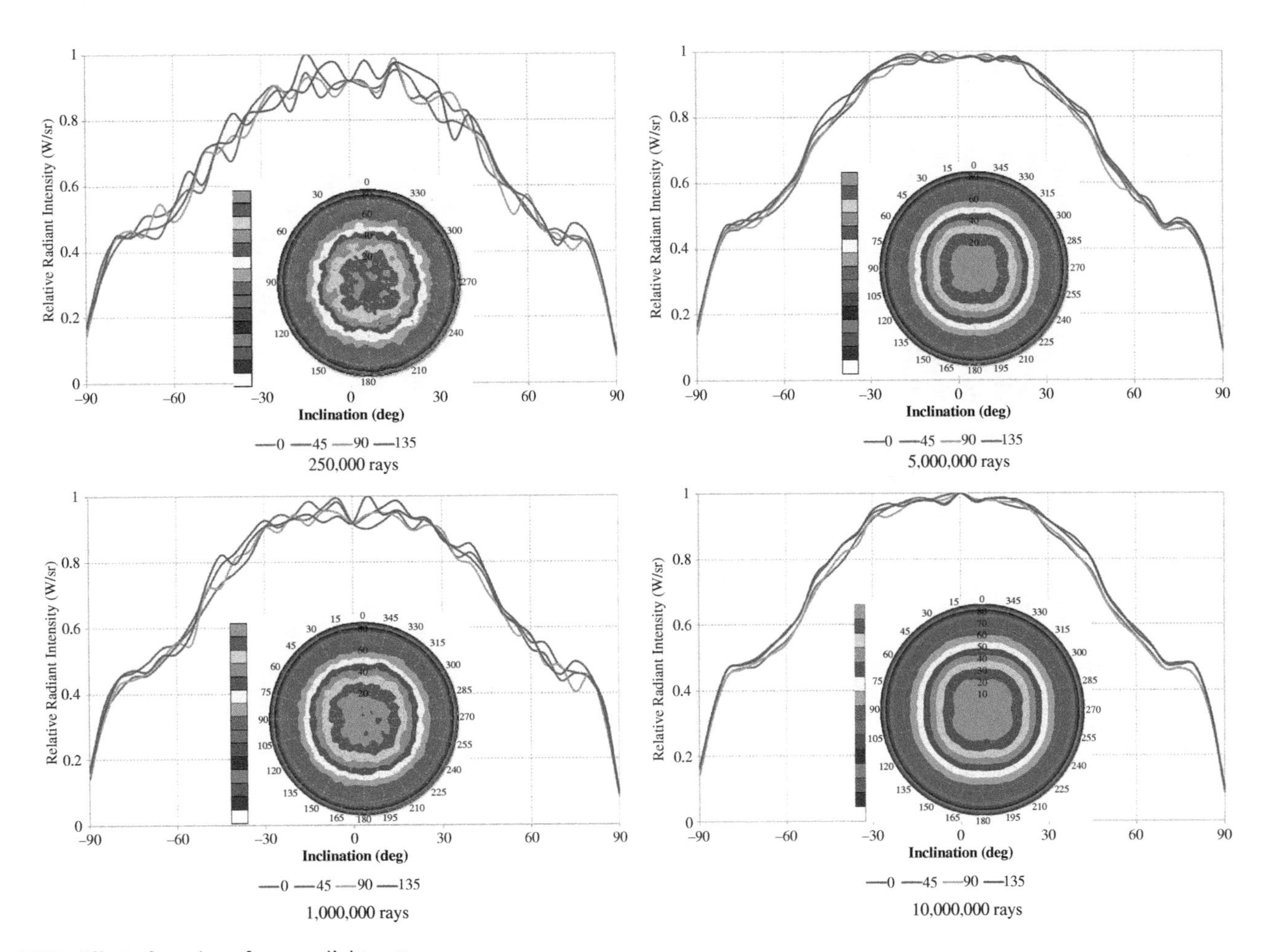

Figure 2.53 Effect of number of rays on light pattern.

After defining the geometry, light source, and material properties, the next step is ray tracing. Rays are randomly generated from the light source according to the predefined light pattern. Each ray generated has its own optical power and wavelength. The final results are calculated based on the rays hitting the user-defined light receiving plane. The number of rays used in the simulation is an important factor. In general, more rays are preferable as the process involves random variables. Nevertheless, it takes a longer computational time and consumes more resources.

The effects of different number of rays are demonstrated by a case study. A lens type LED was simulated. Figure 2.52 shows that the simulated radiant power dropped when the number of rays increased. The results started converging when 250 000 rays were used, indicating that 250 000 rays were enough to calculate the radiant power for this particular case.

Figure 2.53 shows that more rays are required to analyze the light pattern. The light pattern obtained by tracing 250 000 rays was not smooth, and at least 500 000 rays were required. Tracing 10 000 000 rays showed a better result, but it took much longer to complete the simulation.

It is always important to validate the optical model and the simulation results with experiment measurements. Optical simulation generates results and plots no matter whether the geometry, light source, and optical properties are properly defined or not. In some cases, assumptions are set for simplification. Experimental validation can help trace the errors. A validated model is a powerful tool for parametric studies and design optimization.

References

1 Manikam, V.R. and Cheong, K.Y. (2011). Die attach materials for high temperature applications: a review. *IEEE Transactions on Components, Packaging and Manufacturing Technology* 1 (4): 457–478.

2 You, J.P., He, Y., and Shi, F.G. (2007). Thermal management of high power LEDs: impact of die attach materials. In: *Proceedings of the 2nd International Microsystems, Packaging, Assembly and Circuits Technology Conference (IMPACT)*, Taipei (1–3 October 2007), 239–242. IEEE.

3 Yin, L., Yang, L., Xu, G. et al. (2011). Effects of die-attach materials on the optical durability and thermal performances of HP-LED. In: *Proceedings of the 12th International Conference on Electronic Packaging Technology and High Density Packaging (ICEPT-HDP)*, Shanghai, China (8–11 August 2011), 1116–1119. IEEE.

4 Muriel, T. (2007). Die-attach materials and process. *Solid State Technology*, April, https://sst.semiconductor-digest.com/2007/04/die-attach-materials-and-processes/ (accessed 19 March 2021).

5 Thang, T.S., Sun, D., Koay, H.K. et al. (2005). Characterization of Au-Sn eutectic die attach process for optoelectronics device. In: *Proceedings of the 7th International Symposium on Electronics Materials and Packaging (EMAP2005)*, Tokyo, Japan (11–14 December 2005), 118–124. IEEE.

6 Wang, Q., Choa, S.H., Kim, W. et al. (2006). Application of Au-Sn eutectic bonding in hermetic radio-frequency microelectromechanical system wafer level packaging. *Journal of Electronic Materials* 35 (3): 425–432.

7 Ivey, D.G. (1998). Microstructural characterization of Au/Sn solder for packaging in optoelectronic applications. *Micron* 29 (4): 281–287.

8 Elger, G., Hutter, M., Oppermann, H. et al. (2002). Development of an assembly process and reliability investigations for flip-chip LEDs using the AuSn soldering. *Microsystem Technologies* 7 (5–6): 239–243.

9 John, C.M. (2008). Processing and reliability issues for eutectic AuSn solder joints. In: *Proceedings of the 41st International Symposium on Microelectronics (IMAPS 2008)*, Providence, RI, 909–916. IMAPS.

10 Tsai, J.Y., Chang, C.W., Ho, C.E. et al. (2006). Microstructure evolution of gold-tin eutectic solder on Cu and Ni substrates. *Journal of Electronic Materials* 35 (1): 65–71.

11 Evans, D.D. Jr., and Bok, Z. (2010). AuSi and AuSn Eutectic die attach studies from small (12 mil) to large (453 mil) die. In: *Proceedings of the 43rd International Symposium on Microelectronics (IMAPS 2010)*. IMAPS, Raleigh, NC (31 October to 4 November 2010).

12 Park, J.W., Yoon, Y.B., Shin, S.H. et al. (2006). Joint structure in high brightness light emitting diode (HB LED) packages. *Materials Science and Engineering A* 441: 357–361.

13 Li, Z.T., Tang, Y., Ding, X. et al. (2014). Reconstruction and thermal performance analysis of die-bonding filling states for high-power light-emitting diode devices. *Applied Thermal Engineering* 65: 236–245.

14 van Driel, W.D., van Silfhout, R.B.R., Zhang, G.Q. et al. (2008). Reliability of wirebonds in micro-electronic packages. *Microelectronics International* 26 (2): 15–22.

15 Liu, Y., Allen, H., Luk, T. et al. (2006). Simulation and experimental analysis for a ball stitch on bump wire bonding process above a laminate substrate. In: *Proceedings of the 56th Electronic Components and Technology Conference (ECTC 2006)*, San Diego, CA (30 May to 2 June 2006), 1918–1923. IEEE.

16 Tan, B.W., Wang, L., and Lu, H.L. (2008). Application of ultra low loop gold wire bonding technique in super thin (JEDEC package profile height sub code "X2") quad flat no Lead package (QFN). In: *Proceedings of the 33rd IEEE/CPMT International Electronic Manufacturing Technology Symposium (IEMT)*, Penang, Malaysia, 4–6 November 2008, 1–6.

17 Oda, Y., Kiso, M., Kurosaka, S., et al. (2008). Study of suitable palladium and gold thickness in ENEPIC Deposits for lead free soldering and gold wire bonding. Proceedings of the 41st International Symposium on Microelectronics (IMAPS 2008), Providence, RI (2–6 November 2008). https://www.semanticscholar.org/paper/Study-of-Suitable-Palladium-and-Gold-Thickness-in-Oda-Kiso/22153c7fcbf0ea37bc19f29fca52521df9731211 (accessed 19 March 2021).

18 Kim, H., Lee, S.N., and Cho, J. (2010). Electrical and optical characterization of GaN-based light-emitting diodes fabricated with top-emission and flip-chip structures. *Materials Science in Semiconductor Processing* 13 (3): 180–184.

19 Philips Lumileds Press. (2020). Information PR77. http://www.research. http://philips .com/technologies/lumaramic.html (accessed 3 September 2020).

20 Chen, X., Kong, F., Li, K. et al. (2014). Study of light extraction efficiency of flip-chip GaN-based LEDs with different periodic arrays. *Optics Communications* 314: 90–96.

21 Qin, P., Li, Q., and Chan, Y.C. (2011). Thermal analysis of high brightness flip-chip LED packages. In: *Proceeding of the 13th Electronics Packaging Technology Conference (EPTC)*, Singapore (7–9 December 2011), 722–725. IEEE.

22 Tang, C.Y., Tsai, M.Y., Lin, C.C. et al. (2010). Thermal measurements and analysis of flip-chip LED packages with and without underfills. In: *Proceedings of the 5th International Microsystems Packaging Assembly and Circuits Technology Conference (IMPACT)*, Taipei (20–22 October 2010), 1–4. IEEE.

23 Tang, C.Y., Tsai, M.Y., Yen, C.Y. et al. (2011). Characterization of thermal and optical behaviors of flip-chip LED packages with various underfills. In: *Proceedings of the 6th International Microsystems, Packaging, Assembly and Circuits Technology Conference (IMPACT)*, Taipei (19–21 October 2011), 327–331. IEEE.

24 Lin, C.C., Chang, L.B., Jeng, M.J. et al. (2010). Fabrication and thermal analysis of flip-chip light-emitting diodes with different numbers of Au stub bumps. *Microelectronics Reliability* 50 (5): 683–687.

25 Kim, S.C. and Kim, Y.H. (2013). Review paper: flip chip bonding with anisotropic conductive film (ACF) and nonconductive adhesive (NCA). *Current Applied Physics* 13: S14–S15.

26 Wang, Y.H., Nishida, K., Hutter, M. et al. (2007). Low-temperature process of fine-pitch Au-Sn bump bonding in ambient air. *Japanese Journal of Applied Physics* 46 (4B): 1961–1967.

27 Yum, J., Seo, S., Lee, S. et al. (2003). $Y_3Al_5O_{12}:Ce_{0.05}$ phosphor coatings on gallium nitride for white light emitting diodes. *Journal of the Electrochemical Society* 150 (2): H47–H52.

28 Smet, P.F., Parmentier, A.B., and Poelman, D. (2011). Selecting conversion phosphors for white light-emitting didoes. *Journal of the Electrochemical Society* 158 (6): R37–R54.

29 Xie, R.J., Hirosaki, N., Kimura, N. et al. (2007). 2-phosphor-converted white light-emitting didoes using oxynitride/nitride phosphors. *Applied Physics Letters* 90: 191101-1–191101-3.

30 Won, Y.H., Jang, H.S., Cho, K.W. et al. (2013). Effect of phosphor geometry on the luminous efficiency of high-power white light-emitting diodes with excellent color rendering property. *Optics Letters* 34 (1): 1–3.

31 Sommer, C., Krenn, J.R., Hartmann, P. et al. (2009). The effect of the phosphor particle sizes on the angular homogeneity of phosphor-converted high-power white LED light sources. *IEEE Journal of Selected Topics in Quantum Electronics* 15 (4): 1181–1188.

32 Tran, N.T., You, J.P., and Shi, F.G. (2009). Effect of phosphor particle size on luminous efficacy of phosphor-converted white LED. *Journal of Lightwave Technology* 27 (22): 5145–5150.

33 Liu, Z., Liu, S., Wang, K. et al. (2008). Optical analysis of color distribution in white LEDs with various packaging methods. *IEEE Photonic Technology Letters* 20 (24): 2027–2029.

34 Tran, N.T. and Shi, F.G. (2008). Studies of phosphor concentration and thickness for phosphor-based white light-emitting-diodes. *Journal of Lightwave Technology* 26 (21): 3556–3559.

35 Shuai, Y., He, Y., Tran, N.T. et al. (2011). Angular CCT uniformity of phosphor converted white LEDs: effects of phosphor materials and packaging structures. *IEEE Photonics Technology Letters* 23 (3): 137–139.

36 Sun, C.C., Chen, C.Y., Chen, C.C. et al. (2012). High uniformity in angular correlated-color-temperature distribution of white LEDs from 2800K to 6500K. *Optics Express* 20 (6): 6622–6630.

37 Sommer, C., Hartmann, P., Pachler, P. et al. (2009). A detailed study on the requirements for angular homogeneity of phosphor converted high power white LED light sources. *Optical Materials* 31: 837–848.

38 Sommer, C., Krenn, J.R., Hartmann, P. et al. (2009). On the requirements for achieving angular homogeneity in phosphor converted high power flip-chip light emitting diodes. *Japanese Journal of Applied Physics* 48: 070208-1–070208-3.

39 Choi, J.K., Sluzky, E., Anc, M. et al. (2012). EPD of phosphors for display and solid state lighting technologies. *Key Engineering Materials* 507: 149–153.

40 Chen, W., Bai, J., Meen, T. et al. (2009). Electrophoretic deposition of YAG phosphor on SMD LED. *ECS Transactions* 19 (12): 27–35.

41 Collins, III, W.D., Krames, M.R., Verhoecks, C.J., et al.(2003). Using electrophoresis to produce a conformally coated phosphor-converted light emitting semiconductor. US Patent, US6576488B2, filed 11 June 2001 and granted 10 June 2003.

42 Dickerson, J.H. and Boccaccini, A.R. (2012). *Electrophoretic Deposition of Nanomaterial, Nanostructure Science and Technology*. New York: Springer.

43 Yang, L., Wang, S., Lv, Z. et al. (2013). Color deviation controlling of phosphor conformal coating by advanced spray painting technology for white LEDs. *Applied Optics* 52 (10): 2075–2079.

44 Chen, K.J., Han, H.V., Lin, B.C. et al. (2013). Improving the angular color uniformity of hybrid phosphor structures in white light-emitting didoes. *IEEE Electron Device Letters* 34 (10): 1280–1282.

45 Huang, H.T., Tsai, C.C., and Huang, Y.P. (2010). Conformal phosphor coating using pulsed spray to reduce color deviation of white LEDs. *Optics Express* 18 (S2): A201–A206.

46 Cannon, N.O., Jackson, M. (2012). Phosphor coating systems and methods for light emitting structures and packaged light emitting diodes including phosphor coating. US Patent, US20090179213A1, filed 15 January 2008 and granted 15 November 2011.

47 Yu, S., Yang, L., Bin, C. et al. (2011). Development process of phosphor coating with screen printing for white LED packaging. In: *Proceedings of the 12th International Conference on Electronic Packaging Technology and High Density Packaging (ICEPT-HDP)*, Shanghai (8–11 August 2011), 103–107. IEEE.

48 Lee, K.H. and Lee, S.W.R. (2006). Process development for yellow phosphor coating on blue light emitting diodes (LEDs) for white light illumination. In: *Proceedings of the 8th Electronics Packaging Technology Conference (EPTC)*, Singapore (6–8 December 2006), 379–384. IEEE.

49 Lo, J.C.C., Lee, S.W.R., Zhang, R. et al. (2012). Reverse wire bonding and phosphor printing for LED wafer level packaging. In: *Proceedings of the 62nd Electronic Components and Technology Conference (ECTC)*, San Diego, AA (29 May to 1 June 2012), 1814–1818. IEEE.

50 Fujita, S., Sakamoto, A., and Tanabe, S. (2008). Luminescence characteristics of YAG class-ceramic phosphor for white LED. *IEEE Journal of Selected Topics in Quantum Electronics* 14 (5): 1387–1391.

51 Raukas, M., Kelso, J., Zheng, Y. et al. (2013). Ceramic phosphors for light conversion in LEDs. *ESC Journal of Solid State Science and Technology* 2 (2): R3168–R3176.

52 Rao, H., Wang, W., Wan, X. et al. (2013). An Improved slurry method of self-adaptive phosphor coating for white pc-LED packaging. *Journal of Display Technology* 9 (6): 453–458.

53 Hou, B., Rao, B., and Li, J. (2009). Methods of increasing luminous efficiency of phosphor-converted LED realized by conformal phosphor coating. *Journal of Display Technology* 5 (2): 57–60.

54 Lee, S.W.R., Guo, X., Niu, D. et al. (2013). Quasi-conformal phosphor dispensing on LED for white light illumination. In: *Proceedings of the 63rd Electronic Components and Technology Conference (ECTC 2013)*, Las Vegas, NV (28–31 May 2013), 563–567. IEEE.

55 Zheng, H., Wang, Y., Li, L. et al. (2013). Dip-transfer phosphor coating on designed substrate structure for high angular color uniformity of white light emitting diodes with conventional chips. *Optics Express* 21 (S6): A933–A941.

56 Zheng, H., Liu, S., and Luo, X. (2013). Enhancing angular color uniformity of phosphor-converted white light-emitting didoes by phosphor dip transfer coating. *Journal of Lightwave Technology* 31 (12): 1987–1993.

57 Meneghini, M., Lago, M.D., Trivellin, N. et al. (2013). Thermally activated degradation of remote phosphors for application in LED lighting. *IEEE Transactions on Device and Materials Reliability* 13 (1): 316–318.

58 Chen, K.J., Chen, H.C., Lin, C.C. et al. (2013). An investigation of the optical analysis in white light-emitting diodes with conformal and remote phosphor structure. *Journal of Display Technology* 9 (11): 915–920.

59 Narendran, N. (2005). Improved performance white LED. In: *Proceedings. of the SPIE 5941, 5th International Conference on Solid State Lighting*. SPIE. https://doi.org/10.1117/12.625921.

60 Narendran, N., Gu, Y., Freyssinier-Nova, J.P. et al. (2012). Extracting phosphor-scattered photons to improve white LED efficiency. *Physica Status Solidi (A)* 202 (6): R60–R62.

61 Zhu, Y. and Narendran, N. (2008). Optimizing the performance of remote phosphor LEDs. *Journal of Light & Visual Environment* 32 (2): 115–119.

62 Yu, R., Jin, S., Cen, S. et al. (2010). Effect of the phosphor geometry on the luminous flux of phosphor-converted light-emitting didoes. *IEEE Photonics Technology Letters* 22 (23): 1765–1767.

63 Kim, J.K., Luo, H., Schubert, E.F. et al. (2005). Strongly enhanced phosphor efficiency in GaInN white light-emitting didoes using remote phosphor configuration and diffuse reflector cup. *Japanese Journal of Applied Physics* 44 (21): L649–L651.

64 Luo, H., Kim, J.K., Schubert, E.F. et al. (2005). Analysis of high-power packages for phosphor-based white-light-emitting diodes. *Applied Physics Letters* 86: 243505-1–243505-3.

65 Luo, H., Kim, J.K., Schubert, E.F. et al. (2005). Trapped whispering-gallery optical modes in white light-emitting diode lamps with remote phosphor. *Applied Physics Letters* 89: 041125-1–041125-3.

66 Xiao, H., Lu, Y.J., Shih, T.M. et al. (2014). Improvements on remote diffuser-phosphor-packaged light-emitting diode systems. *IEEE Photonics Journal* 6 (2): 8200108.
67 Hoelen, C., Borel, H., Graaf, J. et al. (2008). Remote phosphor LED modules for general illumination – towards 200 lm/W general lighting LED light sources. In: *Proc. SPIE 7058, 8th International Conference on Solid State Lighting*. SPIE. https://doi.org/10.1117/12.799502.
68 Schiel, M. (2011). Remote-phosphor technology can deliver a more uniform and attractive light output from LED lamps. *LEDs Magazine*: 37–40.
69 Huang, K.C., Lai, T.H., Chen, C.Y., and Improved, C.C.T. (2013). Uniformity of white LED using remote phosphor with patterned sapphire substrate. *Applied Optics* 52 (30): 7376–7381.
70 Lin, M.T., Ying, S.P., Lin, M.Y. et al. (2010). Design of the ring remote phosphor structure for phosphor-converted white-light-emitting didoes. *Japanese Journal of Applied Physics* 49: 072101-1–072101-6.
71 Lin, M.T., Ying, S.P., Lin, M.Y. et al. (2010). Ring remote phosphor structure for phosphor-converted white LEDs. *IEEE Photonics Technology Letters* 22 (8): 574–576.
72 Kuo, H.C., Hung, C.W., Chen, H.C. et al. (2011). Patterned structure of remote phosphor for phosphor-converted white LEDs. *Optics Express* 19 (S4): A930–A936.
73 Allen, S.C. and Steckl, A.J. (2008). A nearly ideal phosphor-converted white light-emitting diode. *Applied Physics Letters* 92: 143309-1–143309-3.
74 Chen, K.J., Chen, H.C., Shih, M.H. et al. (2013). Enhanced luminous efficiency of WLEDs using a dual-layer structure of remote phosphor package. *Journal of Lightwave Technology* 31 (12): 1941–1945.
75 He, S.G., Li, L., and Qi, E.S. (2007). Study on the quality improvement of injection molding in LED packaging processes based on DoE and data mining. In: *Proceedings of the International Conference on Wireless Communications, Networking and Mobile Computing (WiCom 2007)*, Shanghai (21–25 September 2007), 6625–6628. IEEE.
76 Chen, W.C., Lai, T.T., Wang, M.W. et al. (2001). An optimization system for LED lens design. *Expert Systems with Applications* 38: 11976–11983.
77 Wang, K., Liu, S., Chen, F. et al. (2009). Effect of manufacturing defects on optical performance of discontinuous freeform lenses. *Optics Express* 17 (7): 5457–5465.
78 Chen, H.C., Lin, J.Y., and Chiu, H.Y. (2013). Rectangular illumination using a secondary optics with cylindrical lens for LED street light. *Optics Express* 21 (3): 3201–3212.
79 Chang, Y.C., Ou, C.J., Tsai, Y.S. et al. (2009). Nonspherical LED packaging lens for uniformity improvement. *Optical Review* 16 (3): 323–325.
80 Zheng, Z., Hao, X., and Liu, X. (2009). Freeform surface lens for LED uniform illumination. *Applied Optics* 48 (35): 6627–6634.

3

Chip Scale and Wafer Level Packaging of LEDs

3.1 Introduction

A typical LED has a chip size of 0.2–1.0 mm. Such a tiny device needs to be turned into a packaged component before it can be used for its intended applications. At present, most of the LED packaging processes are adopted from conventional integrated circuit (IC) packaging. The packaging processes of LED components (from a single chip to a final LED component) are discussed in detail in Chapter 2. Militarization and cost reduction are common objectives in the electric packaging industry. From chip to component involves many processes and materials. This significantly adds both cost and size to the final product. A small package usually consists of less packaging material and hence has a lower cost. The smallest package size that can be achieved will be of the same size as a chip. This particular package category is called a chip scale package (CSP). This package type also offers various advantages for LED packaging. Figure 3.1 shows the militarization trend of LEDs. The optical property of this small package is close to a point light source. This small form factor package is also suitable for fine pitch LED array panels. This chapter briefly discusses this specific package type in LED packaging.

In the past decade, the IC packaging industry has been migrating to wafer level packaging (WLP) [1], which means there is an emerging need for LED packaging to catch up. The basic concept of WLP is to package the whole wafer instead of individual chips. This packaging technology is suitable for low input/output (I/O) devices. WLP has the twin merits of high throughput and low cost [2, 3]. It has been evidenced by many case studies in the IC and microelectromechanical systems (MEMS) packaging industries that WLP technologies may lead to a 20%–30% cost reduction in components at high volume productions [4]. Typical applications of WLP include mobile phones, digital cameras, laptop computers, image sensors, dynamic random access memory (DRAM), and integrated passive devices. According to a 2010 industrial survey [5], WLP reached an amazing compound annual growth rate (CAGR) of 20%. This in turn prompted the industry-wide adoption of WLP technologies.

The industry also started to become aware of the importance of WLP technologies for LEDs. One of the major industrial leaders, TSMC, announced an investment of $80 million to build an LED lab in 2009. On their website [6], it states: "TSMC's LED chip and packaging processes are executed at the wafer level, rather than the individual LED chip level to create significant potential cost reduction … silicon-based manufacturing to deliver a fast ramp, high yield and narrow bin distribution". It is obvious that "low cost, fast throughput

From LED to Solid State Lighting: Principles, Materials, Packaging, Characterization, and Applications, First Edition.
Shi-Wei Ricky Lee, Jeffery C. C. Lo, Mian Tao, and Huaiyu Ye.

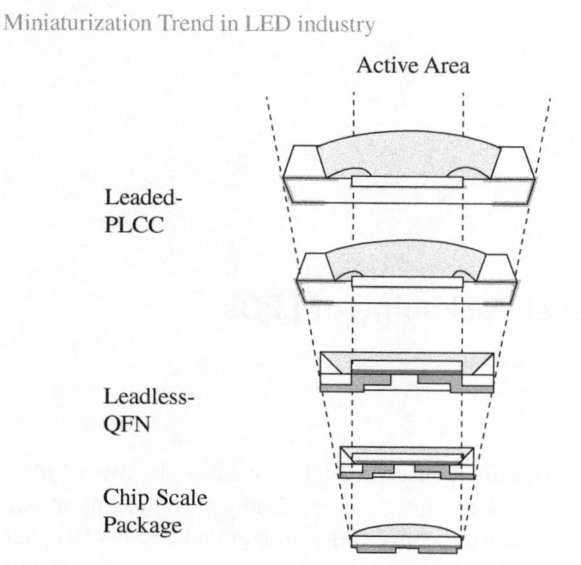

Figure 3.1 Miniaturization trend in the LED industry. Source: Luminleds.

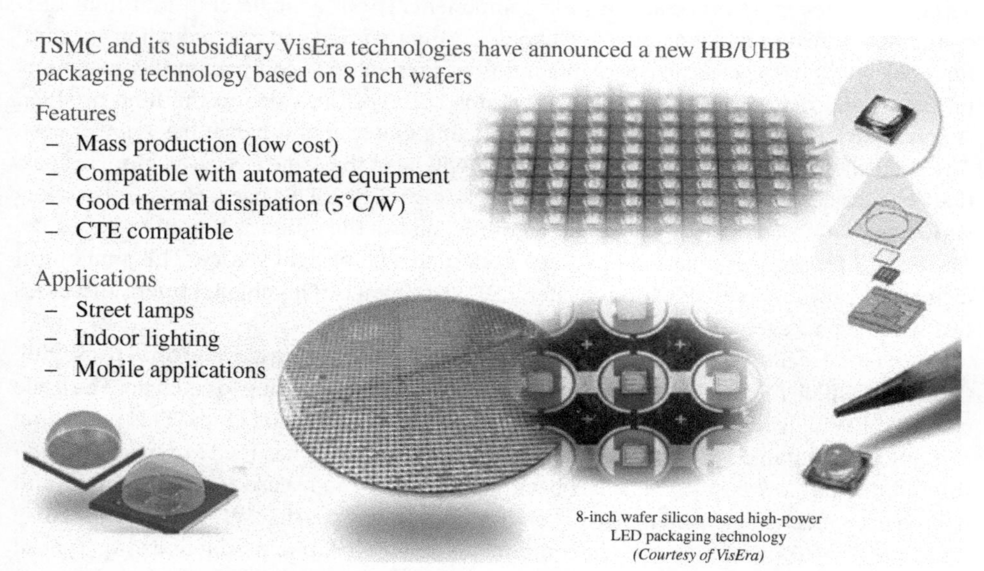

Figure 3.2 Current industrial practice of LED WLP [6]. Source: DIGITIMES. September 2009. VisEra announces new LED packaging technology. https://www.digitimes.com/news/a20090917PD207.html (cited date 3 September 2020).

and high yield" are the incentives for the LED industry to move toward WLP. TSMC and its subsidiary VisEra then announced their first version of LED WLP, as shown in Figure 3.2.

In LED WLP, a silicon submount is used as a carrier. The basic functions of the silicon submount are similar to a metal lead frame or a ceramic substrate. It provides mechanical support, electrical circuitry, and a heat dissipation path. On the other hand, a silicon

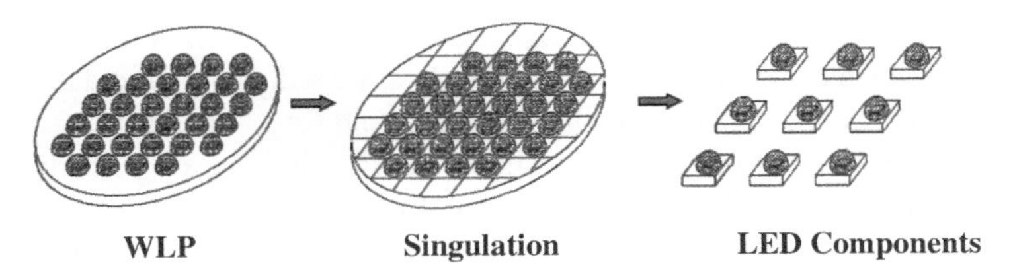

Figure 3.3 LED wafer level packaging process.

wafer is extensively used as the substrate of IC devices. This provides a platform for system integration [7, 8]. It is possible to add more functions to the LED package by fabricating different devices on the silicon submount. One common example is integrating a Zener diode for electrostatic protection [9]. Kim and Lee fabricated a photodetector on the silicon submount to monitor the optical output of the LED [10].

An LED WLP offers a better thermal management solution. The thermal conductivity of the silicon wafer is relatively high. Besides, it is common to thin a wafer by grinding and polishing. The package has a low thermal resistance because of the high thermal conductivity of the silicon and the short heat dissipation path. As discussed in Chapter 2, there are tough requirements on the quality of the substrate in packaging flip-chip LEDs. The silicon submount offers a solution to these technical challenges, as, because it is flat, fine features can be defined on it easily by a standard photolithographic process.

An LED WLP can be divided into two parts: silicon submount fabrication and LED packaging at the wafer level. Fabricating silicon submounts involves some fundamental processes: photolithography, wafer etching, through silicon vias (TSVs) etching and filling, and bond pad metallization. The packaging process includes chip mounting and interconnections, phosphor deposition, and encapsulation. All these are completed at the wafer level. The wafer is diced into single components or an array panel after encapsulation. The whole process is illustrated in Figure 3.3. These are the key enabling technologies for LED WLP [11, 12] and are discussed in Section 3.3.

Section 3.5 reviews two types of advanced LED wafer level packages which utilize the technologies described in this chapter. These packages have their unique features and fully utilize the advantages offered by WLP. The first one is a multichip package with a small form factor, while the other one has a cavity for die mounting to achieve a thin package. The application of WLP panels for general lighting is also be covered.

3.2 Chip Scale Packaging

CSP is one advanced packaging method in electronic packaging. Interestingly, except the package size, the name CSP does not provide any specific information on the package configuration. Also, only a few package types, such as flip-chip and fan-in wafer level chip scale packages (FI-WLCSP), have the same size chip. Therefore, IPC standard J-STD-012, "Implementation of Flip Chip and Chip Scale Technology," defines CSP. A CSP is defined as having an area of no more than 1.2 times the area of the original chip and has to be surface mount compatible.

There are numerous package types sorted as CSP according to the definition in the IPC standard. A flip-chip has the smallest form factor, but it is not always the best candidate. The size and pitch of the I/Os on the chip are usually very small and fine. Regular printed circuit board (PCB) technology cannot fabricate such small and fine pitch I/Os to directly support flip-chip assembly. A high-density interconnect substrate is required to redistribute the I/Os on the chip to a coarse area array. This helps improve the robustness of the subsequent PCB assembly process. There are various types of substrate used in CSPs. Some use a small lead frame to support the chip, and molding is required for protection. Others, such as organic, ceramic, or flexible substrates, are also widely used in CSPs. To maintain the package size no larger than 1.2 times of the chip size after encapsulation, a flip-chip is usually used as the interconnect between the chip and the substrate.

The high-density interconnect substrate and the related packaging processes are expensive, and will substantially increase the CSP cost. WLCSPs package the chips in a wafer level and eliminate the substrate by a redistribution layer (RDL) to address this issue. An RDL consists of a dielectric layer and a metal layer. Both the dielectric layer and the metal layer are deposited and patterned on a full wafer by a standard photolithography process, which is described in Section 3.3.1. After all packaging processes, the wafer is diced into packages. The package size is the same size as a chip. WLCSP has a high throughput and low packaging cost, thus it has been widely adopted in the industry.

LED CSPs are used in many applications, such as general lighting, flashlights, etc. A uniform back light panel is also fabricated by fine pitch LED CSPs. Similar to conventional IC packages, LED CSPs can be divided into two categories: with and without substrates.

Thermal management is one key consideration in LED packaging. The thermal conductivity of the substrate used in LED CSPs has to be high enough to provide an efficient thermal dissipation path, thus a ceramic substrate is commonly used. To maintain a small package size, the LED chip is bonded on the substrate by a flip-chip. Compared to conventional wire bond type LEDs, LED CSPs have a better thermal and optical performance.

In a typical fabrication process of LED CSPs with a ceramic substrate, flip-chip bumps are first fabricated on an Epi wafer. The bumping process is discussed in Chapter 2. The bumped Epi wafer is then bonded on a ceramic substrate. There are vias in the substrate to provide electrical connections between the top side and the surface mounted technology (SMT) pads on the bottom side. The substrate in then singulated and then bonded on a temporary panel. There is a gap between each singulated module on the temporary panel. A phosphor layer is deposited by dispensing or molding. Phosphor materials fill the gap so that the sidewalls of the LED chip are also covered by phosphor. This enhance the angular color uniformity of the package. Finally, the temporary panel is removed, and LED CSPs are fabricated after singulation.

The fabrication process of LED CSPs without a substrate is similar. Instead of bonding the bumped Epi wafer onto a substrate, the wafer is bonded onto a temporary panel. After phosphor deposition and singulation, LED CSPs without a substrate are fabricated.

Samsung first introduced their LED CSPs in 2014. Figure 3.4 shows one Samsung LED CSP package (LM101A) with conformal phosphor coating. They called this PoC CSP. All five surfaces of the sapphire are conformally covered by a uniform phosphor layer. This yields a uniform white light in a larger beam angle range. It does not have a substrate and there are large bond pads on the chip for regular SMT process. Figure 3.5 shows a schematic

Figure 3.4 Samsung PoC LED, LM101A. Source: Samsung.

Figure 3.5 Schematic diagram of PoC LED. Source: Samsung.

diagram of a PoC CSP. APT Electronics Co. Ltd. also launched conformal phosphor coating CSP, as shown in Figure 3.6.

Although conformal phosphor coating CSPs deliver excellent optical performance, the packaging process is costly and complicated. Each chip has to be separated evenly to maintain a uniform gap for phosphor coating on the sapphire sidewalls. Also, the efficacy level is not promising compared to conventional LED packages. Samsung announced FEC LEDs, which have a higher light efficacy than other CSPs.

Instead of five surfaces, only the top surface of the chip is covered by a uniform phosphor layer. The sidewalls of the chip are covered by titanium dioxide. The titanium dioxide prevents blue light leakage from the sidewalls (Figure 3.7). It directs the blue light upward to the top surface, which is covered by a phosphor layer for light conversion. Since there is no phosphor layer on the sidewall, the FEC LED is the same size as the LED chip. Figure 3.8 shows one Samsung FEC LED package (LH181B) which has titanium dioxide on its sidewall. The efficacy of this package is up to 190 lm/W (350 mA, 5000 K).

Figure 3.6 APT white LED chip CSP. Source: APT Electronics Co., Ltd.

Figure 3.7 Schematic diagram of FEC LED. Source: Samsung.

Figure 3.8 Samsung FEC LED, LH181B. Source: Samsung.

The LED CSP has a small footprint and better thermal performance than conventional lead frame type LED packages. However, it is not always cheap. The equipment and process costs of flip-chip bumping and bonding are much higher than that of conventional die bonding and wire bonding. Also, ceramic substrates are more expensive than lead frames.

3.3 Enabling Technologies for Wafer Level Packaging

The overall LED WLP process can be divided into two parts: silicon submount fabrication and the packaging process at the wafer level. This section briefly discusses the key enabling technologies for LED WLP.

3.3.1 Photolithography

A bare silicon wafer is a flat, brittle, and thin substrate. It cannot perform any functions without having deposited and patterned different layers of materials thereon. In LED WLP, silicone oxide or nitride is used as the passivation and electrical isolation, while metals such as aluminum are used for circuitry. In order to improve reflectivity, a silver layer is used on occasion. There are different thin film material deposition methods: thermal diffusion, sputtering, chemical vapor deposition (CVD), plasma enhanced chemical vapor deposition (PECVD), etc. The thin film material will cover the entire wafer surface after deposition. Patterns and features of the layers are then defined by photolithography.

In a regular photolithographic process, a thin film is first deposited onto the silicon wafer. A photosensitive polymer material, photoresist (PR), is then added thereon. A photochemical process occurs when PR is exposed to light with a certain wavelength, say ultraviolet (UV). There are two types of PR: positive PR and negative PR. Upon UV exposure, the former is soluble to the developing agent while the latter is cured and insoluble to the developing agent.

There are different methods to coat the PR material on the wafer. A uniform PR layer on a flat wafer can be easily obtained by spin coating. In this process, PR is applied to a wafer which is spinning at a high speed (up to 4000 r/min). The final thickness of the PR (normally 1–2 μm) is controlled by its viscosity and the wafer spinning speed. If the wafer is not flat, the final thickness of the PR on the wafer will be very nonuniform. The PR layer is usually found thicker in the region close to the cavity, which requires a longer exposure time to complete the photochemical process. Such long exposure times will result in overexposure for the region with a thinner PR layer, which will have an adverse impact on the subsequent PR development process. As a result, fine and small features cannot be patterned on the entire wafer successfully. Figure 3.9 shows the nonuniform PR thickness by spin coating on a wafer with cavities. To achieve a better result, PR is therefore deposited on a nonflat wafer by spray coating.

The wafer is aligned with a mask after applying PR. The mask is usually made of glass with a designed pattern for blocking UV light during the exposure process. The type of PR used, together with the mask, will define the final pattern. If a positive PR is used, the exact pattern as on the mask will be transferred to the wafer. On the other hand, an opposite pattern will result if a negative PR is used. After mask alignment, the PR is exposed under UV light. The PR, which is soluble to the developing agent after the photochemical process, is then removed. The pattern on the mask is transferred to the PR layer after the developing process.

Subsequently, the thin film is etched by the corresponding etchant. This can be done by either dry etching (by gases) or wet etching (by chemical solutions). The thin film covered by the PR is protected and only the region which is exposed to the etchant is etched. The PR is then stripped off and the thin film is patterned with the designed feature. The wafer is

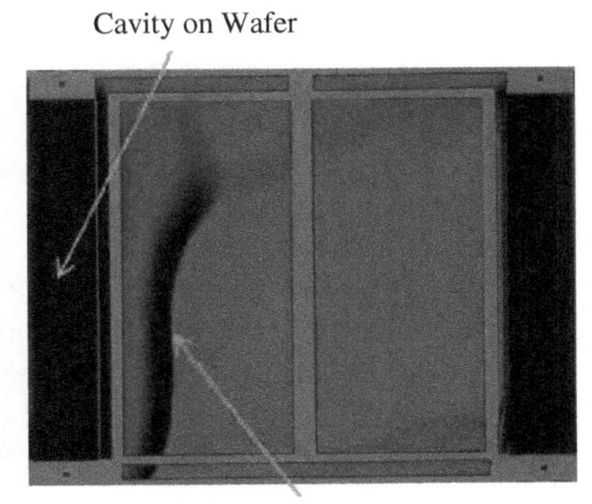

Figure 3.9 Nonuniform PR thickness by spin coating on a wafer with cavities.

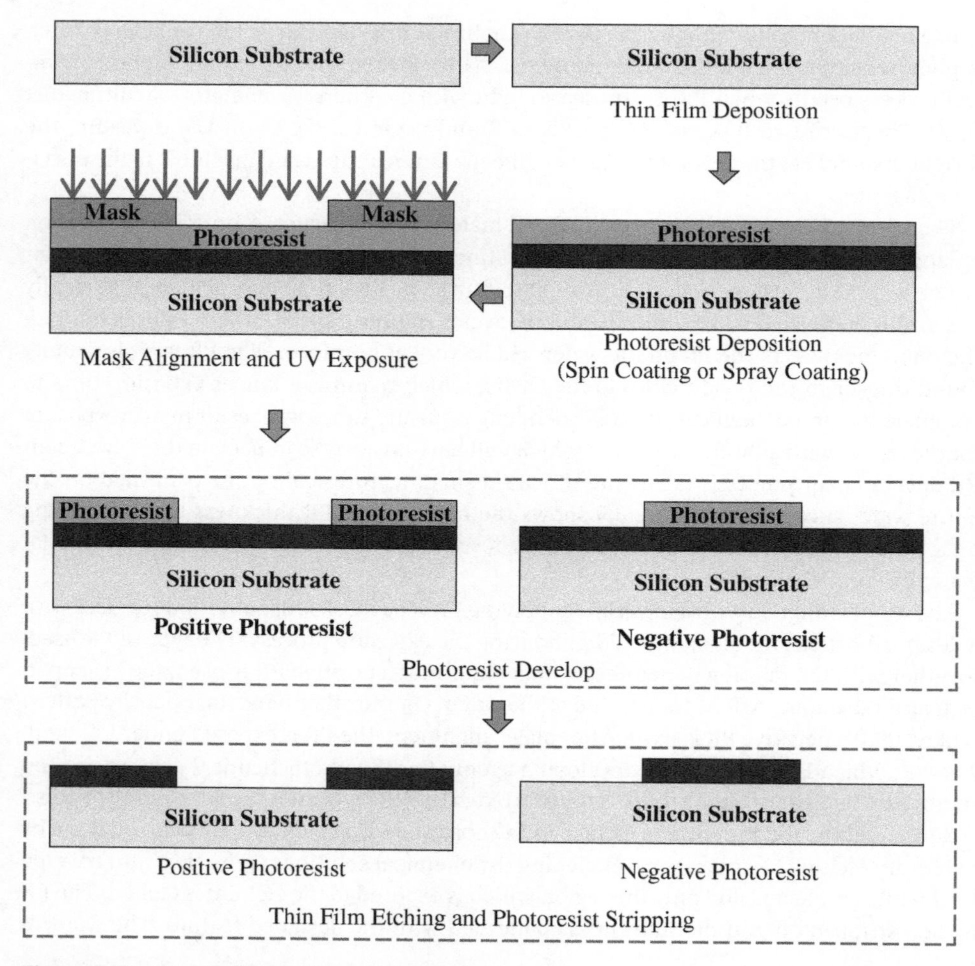

Figure 3.10 Standard photolithography process.

now ready for depositing another thin film layer. The overall photolithographic process is summarized in Figure 3.10.

In the aforementioned procedures, the entire wafer is exposed to the etchant during the thin film etching process. Some thin film materials may require very strong chemical etchants which the PR or other materials on the wafer may not be able to withstand. In this case, a lift-off process is required. The PR is patterned and developed on the wafer before the thin film deposition. The thin film materials on top of the PR are stripped together with the PR. The overall lift-off process is shown in Figure 3.11.

3.3.2 Wafer Etching

Wafer etching is commonly used for fabricating complicated three-dimensional (3D) structures on the silicon wafer, such as micro-channels and v-grooves, for MEMS optoelectronic devices. TSVs for vertical interconnects on silicon wafers are also required wafer etching.

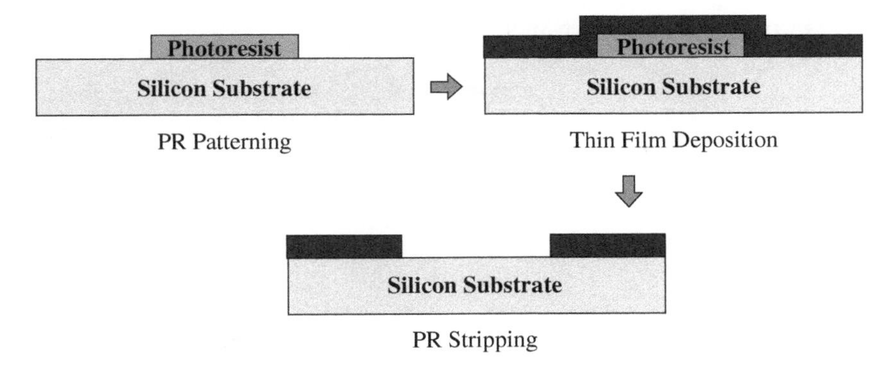

Figure 3.11 Lift-off process.

As a silicon wafer is made by a single silicon crystal, the crystal structure plays an important role in certain etching processes. Silicon atoms are bonded in a tetrahedral pattern to form a diamond-like lattice in the crystal structure. A 3D silicon structure is shown in Figure 3.12.

There are two types of etching process in silicon micromachining, namely dry etching and wet etching. In the wet etching process, the silicon wafer is placed into some etching solutions. The etching process is either isotropic or anisotropic, depending on the type of etchant used. For isotropic etching, the etch rate of silicon is independent of the crystal plane direction. Therefore, a circular and round side wall is obtained. In fact, different crystal planes have different etching rates in anisotropic etching. The crystal planes which have the slowest etching rate are exposed after the etching process. Figure 3.13 illustrates the results obtained by isotropic etching and anisotropic etching [13]. In anisotropic silicon wet etching, {1 1 1} planes have the slowest etch rate. As a result, {1 1 1} planes are obtained after anisotropic silicon etching. Figure 3.14 shows the (1 1 1), (1 1 0), and (1 0 0) crystal planes in a given coordinate. The (1 0 0) wafer is extensively used in the industries and the corresponding {1 1 1} planes have a 54.7° inclination with the wafer top surface.

Figure 3.12 Silicon crystal structure.

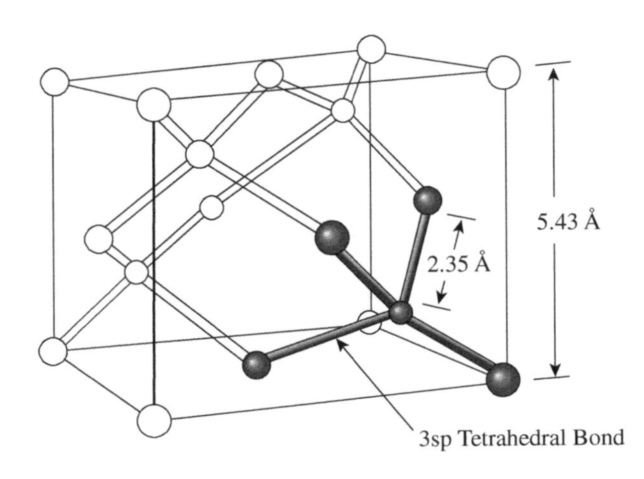

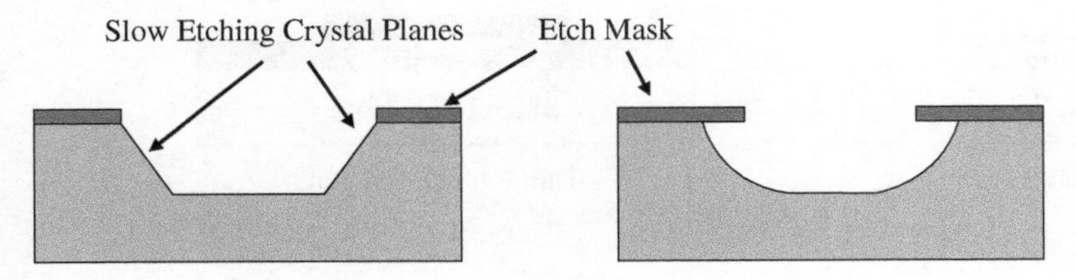

Figure 3.13 Anisotropic vs. isotropic etching.

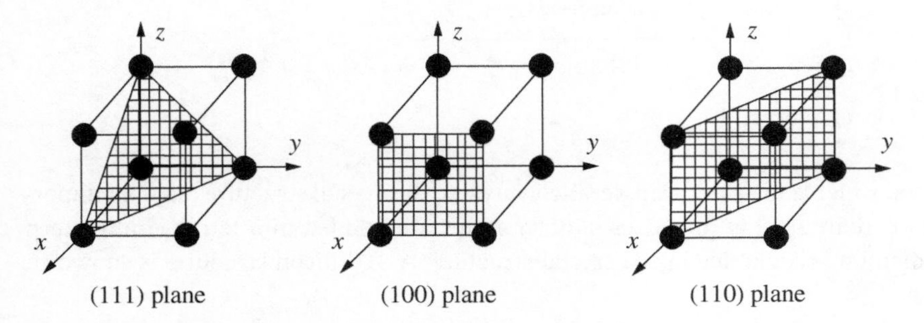

Figure 3.14 Crystal plan orientations.

There are two types of anisotropic silicon etching solution, namely potassium hydroxide (KOH) and tetramethylammonium hydroxide (TMAH). These two etchants have different properties [14]. A higher etch rate is obtained when using KOH. For a 30% KOH solution at 85 °C, the etch rate can reach 80 μm/h, whereas that of a 25% TMAH solution at 85 °C is only 18 μm/h. Besides, the anisotropy of these solutions are different. Anisotropy is the ratio between the etch rate of {1 1 1} planes and that of {1 0 0} planes. A KOH solution has a much higher anisotropy than that of a TMAH solution. A low anisotropy etching will cause a more severe undercut. This will create a larger pattern etched on the wafer than that designed on the mask. Another major difference between the two etchants is the complementary metal oxide semiconductor (CMOS) compatibility. Unlike TMAH, KOH contains potassium ions, which are completely incompatible with the CMOS process.

Since both KOH and TMAH are strongly alkaline, a regular PR cannot be used as the etching mask. Instead, an oxide or a nitride layer is first patterned as the etching mask if TMAH is used. In a KOH solution with a high etching rate of silicon oxide, only a nitride mask is suitable. Table 3.1 compares the two etchants. Figure 3.15 shows the process

Table 3.1 KOH vs. TMAH.

Etchant	KOH	TMAH
Etch rate	80 μm/h	18 μm/h
Anisotropy	1 : 200	1 : 12–1 : 35
Etch mask	Nitride	Nitride/oxide
CMOS compatibility	No	Yes

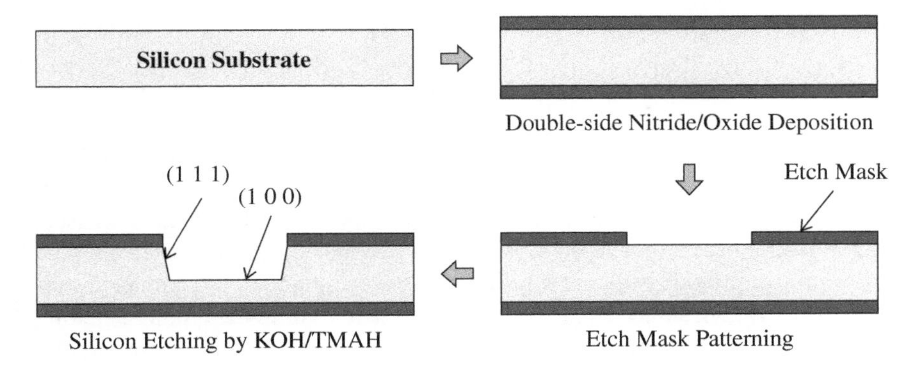

Figure 3.15 Anisotropic silicon wet etching process flow.

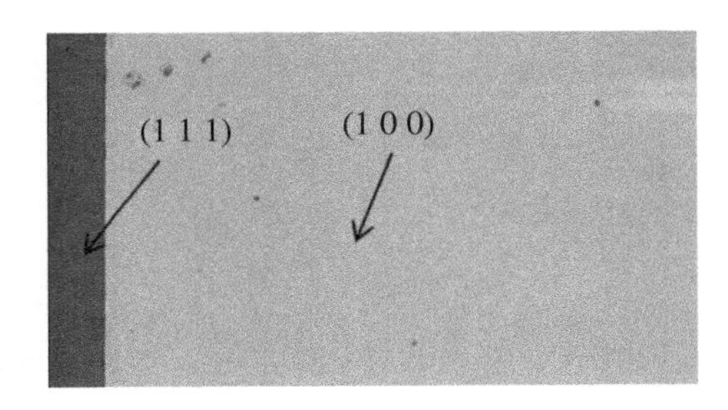

Figure 3.16 Anisotropic silicon wet etching result.

flow of anisotropic silicon etching and Figure 3.16 shows a typical silicon etching result. Anisotropic wet etching can be used to etch through the silicon wafer to form vias. However, the vias are trapezoidal with incline side walls occupying larger areas. It cannot form TSVs with a high aspect ratio.

Both KOH and TMAH are hydroxide and their chemical etching mechanisms are similar [14, 15]. First, the silicon is oxidized by the hydroxyl group in the etchant:

$$Si + 2OH^- \rightarrow Si(OH)_2{}^{2+} + 4e^-$$

Water is then reduced to form hydrogen bubbles and the hydroxyl group:

$$4H_2O + 4e^- \rightarrow 4OH^- + 2H_2$$

Silicate further reacts with hydroxyls to form a water-soluble complex:

$$Si(OH)_2{}^{2+} + 4OH^- \rightarrow SiO_2(OH)_2{}^{2-} + 2H_2O$$

The overall redox reaction of the silicon etching process by hydroxide is described as:

$$Si + 2OH^- + 4H_2O \rightarrow Si(OH)_2{}^{2+} + 2H_2 + 4OH^-$$

From these chemical reactions, it can be noted that hydrogen bubbles are generated in the etching process. The bubbles may adhere on the wafer surface, which will affect the quality of the final etch plane. Therefore, it is necessary to have a good agitation to remove these hydrogen bubbles during the etching process.

In some applications, vertical side walls are required. One common example is drilling high aspect ratio TSVs on the silicon wafer for vertical interconnections. These complicated small features cannot be fabricated on a silicon wafer by regular anisotropic wet etching. In this case, dry etching or laser drilling is used.

The Bosch deep reactive ion etching (DRIE) process is extensively used to etch silicon wafer to form TSVs, blind vias, or other features with vertical side walls [16, 17]. The whole process involves two key steps, namely isotropic etching of silicon and passivation deposition. The silicon is first etched by an isotropic etching gas, SF_6. SF_6 is a strong etchant which also etches regular PRs. Similar to wet etching, an oxide or nitride etch mask is used. After etching the wafer to a certain depth, a passivation polymer C_4F_8 is deposited onto the entire wafer to protect the side wall and top surface. One etching cycle is completed subsequently. The wafer is then subjected to multiple etching cycles until the required etching depth is reached. Figure 3.17 shows the schematic diagram of a typical DRIE process and Figure 3.18 shows a cross-section view of a TSV.

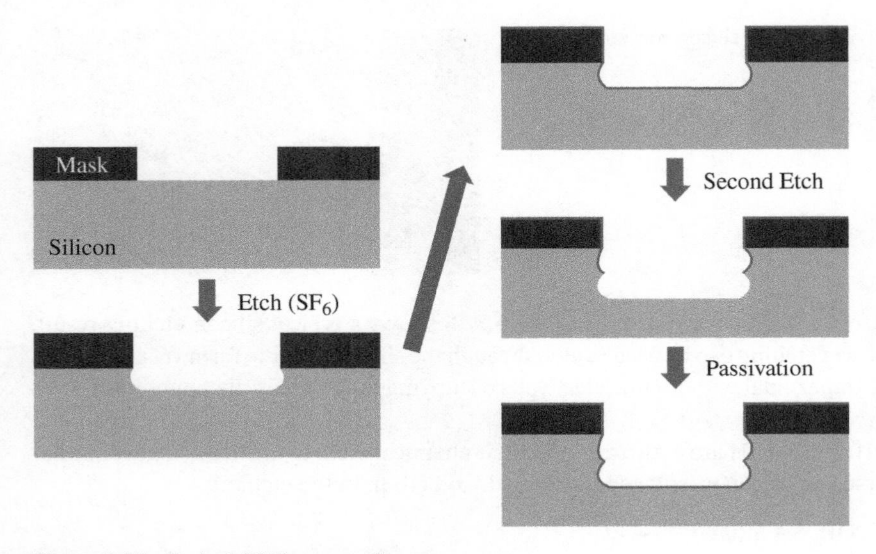

Figure 3.17 Typical DRIE process flow.

Figure 3.18 Through silicon via by DRIE.

3.3.3 TSV Filling

As discussed in Section 3.3.2, vias can be formed on a silicon wafer by either wet etching or dry etching. The vias formed by wet etching are trapezoidal in shape, which occupies a large area. Also, high aspect ratio vias with straight side walls can be obtained by dry etching.

In order to obtain a vertical interconnection, the vias are required to be filled with electrically conductive materials. The TSVs can be filled with different materials, such as copper, gold wire, conductive polymer, etc. Among these, copper-filled TSVs is most commonly used in LED WLP packaging. They have a good electrical conductivity and the thermal conductivity of copper is very high. Therefore, copper-filled TSVs can also serve as thermal vias in LED WLP. Any heat generated from the chip can be dissipated effectively through the copper-filled TSVs.

Figure 3.19 shows a typical process flow of filling TSVs with copper by electroplating. Through vias or blind vias are first formed by DRIE. A passivation layer is then deposited on the top surface of the wafer and the side walls of the vias for electrical isolation purposes. An adhesion layer – normally titanium tungsten (TiW) – and a copper seed layer, which are usually very thin with a thickness of less than 1 μm, are sputtered onto the wafer. The vias are then filled with copper by electroplating. Wafer grinding and chemical mechanical polishing (CMP) are then carried out to remove the excess copper on top of the wafer and expose the vias to the bottom side of the wafer. Figure 3.20 shows 25 μm diameter blind vias with 125 μm depth filled with copper by an electroplating process.

3.3.4 Bond Pad Metallization

Metal thin films are deposited onto the wafer submount for circuit routing. Bond pads are then defined on the metal layer for interconnects between the submount and other

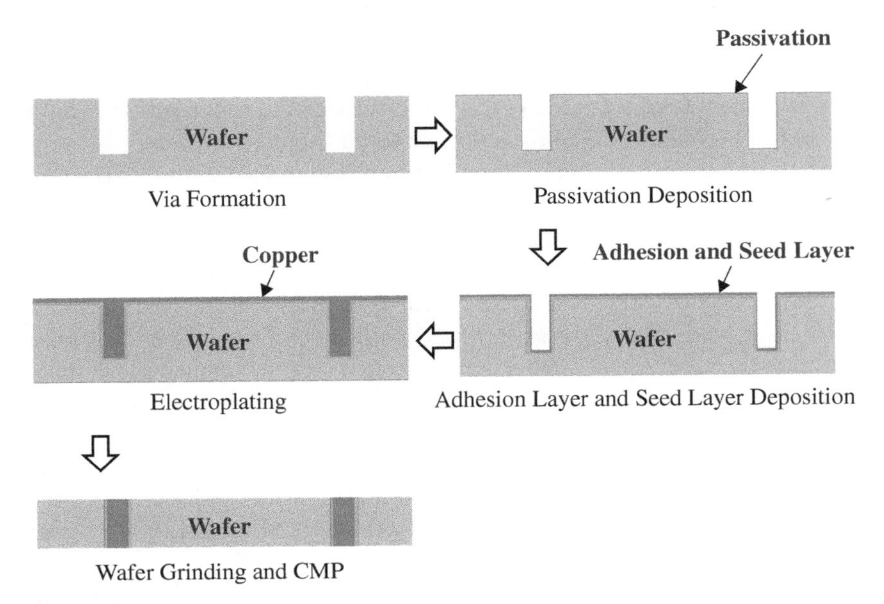

Figure 3.19 Via filling process.

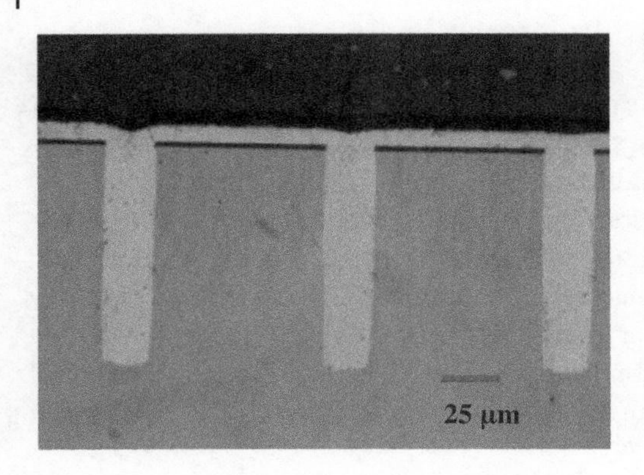

Figure 3.20 Copper-filled TSV.

components. In LED WLP, wire bonds and flip-chip bumps are commonly used interconnection methods between the chip and the submount. During board level assembly, the package or panel is usually connected to the board by soldering. It is necessary to have a proper bond pad metallization for different interconnection methods.

Aluminum is a widely used material for circuit routing and light reflection on the wafer in LED WLP. A 1–2 μm thick aluminum thin film is deposited on the wafer by either sputtering or CVD. Gold wire and aluminum wire bonding are compatible with aluminum. A thin layer of immersion gold is usually deposited on top of the aluminum bond pad for protection. Aluminum bond pads are also compatible with thermionic bonding of flip-chip with gold stud bumps, as its bonding mechanism is similar to that of gold wire bonding.

As discussed in Chapter 2, Au20Sn solder is one widely used solder material in flip-chip LED and eutectic die bonding. In board level assembly, lead free solder such as SAC 305 is extensively used. However, aluminum bond pads on the silicon submount are not compatible with solder materials. Intermetallic compound (IMC) cannot be formed between aluminum and the solder. It is necessary to deposit another metal on top of the aluminum for soldering. Common solder bond pad materials are copper and nickel.

During the soldering process, a portion of the bond pad material is consumed to form IMC with the solder. Therefore, it requires a thicker metal layer on the bond pad (normally >5 μm). Although copper and nickel thin film can be deposited onto the wafer by CVD or other methods, it is not cost effective to deposit a thick layer for soldering purposes. A thick bond pad layer can be achieved by plating. There are two common plating methods, namely electro-copper and solder plating, as well as electroless nickel plating.

The process flow of electro-copper and solder plating is shown in Figure 3.21. An adhesion layer (TiW) and a seed layer (copper) are first deposited. The layers cover the entire wafer so that all aluminum bond pads are electrically shorted together temporarily. A thick PR is deposited and patterned according to the bond pad locations. The PR mask serves as the mold for the copper and solder plating. After that, copper is plated on the bond pad followed by solder plating. Since the seed layer on bond pads are exposed to the plating solutions and the remaining parts are covered by the PR mask, copper and solder are only plated on the bond pads. The PR mask is stripped off thereafter. The temporary seed layer and adhesion layer are etched to prevent a short circuit. At this stage, the solder is still in a mushroom shape. The wafer then goes through a typical reflow process. During reflow, the solder melts and solder bumps are formed.

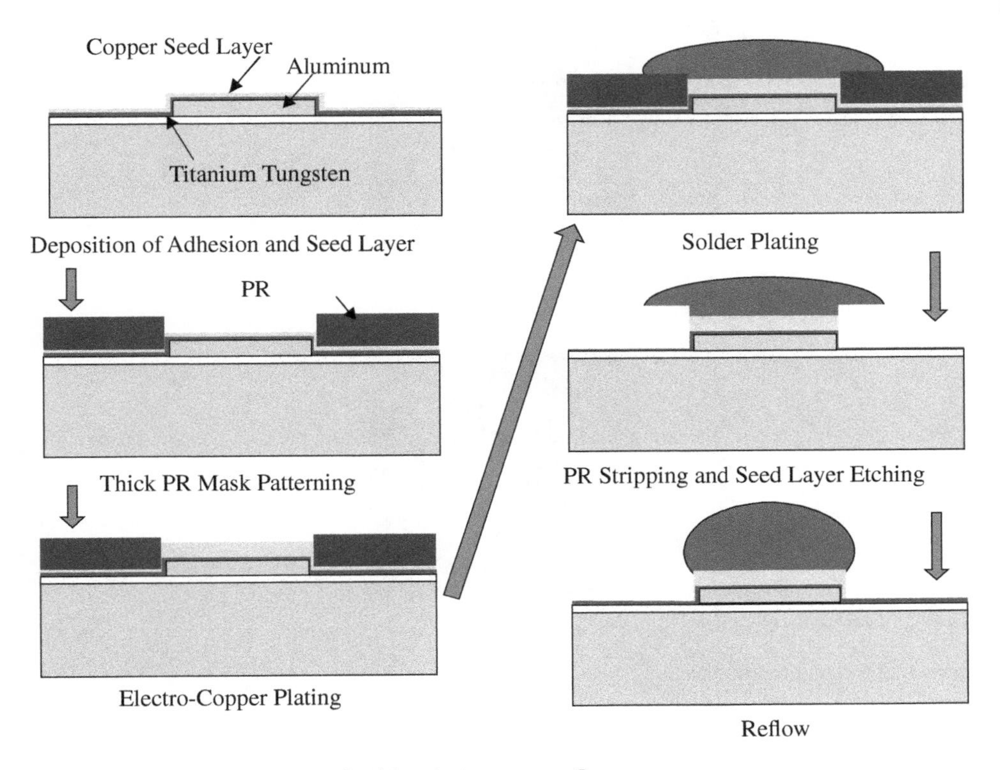

Figure 3.21 Electro-copper and solder plating process flow.

This electroplating process requires an extra mask to define the PR mold. It is not flexible and has a higher processing cost. In contrast, electroless nickel plating does not require an additional mask. In electroless nickel plating, the aluminum bond pads are first cleaned by weak alkaline and acid. Such cleaning removes the contaminants and oxide on the bond pads and exposes a fresh aluminum surface for subsequence processes. The aluminum bond pads are then activated by zincation, namely depositing a thin layer of zinc onto the bond pad. In the zincation process, aluminum is dissolved in the solution in exchange for zinc deposition. Therefore, the zincation time must be well controlled. A short zincation time will result in insufficient zinc activation. Finally, the entire aluminum bond pad is dissolved into the solution if the zincation time is too long. The zincation process should normally be completed in 5–10 seconds. Figure 3.22 shows the bond pads after zincation.

After zincation, the wafer is ready for electroless nickel plating. The plating process is photophobic, hence proper light shielding is required. Besides, the plating solution must be heated up to a temperature above 85 °C. Figure 3.23 shows the electroless nickel plating results.

Most PRs cannot withstand the high temperature of the electroless nickel plating solution and will be peeled off during the plating process. Nickel will then be plated to those areas supposed to be covered by PR. It is necessary to select the correct type of PR for masking purposes.

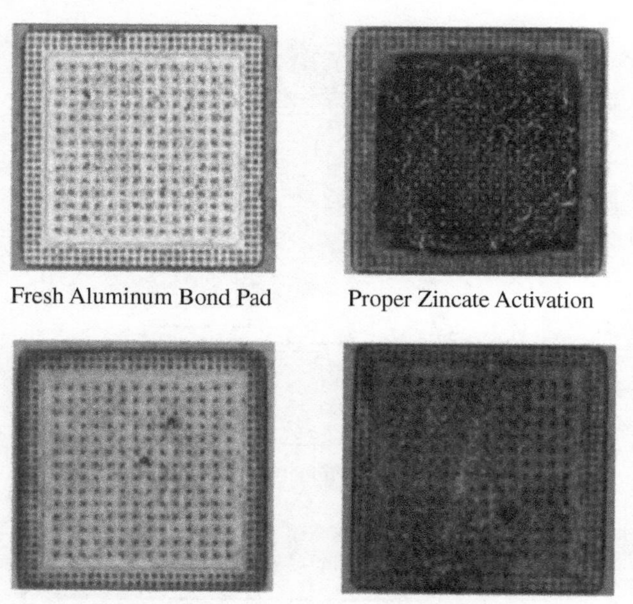

Figure 3.22 Zincation process.

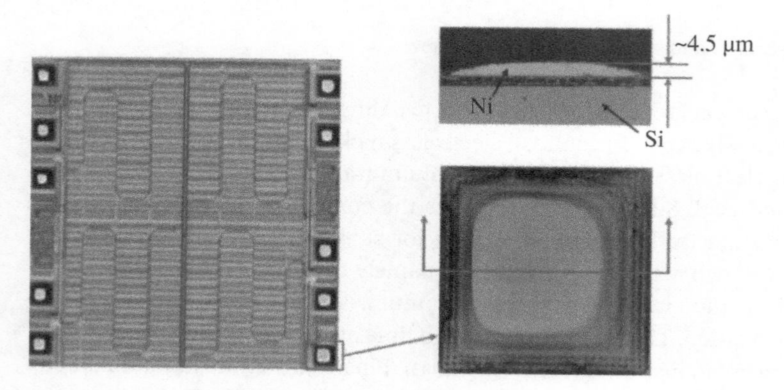

Figure 3.23 Electroless nickel plating results.

3.3.5 Wafer Level Phosphor Deposition Methods

There are different methods of phosphor deposition for LED WLP packaging. This section discusses three innovative methods in detail, namely stencil printing with a silicon stencil, waffle pack remote phosphor film, and remote phosphor dispensing on a pre-encapsulated silicone lens. Other methods, such as spray coating, electrophoretic deposition, and compression molding, are also applicable to LED WLP, and are discussed in Chapter 2.

3.3.5.1 Wafer Level Phosphor Stencil Printing

In this phosphor stencil printing method, a silicon stencil is used. The aperture is defined by a regular photolithographic process and etched by a deep reaction ion etch. The precise aperture size allows for better control of the amount of phosphor coated on the chip's surface and the side wall. Figure 3.24 shows the stencil fabricated on a silicon wafer with a thickness of 200 μm.

The stencil printing method is applicable to both flip-chip [18–20] and wire bond type LEDs [21]. The phosphor printing process of LEDs with wire bonds is more difficult to control than that of the flip-chip LEDs. The interconnects of the flip-chip LEDs are all under the chip. Only the back side of the chip, normally the sapphire substrate, is exposed. The interconnects are protected by the sapphire substrate and will not be damaged by the printing process. In fact, if wire bonds are used, the tiny gold wire interconnects may be damaged during the printing process. Therefore, the stencil design for wire bond type LEDs is crucial. It must be thick enough to cover the LED chip and the wire bonds. The thickness of the stencil also controls the thickness of the phosphor layer on the LED chip. A thicker stencil will give a thicker phosphor layer. In other words, the thinnest phosphor layer which can be printed on the chip surface is solely controlled by the wire bond loop height. The stencil aperture should also be large enough to accommodate the second bonds on the silicon submount. Figure 3.25 shows an appropriate stencil design schematically.

A reserve wire bond process is used to achieve a low loop height. The first bond is on the wafer submount and the second bond is on the LED chip. The reverse wire bond has a very low profile which allows a thinner layer of phosphor to be printed on the chip surface.

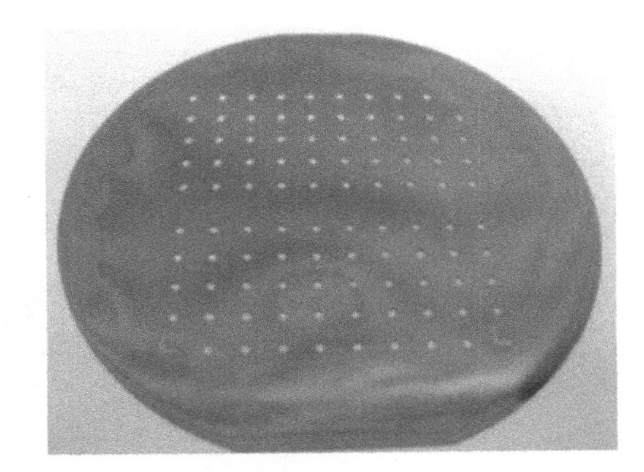

Figure 3.24 Silicon stencil.

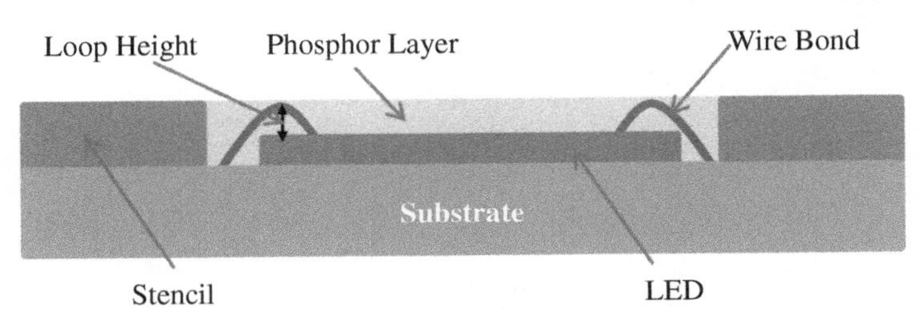

Figure 3.25 Stencil design.

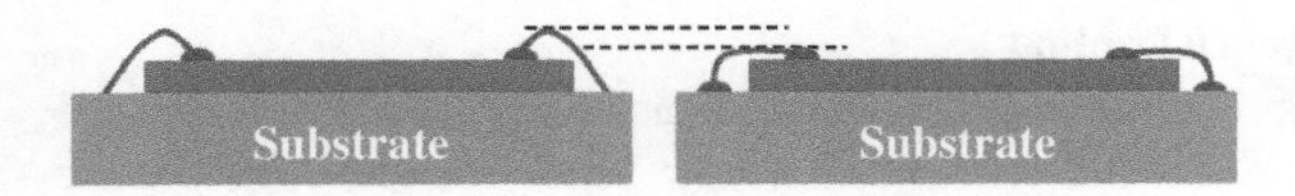

Figure 3.26 Regular and reverse wire bond.

Figure 3.26 compares the loop heights of the regular wire bond and the reverse wire bond schematically.

In order to obtain a low loop height, the first bond is made on the silicon wafer submount and the second bond is made directly on the LED die bond pad. Figure 3.27 shows that the loop height is around 25 μm. However, Figure 3.28 shows that the wire touches the trace on the LED die next to the bond pad. This may result in a short circuit.

In order to prevent a short circuit, a gold stud is first made on the LED die bond pad. This provides a certain standoff between the gold wire and the trace on the LED chip.

Figure 3.27 Reverse wire bond without gold stud (side view).

Figure 3.28 Reverse wire bond without gold stud (close-up at LED die bond pad).

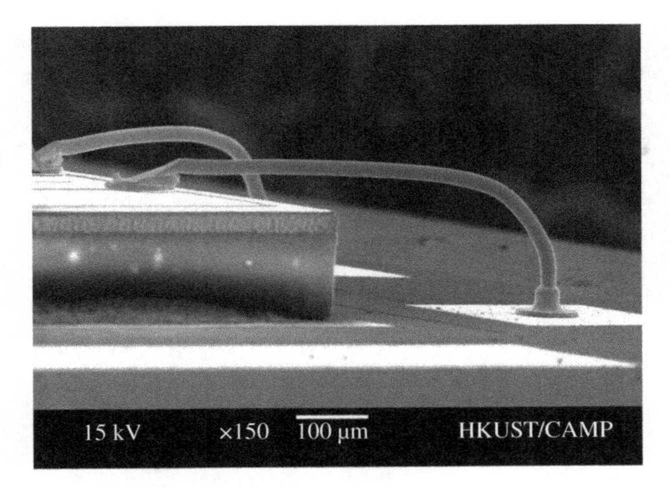

Figure 3.29 Reverse wire bond with gold stud (side view).

Figures 3.29 and 3.30 show the reverse wire bond with a gold stud. The wire bond loop height is then increased from 25 μm to 40 μm. Since the thicknesses of the LED die and the die attach adhesive are 150 μm and 6 μm, respectively, the total thickness is still less than 200 μm even if the wire loop height is 40 μm. This is smaller than the silicon wafer stencil thickness. As a result, the chip and the wire bonds are completely located inside the aperture of the stencil and will not be damaged during the phosphor printing process.

Following that, the silicon wafer stencil is aligned with the submount. The LED chip and the wire bond are located inside the aperture, as shown in Figure 3.31. A small amount of UV curable epoxy is dispensed into the aperture. Dry phosphor particles are then screen printed into the aperture. The mixture is half-cured under a UV light source. The submount and the stencil are separated with the aid of stamp pins. The mixture is fully cured in the end. The printing process is illustrated in Figure 3.32. This method does not require mixing

Figure 3.30 Reverse wire bond with gold stud (close-up at LED die bond pad).

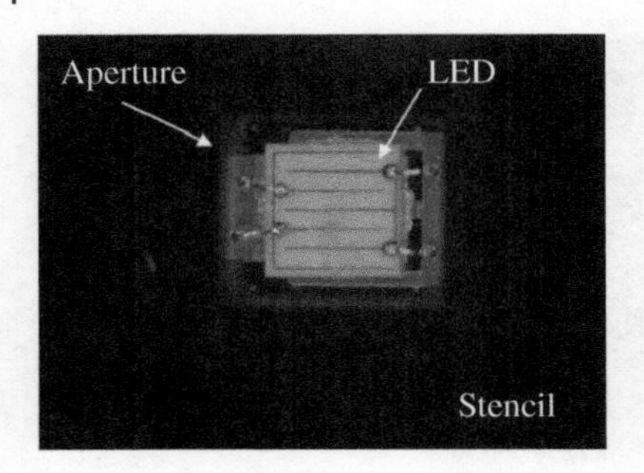

Figure 3.31 Aligned submount and stencil.

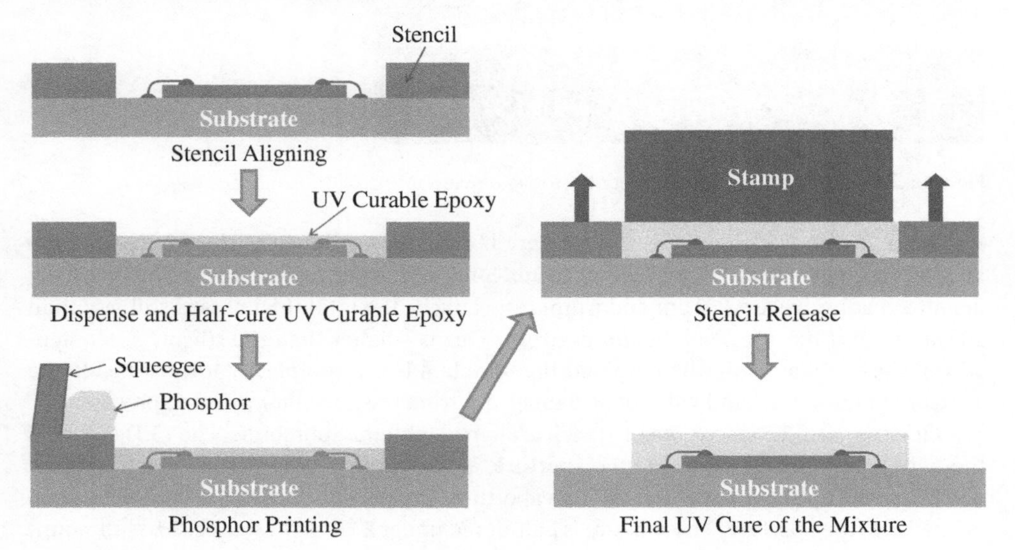

Figure 3.32 Stencil printing process flow.

the phosphor with the epoxy before the printing process. Instead, dry phosphor particles are used during the printing process. After printing, there will not be slurry residue on the stencil surface. The excess phosphor can also be collected and re-used afterward.

The printing results of both flip-chip and wire bond type LEDs are shown in Figure 3.33. A uniform phosphor layer is deposited on the LED chip. Three-dimensional X-ray inspection on the wire bond type package is shown in Figure 3.34. It can be seen that gold wires are not damaged after the printing process.

3.3.5.2 Waffle Pack Remote Phosphor Film

The advantages of remote phosphor are discussed in Chapter 2. In LED WLP, a method called a waffle pack can be used to add a remote phosphor layer to the wafer panel [22, 23].

Unlike the compression molding method, a waffle pack remote phosphor layer is fabricated separately. The film is bonded to the silicon submount on which die mounting and interconnects are already completed. The gap between the die and the film is filled with silicone encapsulant, which protects the die and the interconnects from the environment. It also reduces the reflective index mismatch between the die and air, and hence improves

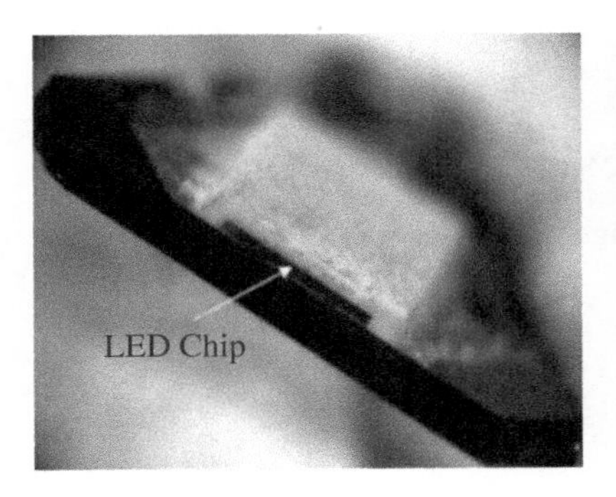

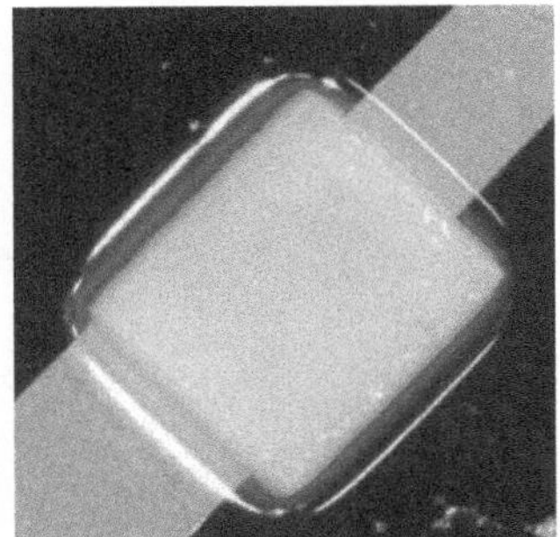

Figure 3.33 Stencil printing results.

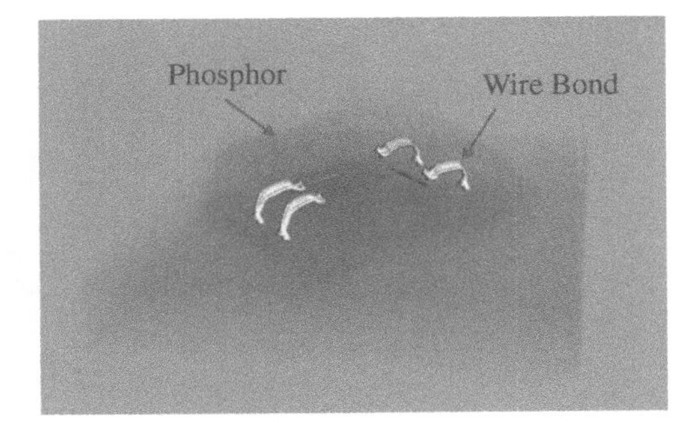

Figure 3.34 Three-dimensional X-ray inspection.

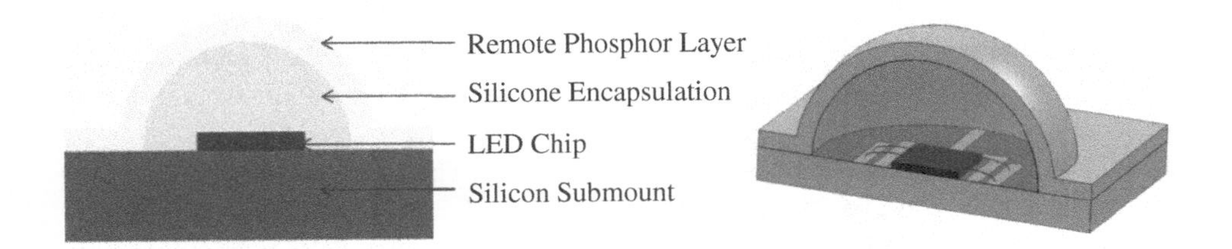

Figure 3.35 Schematic diagram of waffle pack design.

the light extraction efficiency. Figure 3.35 shows the schematic diagram of the LED WLP with a waffle pack.

Flexibility in design and manufacturing is the major merit of this method. Phosphor films with different blending ratios corresponding to different color coordinates are prefabricated. These phosphor films can be characterized and then matched with wafer panels for optical performance optimization.

The light pattern of an LED package is controlled by the lens design. A hemispherical dome shape is used for demonstration purposes. The shape of the waffle pack is determined by the mold design. By changing the mold geometry, it is possible to achieve different light patterns. Figure 3.36 shows the procedures of fabricating the remote phosphor film. First, a thin layer of mold releasing agent is applied to the mold. Premixed phosphor slurry is then dispensed into a female mold (Figure 3.37). Air bubbles in the slurry are removed by

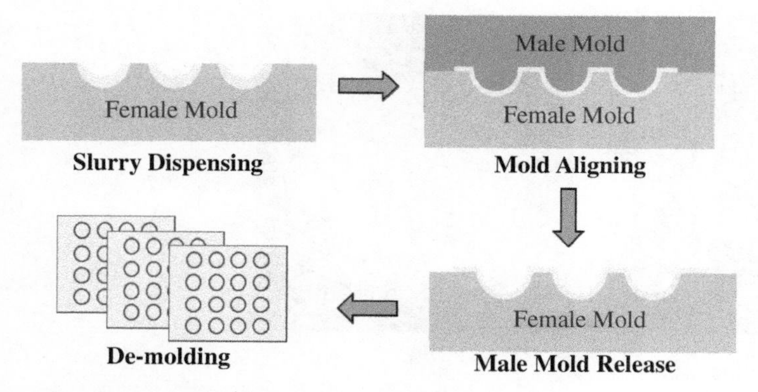

Figure 3.36 Remote phosphor layer fabrication process flow.

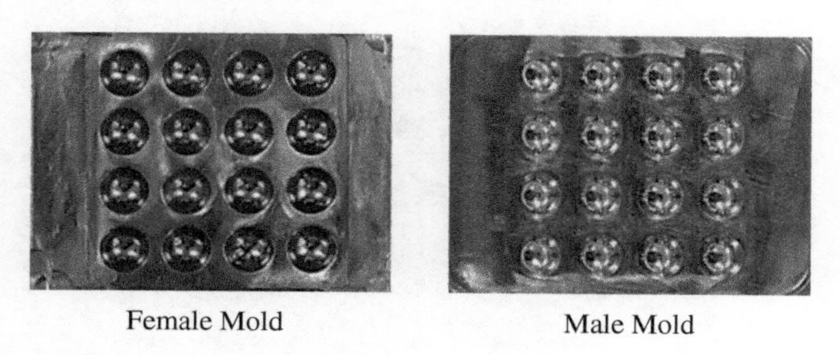

Figure 3.37 Waffle pack molds.

vacuum. A male mold (Figure 3.37) with a matching design is aligned and pressed against the female mold. After curing the slurry, a remote phosphor film is obtained, as shown in Figure 3.38.

A 4×4 wafer panel is used for demonstration purposes. The panel is assembled with the waffle pack remote phosphor film, and the process flow is shown in Figure 3.39. The prefabricated phosphor film is first placed in a mold. Silicone encapsulant is then dispensed into the shell cavity. The wafer panel is aligned with the mold followed by the curing process. A wafer panel with a waffle pack remote phosphor layer is obtained after de-molding, as shown in Figure 3.40.

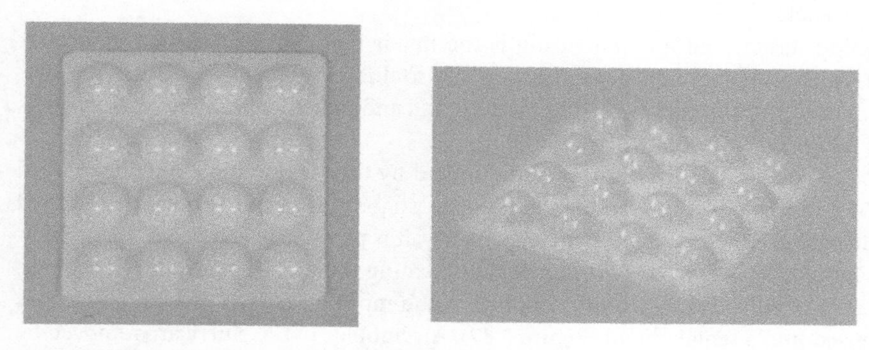

Figure 3.38 Prefabricated remote phosphor film.

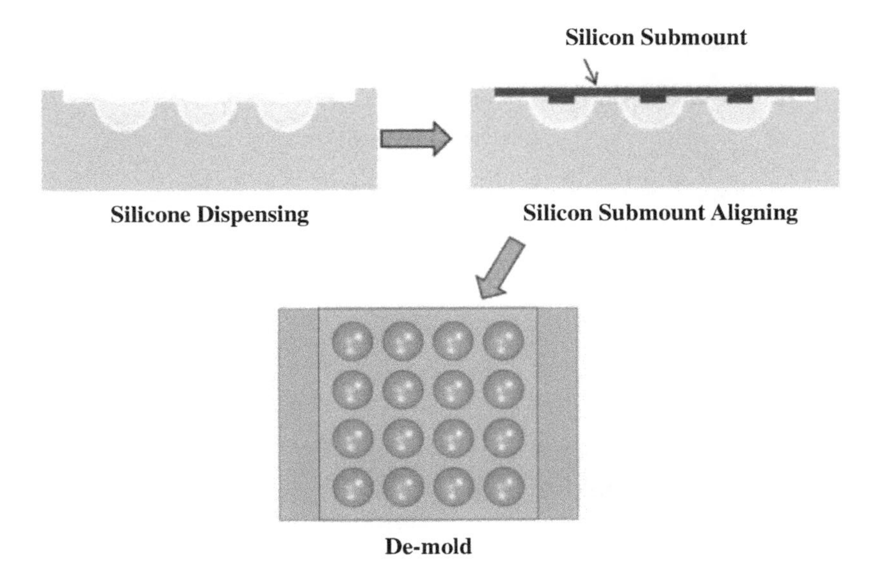

Figure 3.39 Waffle pack assembly process flow.

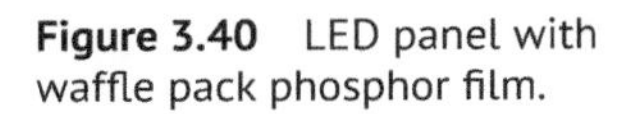

Figure 3.40 LED panel with waffle pack phosphor film.

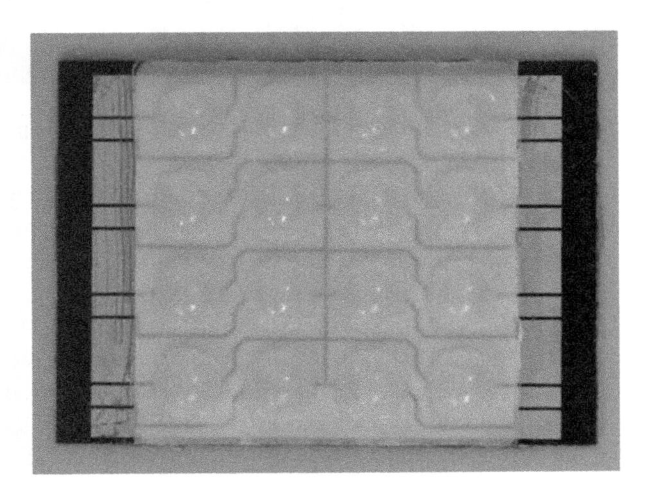

A cross-section inspection is carried out, as indicated in Figure 3.41. Note that phosphor particles are uniformly distributed in the film. No void or delamination is found in the encapsulant. As a result, a uniform white light from each unit on the panel is obtained, as shown in Figure 3.42.

The optical performance of the final wafer panel can be easily adjusted by changing the phosphor-to-silicone ratio in the film. Figure 3.43 shows the chromaticity coordinates of the samples prepared with different blending ratio. The corresponding correlated color temperature (CCT) ranges from 4500 to 8000 K.

3.3.5.3 Remote Phosphor Dispensing on a Pre-Encapsulated Silicone Lens

A remote phosphor layer can be deposited on the silicone lens by a simple dispensing method [24]. A silicone dome with a step is first fabricated on a substrate. The dome serves

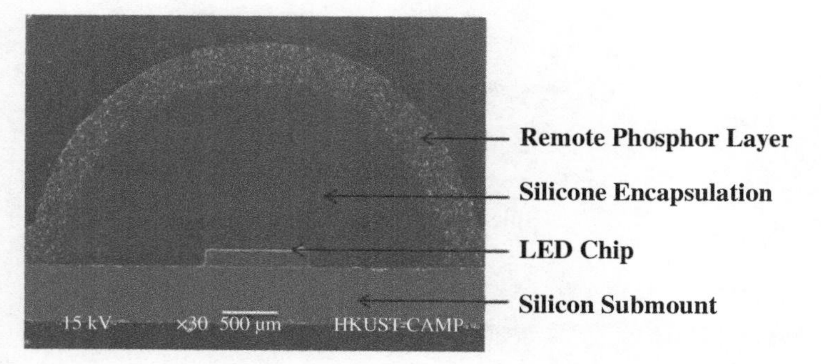

Figure 3.41 Cross-section inspection.

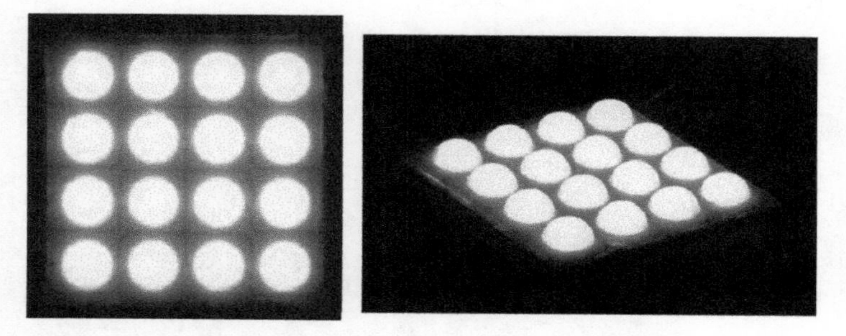

Figure 3.42 Uniform white light from LED WLP with waffle pack.

as the lens in a regular blue LED package. Phosphor slurry is then dispensed on the silicone lens. The slurry flows and spreads on the dome, and stops when it reaches the edge of the step. The unique step feature stops the slurry from overflowing and controls the geometry of the remote phosphor layer. Molding process is not required for applying the remote phosphor layer on top of the curve surface. Figure 3.44 shows the schematic diagram of the proposed dispensing method. The thickness of the remote phosphor layer can be easily adjusted by changing the dispensing volume. It can also be applied to wafer level LED packaging. The special designed lens can be molded onto a silicon wafer substrate using a regular compression molding process. This dispensing method provides a simple and flexible solution which can increase the throughput and reduce overall packaging and processing costs.

After the die bonding and wire bonding processes, the panel is ready for silicone lens molding. An LED array with 16 pieces of 5 mm × 5 mm packages is designed as the test vehicle for demonstration. The base radius of the silicone lens is 4 mm. A unique step feature is introduced in the silicone lens design. The purpose of the step is to confine the flow of the phosphor slurry. The mold design is shown in Figure 3.45. The silicone lenses on the wafer panel are fabricated by a regular compression molding process. The silicone lens and the step are well defined after molding, as shown in Figure 3.46.

The silicone molding process is followed by the remote phosphor deposition process. Phosphor materials are first mixed with silicone to form slurry. The silicone-to-phosphor weight mixing ratio can be adjusted according to the optical performance requirements. The slurry is then dispensed on top of the silicone lens by a fixed volume dispenser. The

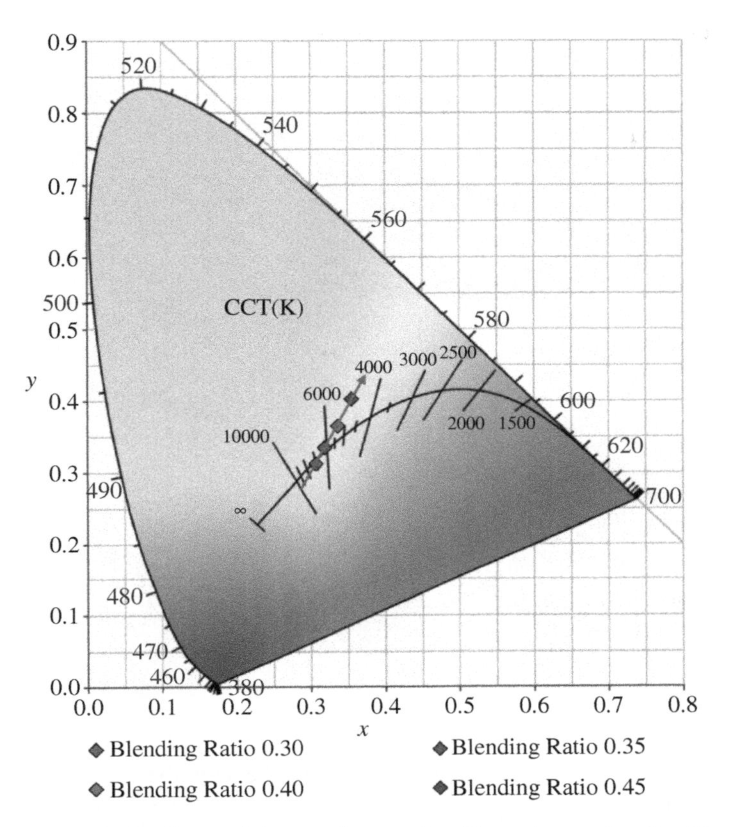

Figure 3.43 Chromaticity coordinates of waffle packs panels with different blending ratio.

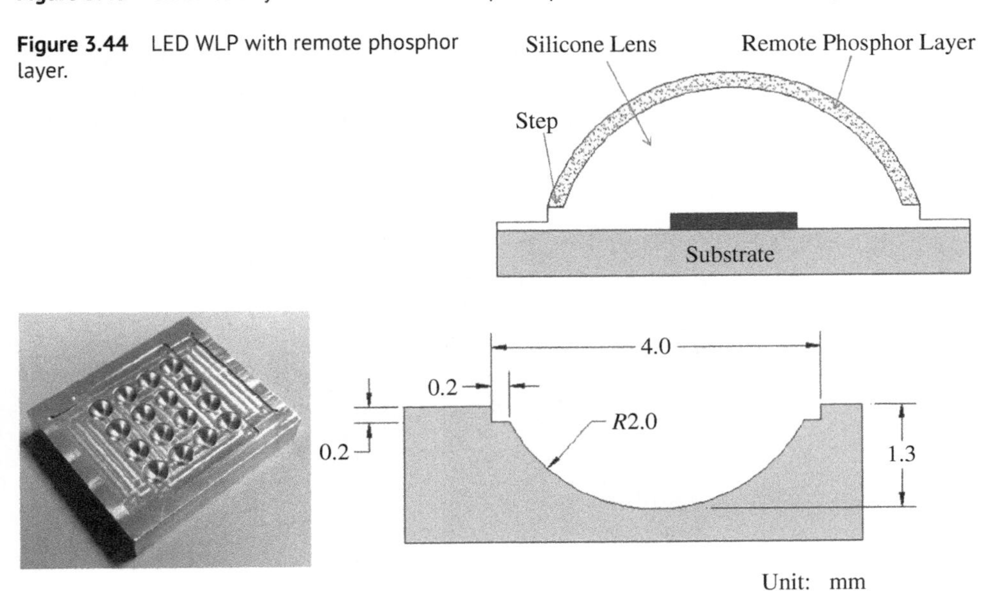

Figure 3.44 LED WLP with remote phosphor layer.

Figure 3.45 Mold design.

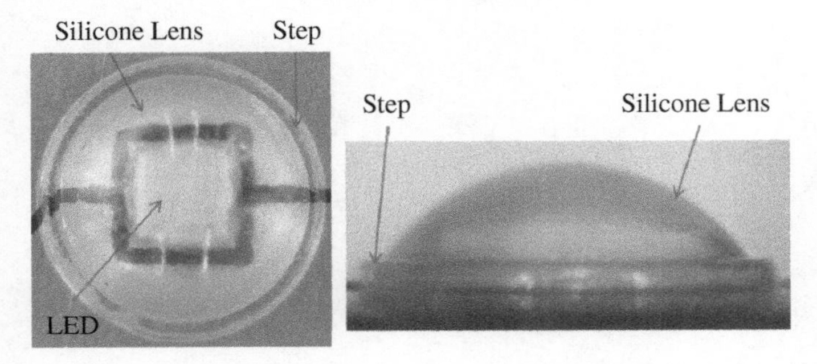

Figure 3.46 Silicone lens with a step.

Figure 3.47 Remote phosphor deposition results.

slurry spreads on the silicone lens. When it reaches the edge of the step, the flow stops. The step provides a boundary which helps confine the flow of the slurry, as shown in Figure 3.47. A uniform remote phosphor layer is formed on top of the silicone lens after curing.

Figure 3.48 shows the LED array panel after the remote phosphor dispensing process. The volume of the phosphor slurry is 0.0050 mL. It is observed that a uniform remote phosphor layer is deposited on each individual unit. Figure 3.49 shows the results, and uniform white light is obtained from the panel.

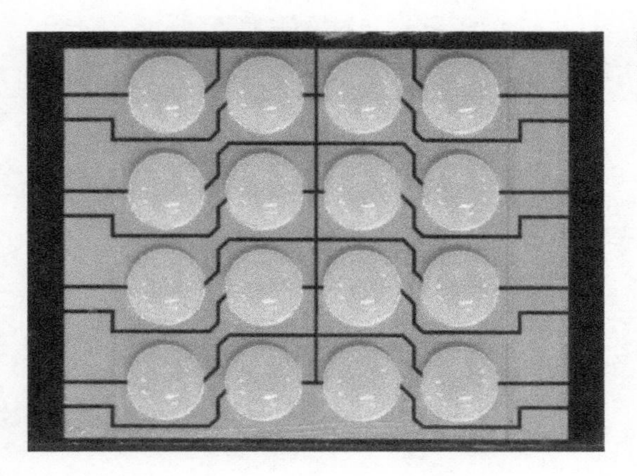

Figure 3.48 LED array panel with remote phosphor.

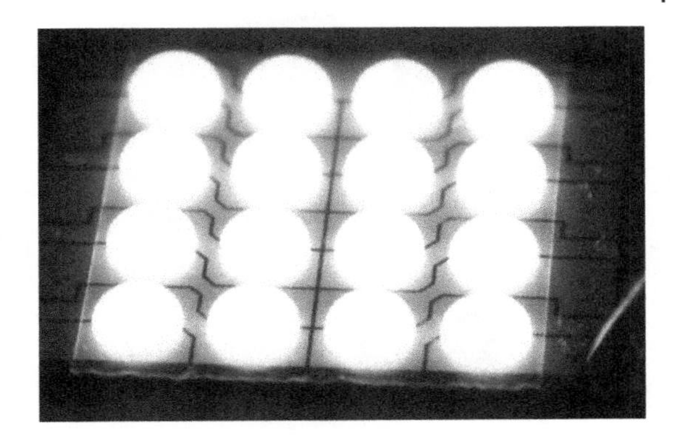

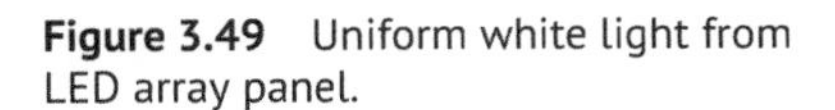

Figure 3.49 Uniform white light from LED array panel.

3.3.6 Moldless Encapsulation

In LED WLP, the encapsulation and the lens are usually formed by compression molding [25, 26]. The compression molding process is not flexible and requires expensive molds and equipment. It is cost effective only when the production volume is high. Zhang et al. propose a moldless encapsulation method [27–29]. In general, the encapsulation lens is simply formed by a regular dispensing process. The geometry of the lens can be easily controlled by the dispensing volume.

When an encapsulant droplet is dispensed on a flat surface, it spreads on the surface. The final geometry is determined by the contact angle between the encapsulant and the surface. It is difficult to control the contact angle precisely to achieve the designed lens shape. Besides, the contact angle is very small in most cases. As a result, a very flat lens which is not suitable for LED packaging is obtained. In the moldless encapsulation method, half cut trenches are fabricated on the wafer either by wafer dicing or DRIE. Those trenches serve as boundaries and constraints which stop the encapsulant flow, as shown in Figure 3.50. Figure 3.51 shows a lens array fabricated by this method. Figure 3.52 shows an example of utilizing moldless the encapsulation method in LED WLP.

The H/L ratio, as illustrated in (Figure 3.53), is an important parameter which controls the optical property of the package. A hemispheric lens which has a 0.5 H/L ratio is more preferable, because less light is trapped by total internal reflection. There are three parameters which control the H/L ratio of the lens fabricated by this moldless method. The first one is the viscosity of the encapsulant. A low viscosity encapsulant is easier to spread on the surface, and hence the trenches cannot stop the flow of the encapsulant efficiently. This results in overflowing and merging of the encapsulant with the adjacent component. Therefore, an encapsulant with a high viscosity is preferred.

The second parameter is the dispensing volume. A higher H/L ratio can be obtained by dispensing more encapsulant onto the wafer. However, if too much encapsulant is dispensed, the trenches cannot stop the flow. This also results in overflowing of the encapsulant. Figures 3.54–3.57 show the relationship between dispensing volume and the H/L on two different package sizes: 2.8 mm × 2.8 mm and 5 mm × 5 mm.

The package size is another parameter which determines the maximum H/L ratio that can be achieved by the moldless encapsulation method. When the package size increases, more encapsulant is required to obtain a high H/L ratio. The effect of gravity becomes more

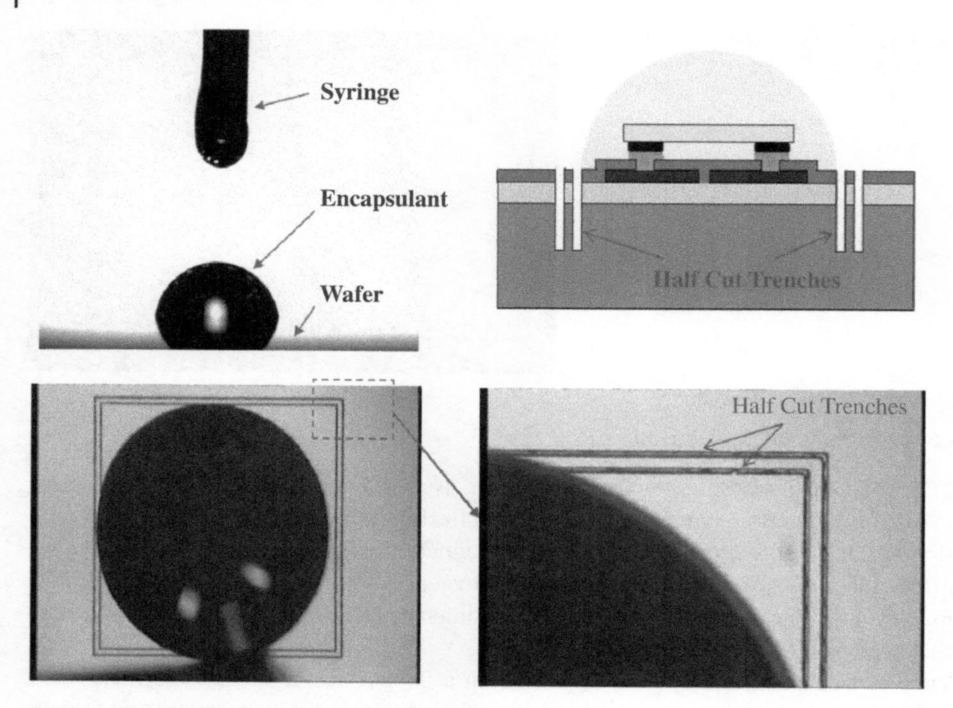

Figure 3.50 Moldless encapsulation method.

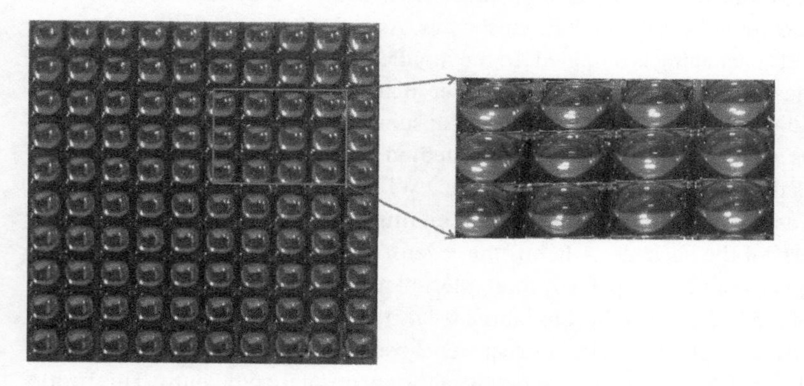

Figure 3.51 Silicone dome array (2.8 mm × 2.8 mm).

significant with the increase in package size. The encapsulant will collapse due to its weight. By comparing the results shown in Figures 3.54 and 3.55, it can be observed that the maximum H/L ratio of a 2.8 mm × 2.8 mm package is much higher than that of a 5 mm × 5 mm package. An H/L ratio of 0.47, which is very close to a hemispherical lens, can be obtained in a 2.8 mm × 2.8 mm package. In contrast, the maximum H/L ratio of a 5 mm × 5 mm package is only 0.36.

Figure 3.52 LED WLP with moldless encapsulation.

Figure 3.53 H/L measurement.

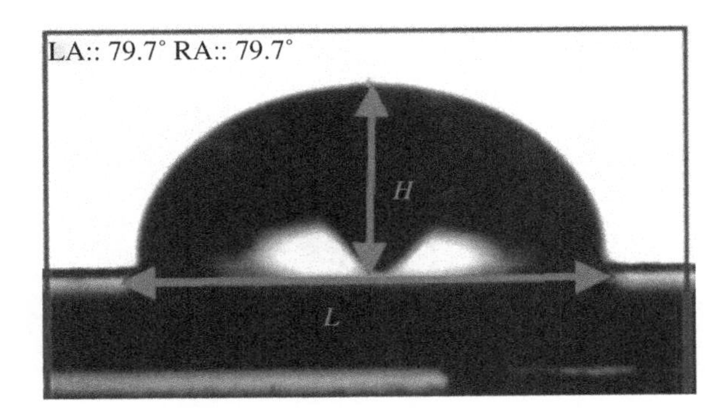

Shot Size = 1.5 mg; H/L = 0.12	Shot Size = 2 mg; H/L = 0.23	Shot Size = 2.5 mg; H/L = 0.31
Shot Size = 3 mg; H/L = 0.35	Shot Size = 3.5 mg; H/L = 0.38	Shot Size = 3.7 mg; H/L = 0.42
Shot Size = 4 mg; H/L = 0.44	Shot Size = 4.2 mg; $H L$ = 0.46	Shot Size = 4.5 mg; H/L = 0.47

Figure 3.54 Silicone dome geometry control (2.8 mm × 2.8 mm).

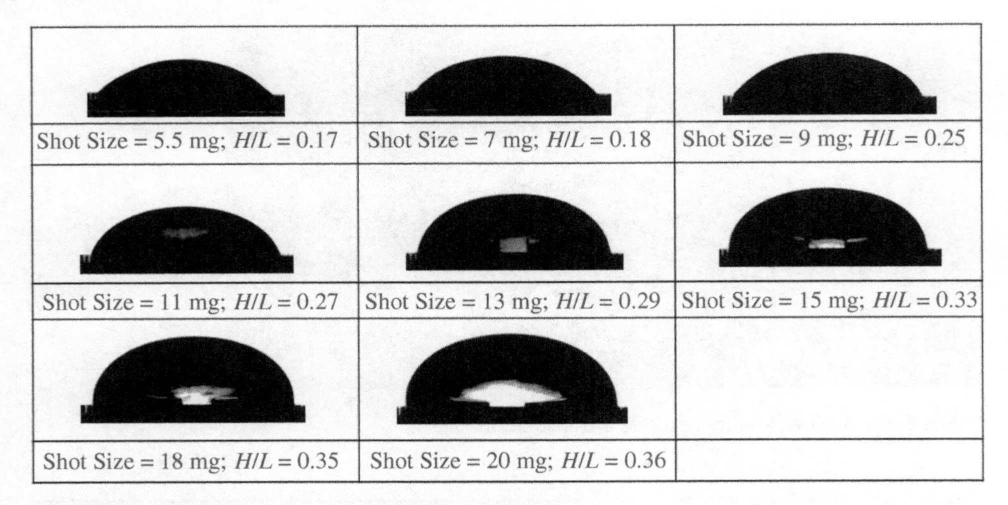

Figure 3.55 Silicone dome geometry control (5 mm × 5 mm).

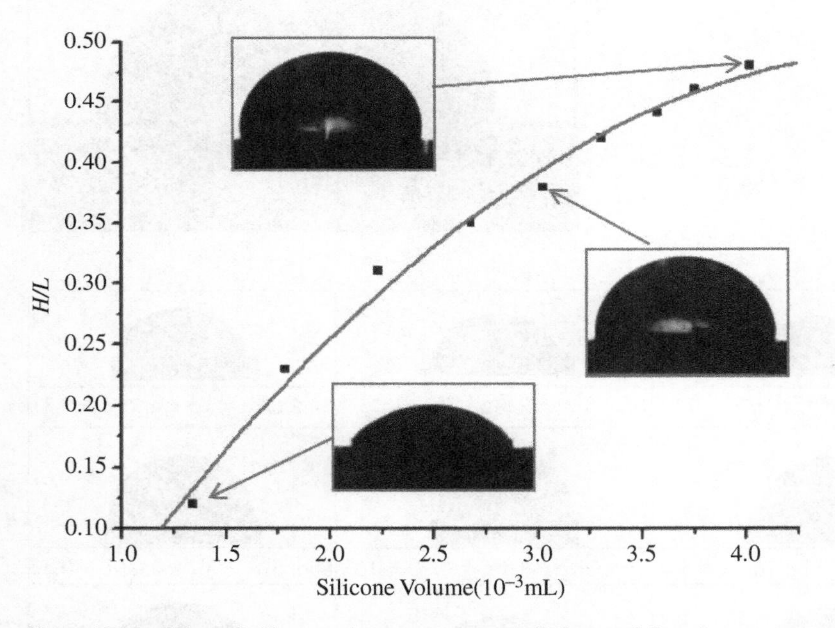

Figure 3.56 Effect of silicone volume on geometry (2.8 mm × 2.8 mm).

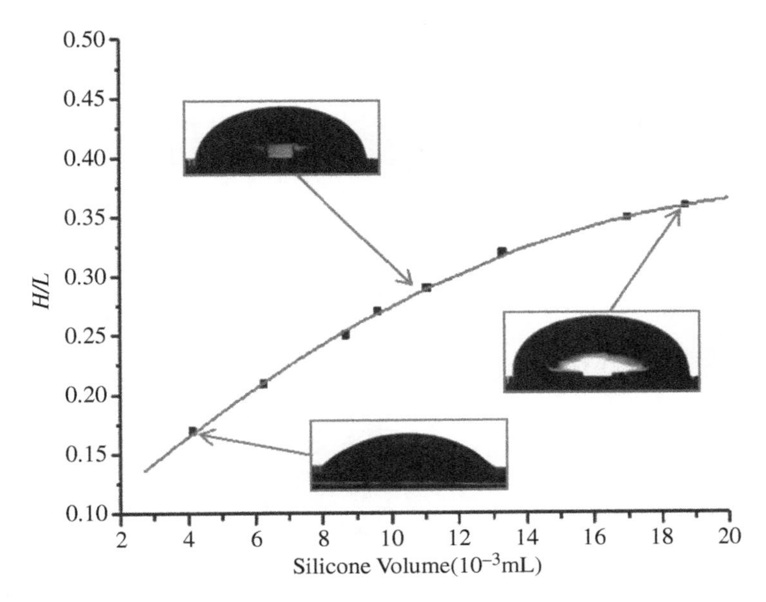

Figure 3.57 Effect of silicone volume on geometry (5 mm × 5 mm).

3.4 Designs and Structures of LED Wafer Level Packaging

With the aforementioned enabling technologies, LED WLP with different structures can be fabricated. In this section, several considerations in designing LED WLP are discussed.

3.4.1 Reflective Layer Design

A flat panel design has the simplest microfabrication process. No cavity will be fabricated on the wafer. The metal layer patterned on the wafer is used to provide an electrical path and reflect light. It is necessary to cover the wafer so that as much light as possible can be reflected. In designing the LED array or a multichip package, the electrical connection method is another concern. The LEDs can be either connected in series, parallel, or with a parallel connection of several serial chains. Each connection method has its pros and cons.

Take an example of a silicon submount for LED WLP. A 4 × 4 LED panel is designed of size of 22 mm × 35 mm. The LEDs are connected in series. Figure 3.58 shows the mask layout and Figure 3.59 shows the corresponding fabrication process. In this design, a 1 μm oxide is first deposited on the wafer. A 2 μm aluminum layer is then deposited and patterned thereon. The aluminum layer serves as a reflective layer and therefore is designed to cover a large area to reflect as much light as possible. It also serves as the bond pad material for a wire bond. Another 1 μm oxide is deposited and patterned on top as a passivation layer, which defines the bond pad locations. LEDs are mounted on the submount, and interconnects are completed by regular die mounting and wire bonding processes, respectively. Figure 3.60 shows the final panel with 16 LEDs.

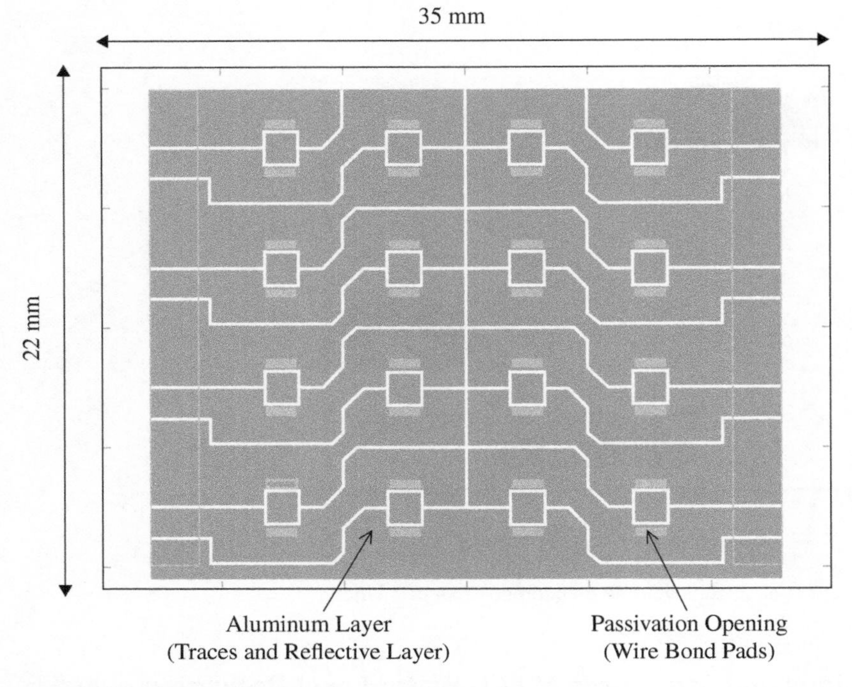

Figure 3.58 4×4 LED panel array mask layout.

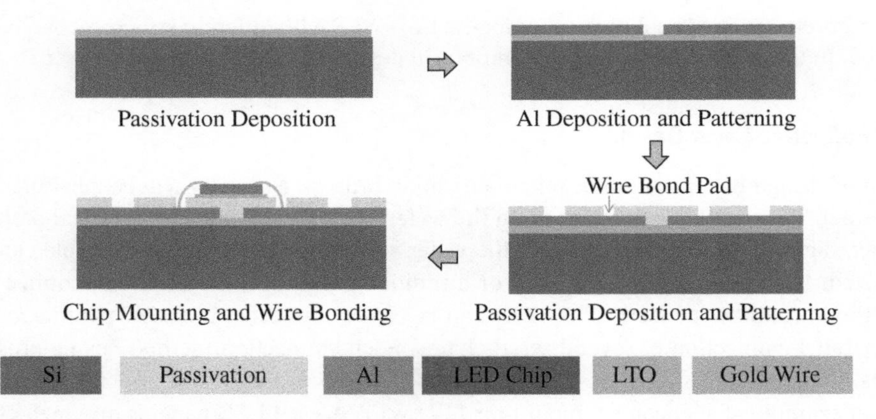

Figure 3.59 4×4 LED panel microfabrication process.

3.4.2 Cavity and Reflective Cup by Wet Etching

In LED WLP, wet etching can be used to form a cavity on the wafer and trapezoidal through holes for vertical interconnections [30, 31]. The (1 0 0) bottom surface is used to hold the LED chip. Wire bond pads may also be fabricated on the surface of the cavity [31]. The incline {1 1 1} side walls are use as the reflective surfaces [32]. Therefore, the quality of the surfaces after etching is critical. It will affect the subsequent packaging process and the final optical performance.

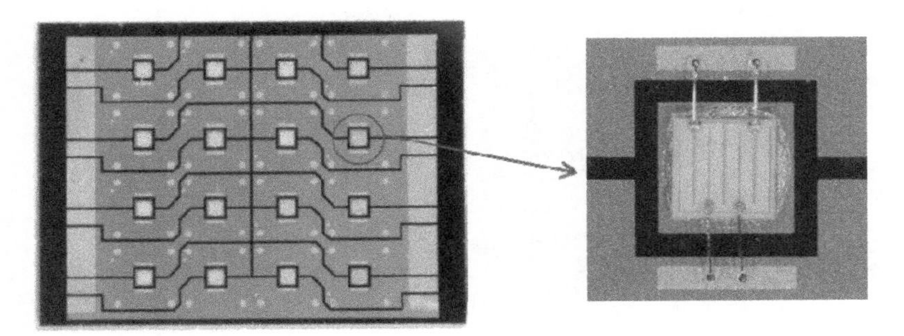

Figure 3.60 4 × 4 LED panel array.

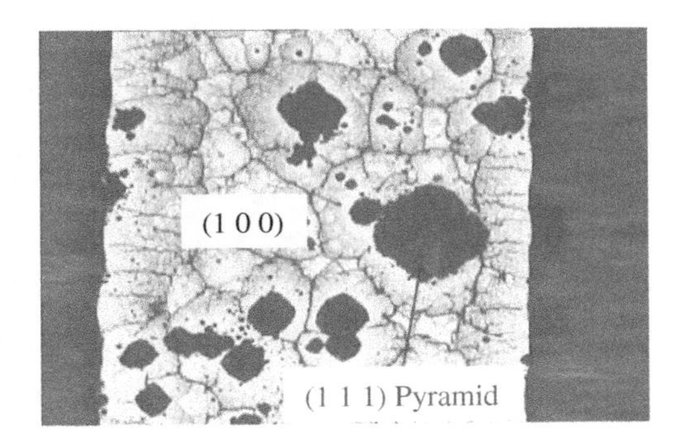

Figure 3.61 Rough (1 0 0) plane.

As illustrated in the chemical equations in Section 3.3.2, hydrogen bubbles will be evolved during the etching process. The bubbles may attach on the wafer surface without proper agitation. They serve as small masks to prevent the etchant from reaching the surface. As a result, tiny pyramids with four {1 1 1} incline planes are formed on the (1 0 0) plane, as shown in Figure 3.61. It cannot provide a flat platform for die mounting and wire bond pads.

As reflective planes, the quality of the {1 1 1} incline planes will affect the optical performance of the package. A mirror-like surface is obtained if an incline plane is formed by a single crystal plane. It can be achieved by having an excellent alignment between the mask pattern and the crystal plane [33–35]. The incline side wall will be formed by multiple {1 1 1} planes should there be a slight angular misalignment between the mask pattern and the crystal plane. This results in the formation of a zigzag feature on the incline plane.

A study has been carried out to evaluate the effect of misalignment between the mask pattern and the crystal plane on the quality of the incline side wall [36]. Figure 3.62 shows the quality of the side walls with different degrees of misalignment. It can be noted that zigzag features are formed even with a 1° misalignment. The roughness of the side walls is measured, and the results are shown in Figure 3.63 and Table 3.2. The roughness (Ra) of the sample with a 3° misalignment is more than 10 times that of the well-aligned sample. The angular misalignment between the mask pattern and the crystal plane will also cause severe undercut, as shown in Figure 3.64. As a result, the final dimensions of the feature obtained after etching are completely different from the mask layout.

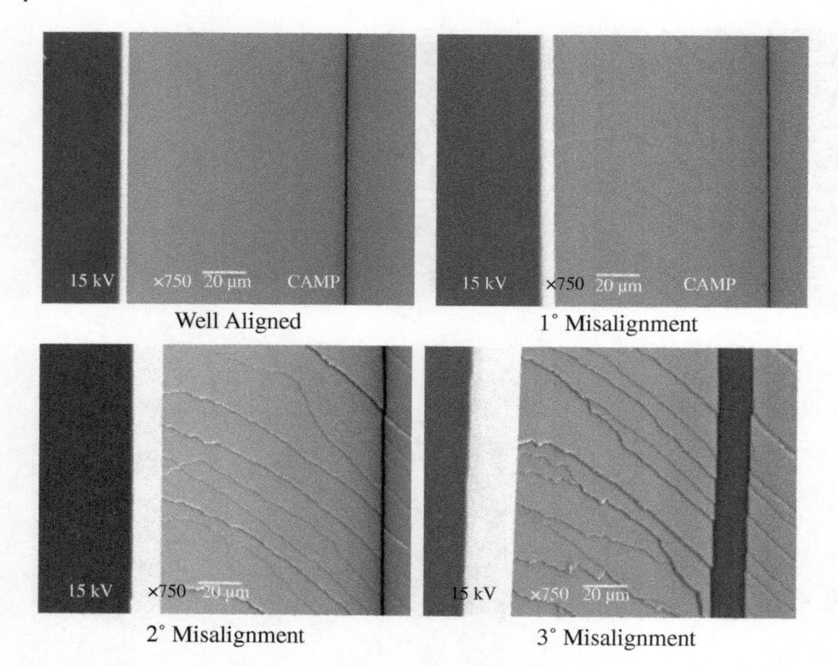

Figure 3.62 Effect of angular misalignment (scanning electron microscope (SEM) microscopy inspection).

Figure 3.63 Effect of angular misalignment on incline plane roughness.

Table 3.2 Effect of angular misalignment on incline plane roughness.

Angular misalignment	Ra (nm)	Rq (nm)	Rt (nm)
0°	11	14	403
1°	57	70	929
2°	127	158	1288
3°	174	218	1569

Note:
Ra: Roughness Average;
Rq: Root Mean Square Roughness;
Rt: Maximum Height of the Profile.

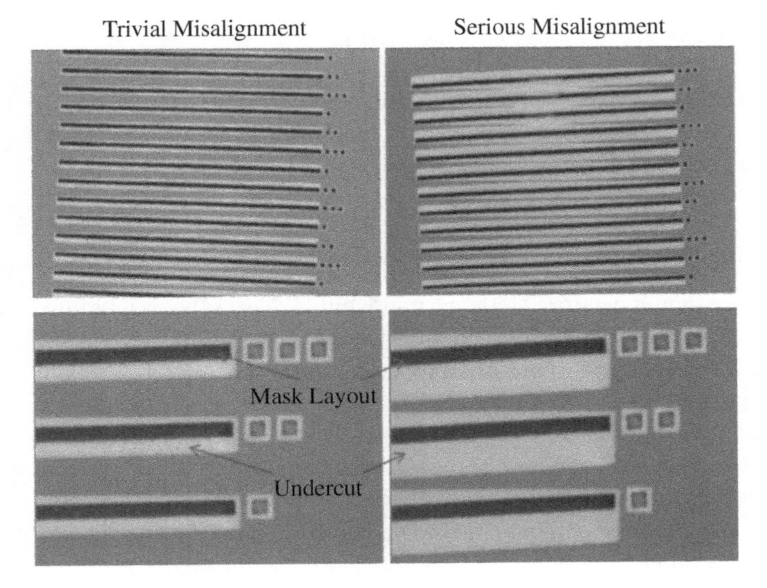

Figure 3.64 Effect of angular misalignment (undercut).

3.4.3 Copper-Filled TSVs for Vertical Interconnection and Heat Dissipation

Copper-filled TSVs in LED WLP are extensively used. They provide vertical interconnection paths and connect the circuit on top of the wafer to the bond pad at the bottom electrically [37, 38]. As copper has a high thermal conductivity, copper-filled TSVs can also serve as thermal vias [39]. With a proper TSV arrangement, the heat dissipation path can be separated from the electrical connection path (Figure 3.65). The thermal vias can then be attached directly to metal heat slugs or heat sinks without electrical insulation. This offers a better and more flexible thermal management solution.

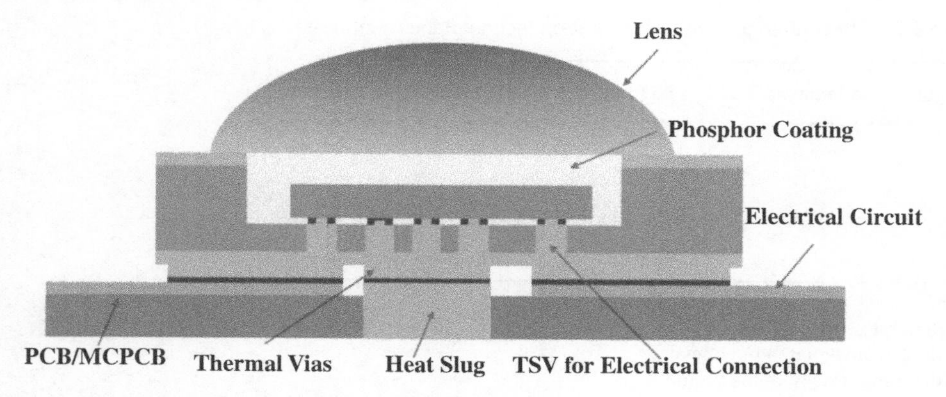

Figure 3.65 LED WLP with copper-filled TSVs.

3.5 Processes of LED Wafer Level Packaging

In this section, three case studies on LED WLP are discussed. The first two cases are "Multichip LED WLP with through Silicon Slots" [40] and "LED WLP with a Cavity" [41–43]. These two cases demonstrate the use of enabling technologies for LED WLP. The remaining case is related to the application of LED WLP panels [44].

3.5.1 Case 1: Multichip LED WLP with Through Silicon Slots

A WLP process for surface mount device (SMD) type multichip LED packages is presented in this case. Two types of packages with different form factors, namely 2.8 mm × 2.8 mm and 5 mm × 5 mm, are fabricated. These packages support different electrical connections of four LED chips. Copper plated through silicon slots are used to direct the electrodes from the top to the bottom side of the wafer. A moldless encapsulation process is applied to form a dome-shaped encapsulation. The geometry of the encapsulation is defined by half cut trenches and can be controlled by adjusting the volume of the encapsulant. After WLP and dicing, the components can be surface mounted to the PCB directly. The schematic diagram of the package is illustrated in Figure 3.66.

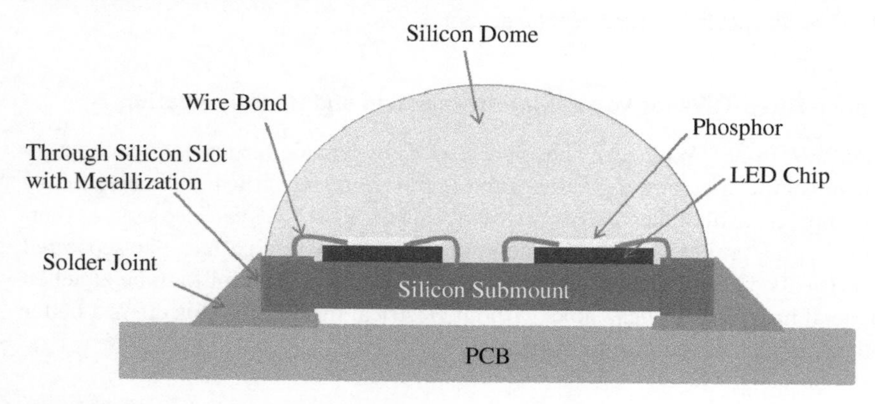

Figure 3.66 Multichip LED WLP with through silicon slots.

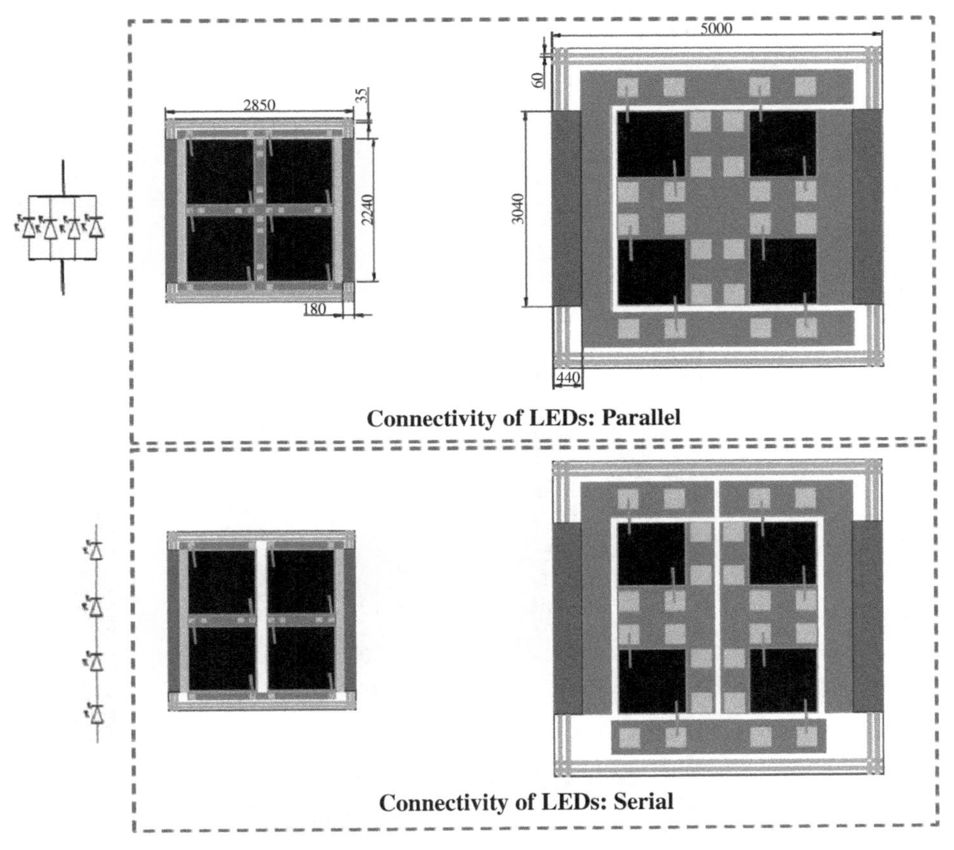

Note:

1. **The pink squares are wire bond pads. Redundant pads are arranged to accommodate possible variations of wire bond pad locations on different LED chips.**
2. **The green rectangles are through silicon slots.**

Units: μm

Figure 3.67 LEDs connection methods.

The connection methods between the four LEDs on the submount can be adjusted by changing the aluminum pattern on top. Figure 3.67 illustrates two possible connection methods for a 2 × 2 LED chip array. The LED chips on submounts can be connected either in parallel or series.

The fabrication process of the silicone submount is described below. The through silicon slot is first etched on a 400 μm thick wafer by DRIE, as shown in Figure 3.68. A 1 μm thick oxide is then deposited on both sides of the wafer and the side walls of the slots for passivation. Aluminum that is 2 μm thick is deposited and patterned as the circuit and reflective layer (Figure 3.69). Another 1 μm thick oxide is deposited and patterned on top of the aluminum layer to define the bond pads and for passivation (Figure 3.70). TiW adhesion layer and copper seed layers are sputtered and patterned on both sides of the wafer and side walls of the slot. Copper, nickel, and gold (by electro-copper plating, electroless nickel plating, and

Figure 3.68 Through silicon slots by DRIE.

2.8 mm × 2.8 mm, Parallel

2.8 mm × 2.8 mm, Serial

Figure 3.69 Aluminum layer deposition and patterning.

2.8 mm × 2.8 mm, Serial

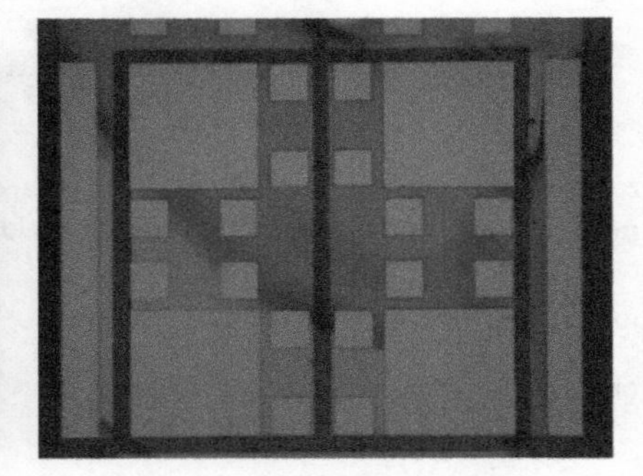

5 mm × 5 mm, Serial

Figure 3.70 Passivation layer deposition and patterning.

immersion gold plating, respectively) metallization is formed on the side walls of the slots and the bond pads at the bottom (Figure 3.71). Half cut trenches are made on the wafer by a regular die sawing process.

Four LED chips (1 W, 1 mm × 1 mm) are mounted on the submount and gold wire bonds are used as interconnects between the chips and the submount. A moldless encapsulation process is then performed to encapsulate the LED chips with a mixture of silicone and phosphor powders. The trenches can limit the spread of silicone, in which a dome-shaped encapsulation is formed. The packages are then singulated from the wafer and can be mounted

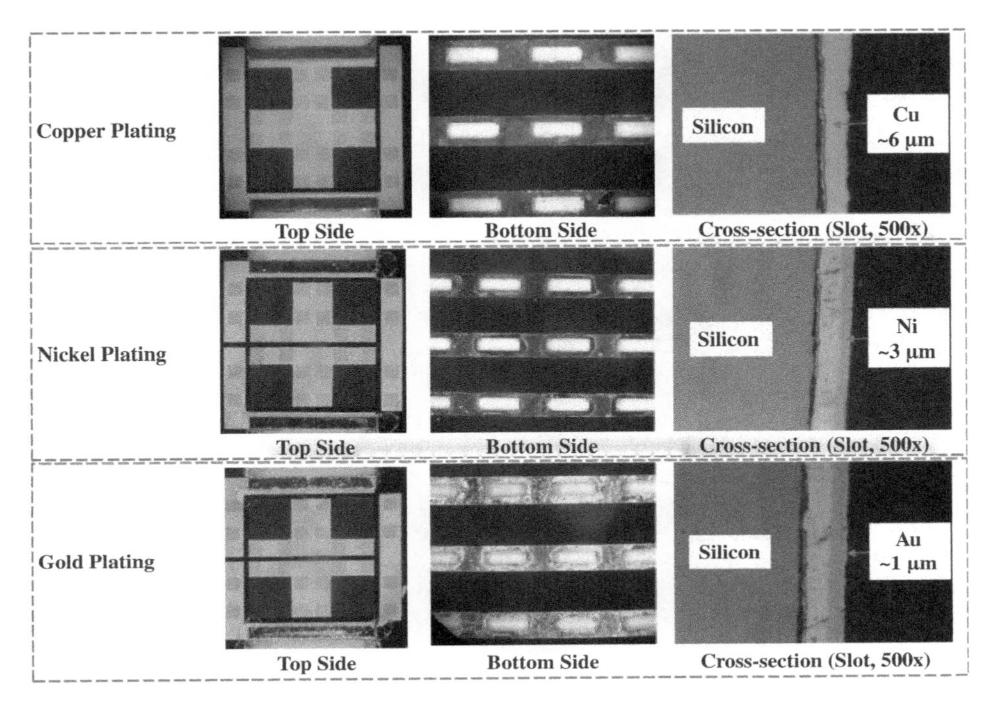

Figure 3.71 Plating results.

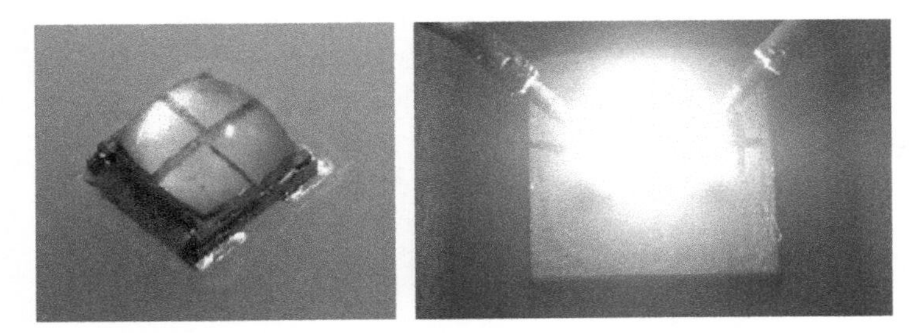

Figure 3.72 Singulated package assembled on a PCB.

on a PCB by the surface mount process. Figure 3.72 shows a singulated prototype package which is powered up after mounting on a PCB. Figure 3.73 shows the cross-section of a package. In this case, the LED WLP is capable of packing multichips in a small form factor.

3.5.2 Case 2: LED WLP with a Cavity

A novel WLP process for phosphor-converted LEDs is presented here. The core of this process is the fabrication of a silicon substrate with cavities for phosphor printing and copper-filled TSVs for vertical interconnection and heat dissipation. The configuration of the proposed LED substrate is shown in Figures 3.74 and 3.75. It consists of a 1.3 mm × 1.3 mm × 0.22 mm cavity on the front side to contain an LED chip and phosphor powders, certain copper pillars plated with solder at the bottom of the cavity for chip mounting,

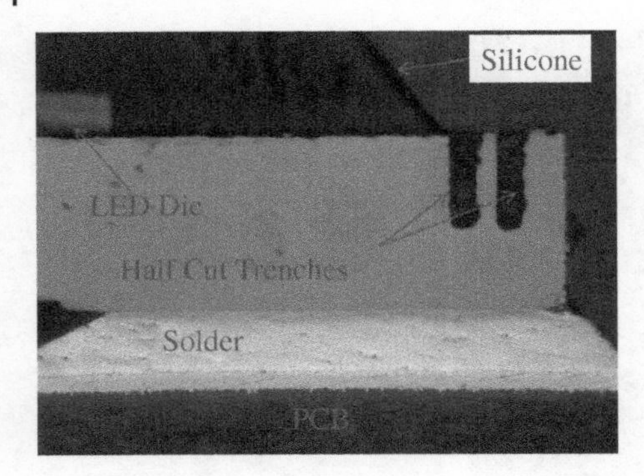

Figure 3.73 Cross-section inspection.

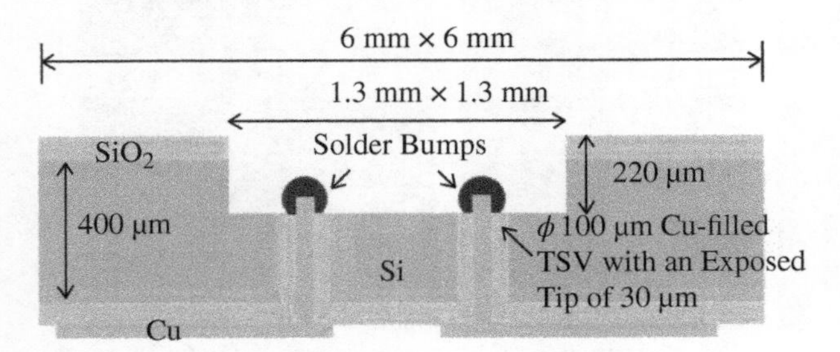

Figure 3.74 LED WLP with a cavity (schematic cross-section view).

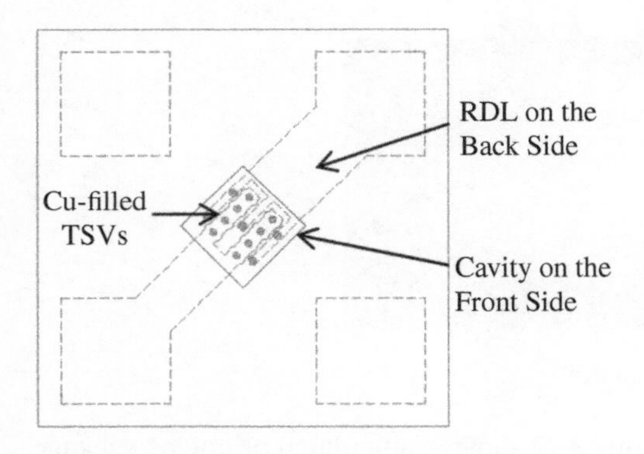

Figure 3.75 LED WLP with a cavity (schematic top view).

and TSVs connecting to the RDL patterned at the back. This structure does not require any thin film deposition or photolithography process at the bottom of the cavity. It offers a solution to the challenge that it is difficult to pattern an RDL in the cavity. Encapsulant and phosphor slurry can be printed directly into the cavity without any molds and stencils.

The overall fabrication process of the submount is shown in Figure 3.76. Blind vias are first etched by DRIE at the back of the wafer. The diameter of the blind vias is 100 μm with a depth of 220 μm. Next, DRIE is performed on the front to create cavities for LED chip mounting and phosphor printing. The cavity is etched to a depth of 150 μm, leaving a 40 μm gap between the bottom of the cavity and that of the blind vias. Copper is then electroplated to fill the blind vias to form copper pillars.

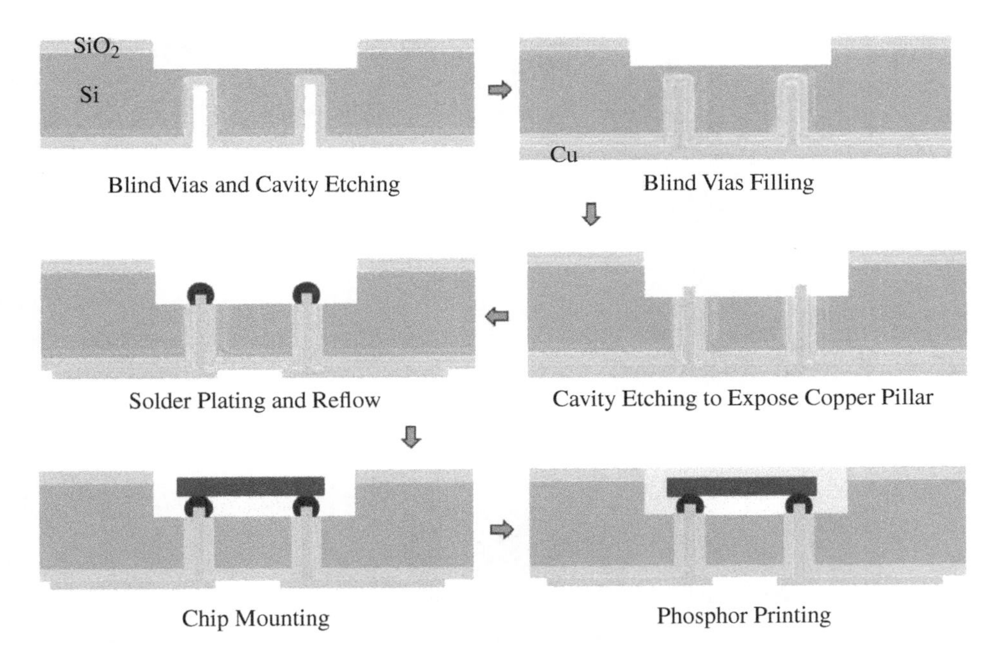

Figure 3.76 Submount microfabrication process flow.

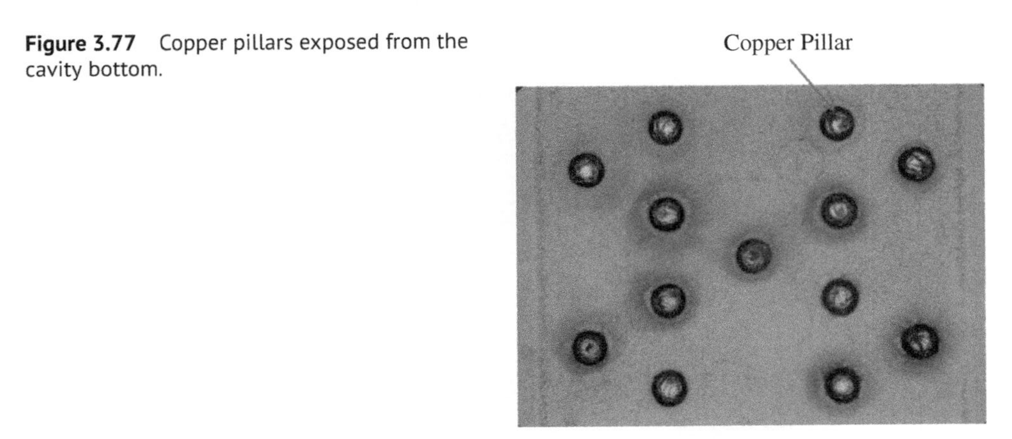

Figure 3.77 Copper pillars exposed from the cavity bottom.

Subsequently, KOH etching is applied on the front to further etch the cavities to expose the embedded copper pillars. Figure 3.77 shows the pillars are exposed at the bottom of the cavity. Solder is then plated on the exposed tips of copper pillars in the cavities. The solder is reflowed and round shape solder bumps are formed (Figure 3.78). Finally, the copper at the back of the wafer is patterned as the RDL for the next level of interconnection. Figure 3.79 shows the top, bottom, and cross-section views of the submount.

Upon fabrication of the silicon submount, blue LEDs are flip-chip mounted on the pre-plated solder bumps in the cavities. Flip-chip blue LED chips with a dimension of 1 mm × 1 mm × 0.07 mm are mounted in the cavity with reflow soldering (Figure 3.80). Following an epoxy dispensing process, yellow phosphor powder is printed into the cavity

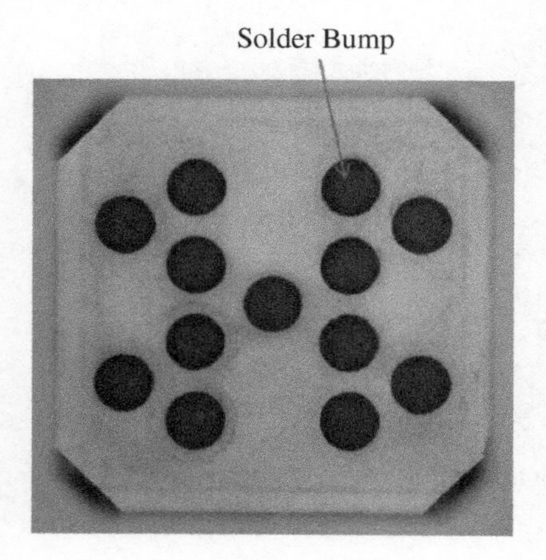

Figure 3.78 Solder bumps on the copper pillars after reflow.

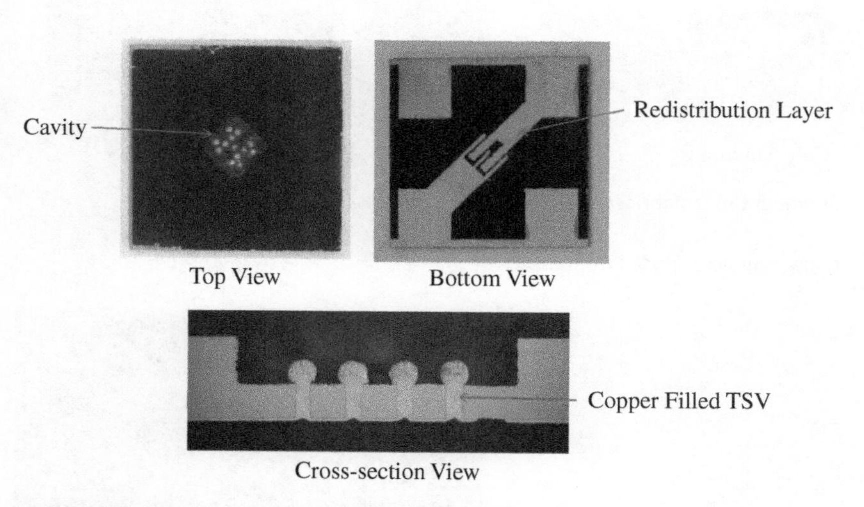

Figure 3.79 Submount with copper-filled TSVs and a cavity.

Figure 3.80 LED chip mounted in the cavity.

Figure 3.81 After phosphor printing.

Figure 3.82 Cross-section inspection of final package.

for color tuning. The pre-dispensed epoxy is then UV-cured to serve as the phosphor powder binder and the LED chip encapsulant (Figure 3.81). The present configuration offers a structure with a low profile and compact footprint for LED WLP.

Figure 3.82 shows the cross-section inspection of the finished LED package. It can be observed that the LED chip is embedded in the cavity and covered by phosphor powder. The encapsulation is formed within the cavity, hence achieving a flat profile. Figure 3.83 shows a lit-up LED package. It should be noted that light is converted from blue to white with yellow phosphor.

3.5.3 Case 3: Applications of an LED WLP Panel

In this case, a 4×4 LED WLP panel with a waffle pack phosphor layer as described in Section 3.5.2 is used as the light source for general lighting. The panel is installed in a

Figure 3.83 Uniform white light from the package.

Figure 3.84 Lightbulb with LED WLP panel.

lightbulb with a diffuser, as shown in Figure 3.84. It is directly attached to the heat sink of the lightbulb by thermal grease. Optical performance of a traditional chip-on-board (COB) package with nine LED chips (Figure 3.85) is also characterized for comparison. As shown in the figure, the LED WLP panel has a larger light-emitting area.

Both the LED WLP panel and COB package are driven by the same condition, 8 W. The corresponding angular radiation power and color temperature distribution are measured by a goniometer. Figure 3.86 compares the angular radiation power distribution of the two packages. The results are normalized by the corresponding maximum value. It can be observed that the LED WLP panel has a wider power distribution, which is more suitable for general lighting.

An LED WLP panel has a promising performance in angular color temperature distribution when compared with a COB package (Figure 3.87). A waffle pack remote phosphor film is applied to the hemispherical lens fabricated on the panel. The phosphor distribution in the film is uniform, as shown in the cross-section inspection in Section 3.5.2. The path length of light traveling in the phosphor layer is similar at different view angles. Therefore, the amount of blue light emitted by the chip being converted to yellow light is similar at different angles. This results in a uniform angular color temperature distribution. In contrast,

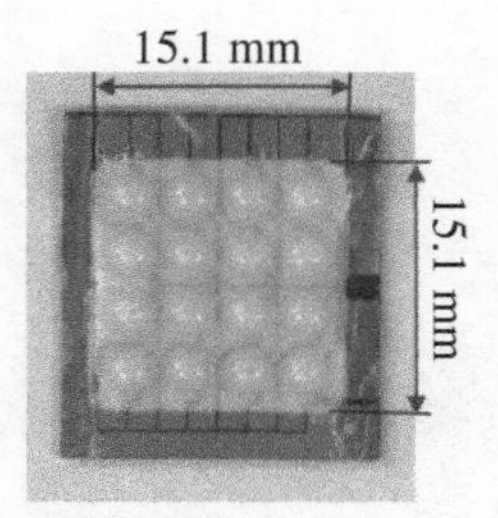

Number of Chips: 16
Emitting Area: 228 mm^2

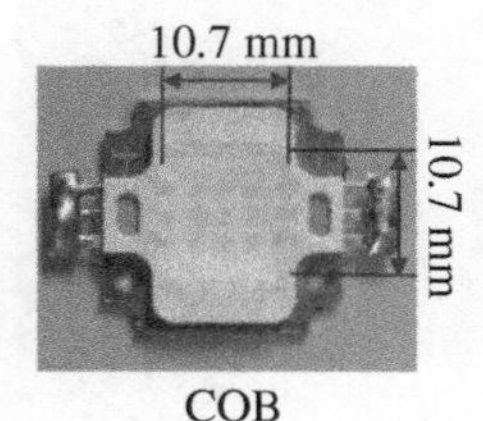

Number of Chips: 9
Emitting Area: 115 mm^2

Figure 3.85 LED WLP panel vs. COB.

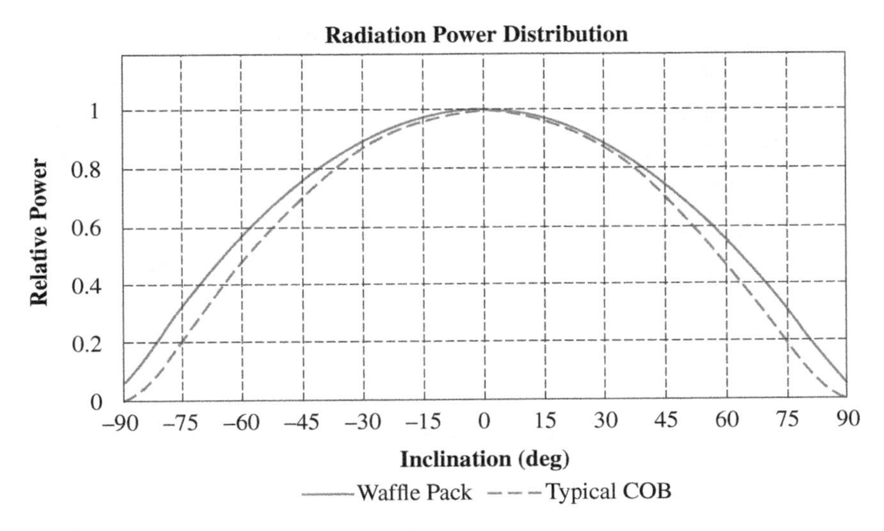

Figure 3.86 Radiation power distribution.

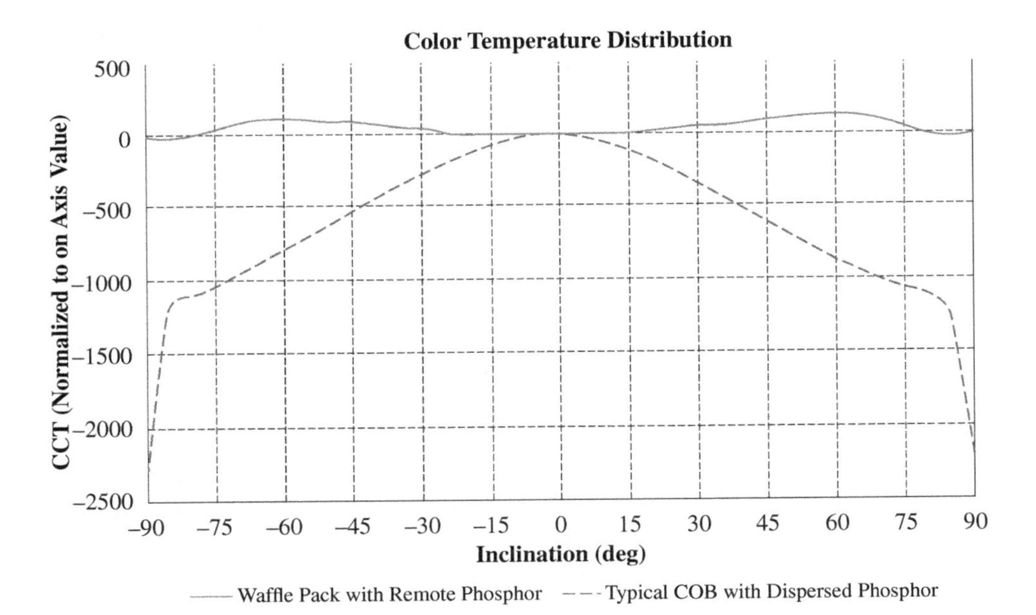

Figure 3.87 Color temperature distribution.

the reflective cup of the COB package is filled with phosphor slurry by a dispersed dispensing method. The path length of light traveling in the phosphor layer varies at different angles. The light emitted to the side has a longer light path, and hence much blue light is converted to yellow light. As a result, the COB package has a much lower CCT at the large inclination angle. This creates a yellow ring, which is not ideal for general lighting applications.

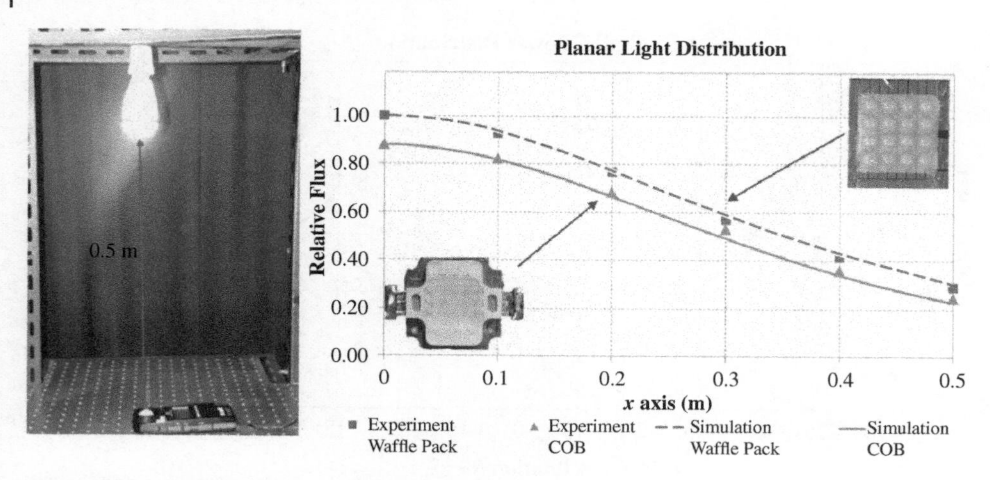

Figure 3.88 Illuminance measurement results.

The illuminance distributions of lightbulbs are measured. The lux meter is placed 0.5 m apart from the tip of the lightbulb. The planner distribution is recorded with a 0.1 m step. From the normalized results, as shown in Figure 3.88, the trends of both lightbulbs are similar. The lightbulb with an LED WLP panel has a better performance than that of the traditional COB package. The LED WLP panel has a higher illuminance and is more efficient for general lighting than the traditional COB lightbulb.

References

1 Lau, J.H. and Lee, S.W.R. (1999). *Chip Scale Package (CSP): Design, Materials, Process, Reliability & Applications*. New York: McGraw-Hill.

2 Kimura, A., Obata, S., Nakayama, T. et al. (2012). Optical characteristics and reliability evaluation of wafer level white LED package. In: *Proceedings of 62nd Electronic Components and Technology Conference (ECTC 2012)*, San Diego, CA (29 May–1 June 2012), IEEE.

3 Korjima, A., Shimada, M., Akimoto, Y.M., et al. (2012). A fully integrated novel wafer-level LED package (WL2P) technology for extremely low-cost solid state lighting devices. In: *Proceedings of International Interconnect Technology Conference (IITC)*, San Jose, CA, (4–6 June 2012), 1–3. IEEE.

4 Techsearch (2008). *Flip Chip and Wafer Level Packaging Trends and Market Forecasts*. Austin, TX: Techsearch.

5 Yole Development (2010). *Wafer Level Packaging: Technologies, Applications and Markets*. Lyon, France: Yole Development.

6 DIGITIMES. (2009). VisEra announces new LED packaging technology. https://www.digitimes.com/news/a20090917PD207.html (accessed 3 September 2020).

7 Fan, X. (2012). Wafer level system packaging and integration for solid state lighting (SSL), Proceedings of the 13th International Conference on Thermal, Mechanical and Multi-Physics Simulation and Experiments in Microelectronics and Microsystems (EuroSimE 2012), Cascais (16–18 April 2012).

8 Lau, J., Lee, S.W.R., Yuen, M. et al. (2010). 3D LED and IC wafer level packaging. *MicroElectronics International* 27 (2): 98–105.
9 Uhrmann, T., Matthias, T., and Lindner, P. (2011). Silicon-based wafer-level packaging for cost reduction of high brightness LEDs. In: *Proceedings of 61st Electronic Components and Technology Conference (ECTC 2011)*, Lake Buena Vista, FL (31 May to 3 June 2011), 1622–1625. IEEE.
10 Kim, J.K. and Lee, H.C. (2013). MSM photodetector on a polysilicon membrane for a silicon-based wafer-level packaged LED. *IEEE Photonics Technology Letters* 25 (24): 2462–2465.
11 Lee, S.W.R., Zhang, R., Chen, K. et al. (2012). Emerging trend for LED wafer level packaging. *Frontiers of Optoelectronics* 5 (2): 119–126.
12 Lee, S.W.R. (2011). Advanced LED wafer level packaging technologies. In: *Proceedings of 6th International Microsystems, Packaging, Assembly and Circuits Technology Conference (IMPACT 2011)*, Taipei (19–21 October 2011), 71–74. IEEE.
13 Bean, K.E. (1978). Anisotropic etching of silicon. *IEEE Transactions on Electron Devices* ED-25: 1185–1193.
14 Campbell, S.A. and Lewerenz, H.J. (1998). *Semiconductor Micromaching*, vol. 2: Techniques and Industrial Applications. Chichester: Wiley.
15 Campbell, S.A. and Lewerenz, H.J. (1998). *Semiconductor Micromaching*, vol. 1: Fundamental Electrochemistry and Physics. Chichester: Wiley.
16 Lau, J. (2013). *Through-Silicon Vias TSV for 3D Integration*. New York: McGraw-Hill Education.
17 Garrou, P., Bower, C., and Ramm, P. (2008). *Handbook of 3D Integration, Technology and Applications of 3D Integrated Circuits*. Weinheim, Germany: Wiley-VCH.
18 Chen, K., Zhang, R., and Lee, S.W.R. (2010). Integration of phosphor printing and encapsulant dispensing processes for wafer level LED array packaging. In: *Proceedings of 11th International Conference on Electronic Packaging Technology & High Density Packaging (ICEPT-HDP)* , Xi'an, China (16–19 August 2010), 1386–1392. IEEE.
19 Lee, K.H. and Lee, S.W.R. (2006). Process development for yellow phosphor coating on blue light emitting diodes (LEDs) for white light illumination. In: *Proceedings of 8th Electronics Packaging Technology Conference (EPTC)*, Singapore (6–8 December 2006), 379–384. IEEE.
20 Lee, K.H., Lee, S.W.R. (2007). Screen printing of yellow phosphor powder on blue light emitting diode (LED) arrays for white light illumination. Proceedings of InterPACK'07, Vancouver, Canada (8–12 July 2007).
21 Lo, J.C.C., Lee, S.W.R., Zhang, R. et al. (2012). Reverse wire bonding and phosphor printing for LED wafer level packaging. In: *Proceedings of 62nd Electronic Components & Technology Conference (ECTC)*, San Diego, CA (29 May to 1 June 2012), 1814–1818. IEEE.
22 Liu, H., Lo, J.C.C., Lee, S.W.R. et al. (2013). LED wafer level packaging with a hemispherical waffle pack remote phosphor layer. In: *Proceedings of International Conference on Electronics Packaging (ICEP 2013)*, Osaka, Japan (10–12 April 2013), 368–373. IEEE.
23 Liu, H., Zhang, R., Lo, J.C.C. et al. (2012). LED wafer level packaging with a remote phosphor cap. In: *Proceedings of 14th International Conference on Electronics Materials and Packaging (EMAP 2012)*, Hong Kong (13–16 December 2012), 251–255. IEEE.

24 Lo, J.C.C., Liu, H., Lee, S.W.R., et al. (2013). Remote phosphor deposition on LED arrays with pre-encapsulated silicone lens. In: *Proceedings of 14th International Conference on Thermal, Mechanical and Multi-Physics Simulation and Experiments in Microelectronics and Microsystems (EuroSimE 2013)*, Wroclaw, Poland (14–17 April 2013).

25 Yu, H., Shang, J., Xu, C. et al. (2011). Chip-on-board (COB) wafer level packaging of LEDs using silicon substrates and chemical foaming process (CFP)-made glass-bubble caps. In: *Proceedings of 12th International Conference on Electronic Packaging Technology and High Density Packaging (ICEPT-HDP 2011)*, Shanghai, China (8–11 August 2011), 133–136. IEEE.

26 Cao, B., Yu, S., Cheng, H. et al. (2012). Silicon-based system in packaging for light emitting diodes. In: *Proceedings of 62nd Electronic Components & Technology Conference (ECTC 2012)*, San Diego, CA (29 May–1 June 2012), 1267–1271. IEEE.

27 Zhang, R. and Lee, S.W.R. (2012). Moldless encapsulation for LED wafer level packaging using integrated DRIE trenches. *Microelectronics Reliability* 52 (5): 922–932.

28 Zhang, R. and Lee, S.W.R. (2008). Wafer level LED packaging with integrated DRIE trenches for encapsulation. In: *Proceedings of 9th International Conference on Electronic Packaging Technology and High Density Packaging (ICEPT-HDP)*, Shanghai, China (28–31 July 2008), 1–6. IEEE.

29 Zhang, R. and Lee, S.W.R. (2007). Wafer level encapsulation process for LED array packaging. In: *Proceedings of 9th International Conference on Electronic Materials and Packaging (EMAP 2007)*, Daejeon, Korea (19–22 November), 1–8. IEEE.

30 Tsou, C. and Huang, Y. (2006). Silicon-based packaging platform for light-emitting diode. *IEEE Transactions on Advanced Packaging* 29 (3): 607–614.

31 Chang, C., Huang, C., Lai, T. et al. (2011). MEMS technology on wafer-lever LED packaging. In: *Proceedings of 13th Electronics Packaging Technology Conference (EPTC)*, Singapore (7–9 December 2011), 84–87. IEEE.

32 Jeung, W.K., Shin, S.H., Hong, S.Y. et al. (2007). Silicon-based, multi-chip LED package. In: *Proceedings of 57th Electronic Components and Technology Conference (ECTC 2007)*, Reno, NV (29 May to 1 June 2007), 722–727. IEEE.

33 Reddy, P.S. and Jessing, J.R. (2004). Pattern alignment effects in through-wafer bulk micromachining of (100) silicon. In: *IEEE Workshop on Microelectronics and Electron Devices*, 89–92. IEEE.

34 Chen, P., Hsieh, C., Peng, H. et al. (2000). Precise mask alignment design to crystal orientation of (100) silicon wafer using wet anisotropic etching. *Micromachining and Microfabrication Process Technology* 4174: 462–466.

35 Li, L., Yi, T., and Kim, C.J. (2000). Effect of mask-to-crystal direction misalignment on fracture strength of silicon microbeam. In: *Proceedings of Microscale Systems: Mechanics and Measurements Symposium, Society for Experimental Mechanics, IX International Congress*, 36–40. Society for Experimental Mechanics.

36 Lam, K.S.J. and Lee, S.W.R. (2007). Effect of V-groove side wall feature on epoxy flow in passive alignment of optical fiber. In: *Proceedings of 6th International Conference on Polymers and Adhesives in Microelectronics and Photonics (Polytronic 2007)*, Odaiba, Tokyo (16–18 January 2007), 202–208. IEEE.

37 Chen, D., Zhang, L., Xie, Y. et al. (2012). A study of novel wafer level LED package based on TSV technology. In: *Proceedings of 13th International Conference on Electronic*

Packaging Technology & High Density Packaging (ICEPT-HDP), Guilin, China (13–16 August 2012), 52–55. IEEE.

38 Xie, Y., Chen, D., Li, Z. et al. (2013). A novel wafer level packaging for white light LED. In: *Proceedings of 14th International Conference on Electronic Packaging Technology (ICEPT 2013)*, Dailian, China (11–14 August), 1170–1174. IEEE.

39 Auersperg, J., Dudek, R., Jordan, R., et al. (2013). On the crack and delamination risk optimization of a Si-interposer for LED packaging. Proceedings of 14th International Conference on Thermal, Mechanical and Multi-Physics Simulation and Experiments in Microelectronics and Microsystems (EuroSimE 2013), Wroclaw, Poland (14–17 April 2013).

40 Zhang, R., Lo, J.C.C., Chen, K. et al. (2011). Multi-chip LED wafer level packaging with through-silicon slots for interconnection. In: *Proceedings of 8th China International Forum on Solid State Lighting (ChinaSSL 2011)*, Guangzhou, China (8–11 November 2011), 208–214. IEEE.

41 Zhang, R., Lo, J.C.C., and Lee, S.W.R. (2012). Design and fabrication of a silicon interposer with TSVs in cavities for 3D IC packaging. *IEEE Transactions on Device and Materials Reliability* 12 (2): 189–193.

42 Zhang, R., SWR, L., Xiao, D.G. et al. (2011). LED packaging using silicon substrate with cavities for phosphor printing and copper-filled TSVs for 3D interconnection. In: *Proceedings of 61st Electronic Components & Technology Conference (ECTC 2011)*, Orlando, FL (31 May–3 June 2011), 1616–1621. IEEE.

43 Zhang, R. and Lee, S.R.W. (2010). Silicon interposer with cavities and copper-filled TSVs for 3D packaging. In: *Proceedings of 12th International Conference on Electronic Materials and Packaging (EMAP 2010)*, 1–5. IEEE.

44 Liu, H., Lo, J.C.C., and Lee, S.W.R. (2013). Waffle pack LED module configured light bulb and simulation validation. In: *Proceeding of 10th China International Forum on Solid State Lighting (ChinaSSL 2013)*, Beijing, China (10–12 November 2013), 186. IEEE.

4

Board Level Assemblies and LED Modules

4.1 Introduction

The chip level packaging and wafer level packaging processes are discussed in detail in Chapter 2 and Chapter 3, respectively. Throughout the packaging process, the tiny chips are either turned into discrete components or LED array panels. After that, the components or panels are assembled to a printed circuit board (PCB) to form a system. The LED components and other integrated circuit (IC) packages are assembled to the boards by soldering. Two types of soldering methods are discussed in this chapter, namely wave soldering for plated-through-hole (PTH) components and reflow soldering for surface mount components.

In traditional IC packaging, the core of the PCB is usually made with polymer materials. The low thermal conductivity of the polymer core cannot provide an effective thermal dissipation path for high-power LED assemblies. Metal core printed circuit boards (MCPCBs), which have a higher thermal conductivity, are commonly used in high-power LED products. MCPCBs have several key functions in the assembly of LEDs. The MCPCB provides mechanical support to the LED components. The circuitry on the MCPCB connects multiple LED components together. It also links the LED components to other essential electronic devices, such as driving ICs, resistors, capacitors, etc. The metal core of the PCB spreads the heat generated from the LED components and transfers it to the heat sink. The manufacturing process of PCBs and MCPCBs is introduced in this chapter. Different methods to reduce the thermal resistance of PCBs and MCPCBs is also discussed.

The aforementioned board level assembly process offers a larger degree of design and manufacturing flexibility. LED system manufacturers can select and acquire LED components which meet their design criteria from the market. The optical performance of the system can be easily adjusted by changing the MCPCB design or adopting different types of LED component. However, this is not the most cost-effective method. If the application of the system is known and well defined, chip-on-board (COB) assemblies offers a cheaper solution.

In COB, the LED chips are directly mounted onto the board. Interconnects are made between the chips and the board, followed by phosphor dispensing and encapsulation. This allows you to put multiple LED chips in a finer pitch. The light-emitting area of COB is much larger and hence is more suitable for general lighting. Since the chips are directly mounted onto the board, no substrate or lead frame is required, which reduces the cost

From LED to Solid State Lighting: Principles, Materials, Packaging, Characterization, and Applications, First Edition.
Shi-Wei Ricky Lee, Jeffery C. C. Lo, Mian Tao, and Huaiyu Ye.

of materials. It also has a shorter thermal dissipation path which offers a better thermal management solution.

The final section of this chapter briefly discusses the design considerations of LED modules. There are multiple chips or packages on an LED module. Those LED components can be connected in series, parallel, or parallel of series strings. Each connection method has its specific requirement on the driver design. The location of the components is another important consideration. It determines the light pattern of the module. It also affects the thermal performance of the entire system.

4.2 Board Level Assembly Processes

In the board level assembly process, discrete LED components or packages are mounted onto a PCB. Other electronic components, such as driving I/Cs, didoes, passive components (resistors and capacitors), etc., are also mounted on the circuit board to form a complete system. This provides a large degree of design and manufacturing flexibility. There are many types of LED components available in the market with different properties (such as color, light pattern, beam angle, light intensity, power consumption, etc.). Manufacturers and product designers can select different types of LED packages and electronic components for their unique products to meet the application requirements. It can be either a flashlight or a toy with different colors, a lightbulb, streetlight, LED display panel, or LED light strips for decoration, etc.

In an LED assembly, PCB provides the mechanical support to the components and defines the locations of the components. This controls the final optical performance of the system, such as the light pattern and intensity uniformity. The circuitry on the PCB also electronically connects the components. The LED components on the board can be connected in series, parallel, or parallel of series strings. The LED's connection method will determine the final electrical requirements (driving voltage and current). For example, if all LED components are connected in series, a high driving voltage will be required. In contrast, if the components are connected in parallel, the driver has to deliver a higher current to power up the system. In addition, the board also has a major role in thermal management. The heat dissipated from the LED components is transferred to the board. Nowadays, most LED assemblies use MCPCBs, which has a better thermal conductivity than traditional FR4 PCBs. The MCPCB allows the heat to spread across the board and to be transferred to the heat sink effectively.

This section discusses different types of MCPCBs. It evaluates the performance of FR4 PCBs with thermal vias in high-power LED applications. The assembly processes of PTH components and surface mount device (SMD) components are also described in detail.

4.2.1 Metal Core Printed Circuit Board

Thermal management is always an important consideration while designing the LED assembly. If the heat generated from the chip cannot dissipate to the environment effectively, a high junction temperature will be generated. This high junction temperature will affect the overall performance of the system, such as color shift, reduction in lumen output, etc. Packaging material aging and lumen decay accelerate at elevated temperatures.

In the long run, the reliability of the system will also be affected. In the worst case, extreme high junction temperatures will cause catastrophic failure after powering up the system in a short time.

In conventional IC packaging, the heat generated from high-power chips can be dissipated in two directions. A portion of the heat is dissipated to the environment through the upper part of the molding compound. A heat sink can be attached to the top of the component to further enhance the heat dissipation. In some cases, a cooling fan or heat pipe is installed to improve the heat transfer efficiency. Another portion of the heat is transferred to the PCB through the molding compound and solders joints. The heat will be further spread across the board and eventually dissipated to the environment. The two heat dissipation paths are illustrated in Figure 4.1.

The heat transfer path of LED assembly is different from conventional IC assemblies. It is not possible to add a heat sink on the top side of the package as it will block all lights emitted from the chip. Transparent encapsulants such as silicones and epoxies are commonly used to protect the LED chip and interconnects. They also serve as the lens in the LED package. However, the thermal conductivity of transparent silicones and epoxies is usually very low (only 0.3–0.5 W/(m · °C)). In addition to low thermal conductivity, the transparent encapsulation is relatively thick as it serves as the lens. The low thermal conductivity of the material and thick encapsulation lead to a very high thermal resistance. Only a small portion of heat generated from the chip is transferred to the ambient through the transparent encapsulation. As a result the major portion of the heat is transferred to the board. The heat is spread in the board and eventually dissipated to the environment. A heat sink is attached to the bottom side of the board to aid heat transfer. Figure 4.2 illustrates the thermal dissipation path of an LED assembly.

Owing to this, the PCB plays an important role in the thermal management of the LED assembly. The core of conventional PCBs is normally made by polymer materials, such as paper phenolic or FR4. The thermal conductivity of this polymer core is very low, normally

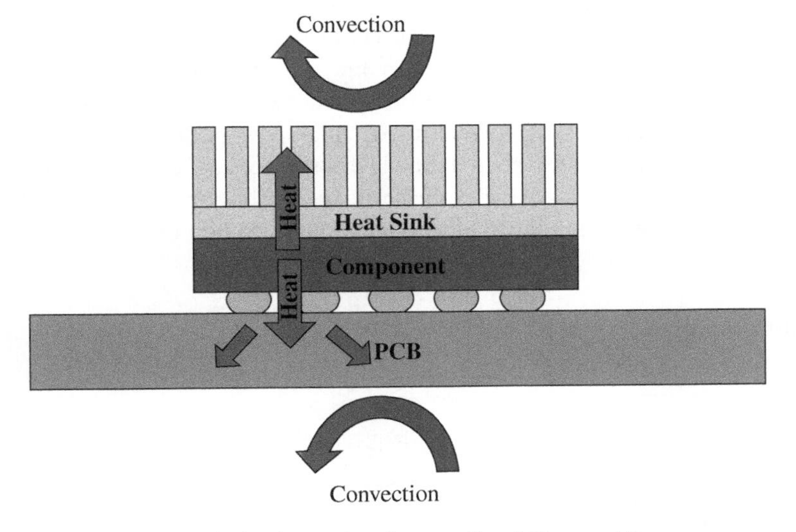

Figure 4.1 Heat dissipation paths of conventional IC assemblies.

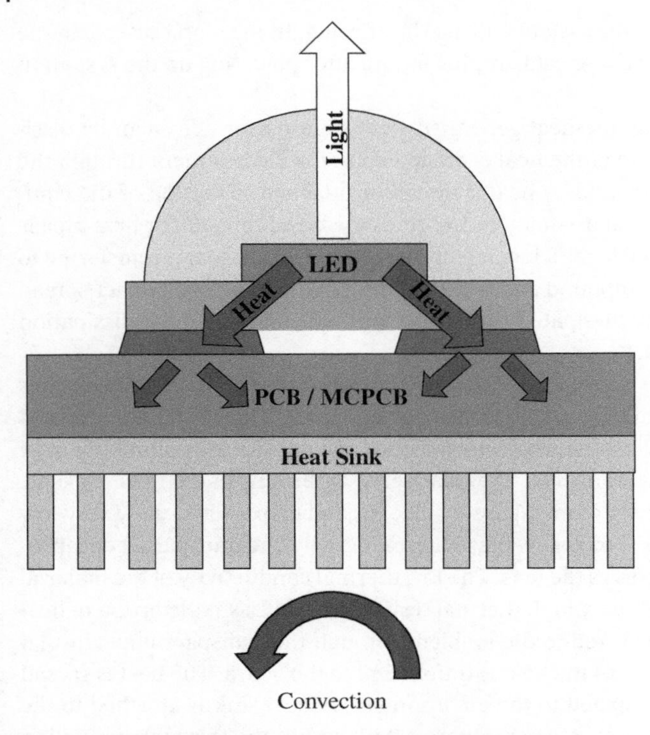

Figure 4.2 Heat dissipation path of LED assemblies.

less than 1 W/(m · °C). The copper trace on the PCB has a high thermal conductivity, which can help to dissipate the heat. However, the overall thermal resistance of PCB is still very high. It cannot effectively spread the heat generated from high-power LED packages and transfer it to the heat sink or environment. In most cases, it can only be used on low-power LED assemblies, for which thermal management is not a critical concern.

In most LED assemblies, there are usually more than one LED packages assembled on the board. The total amount of heat generated by those packages is huge. MCPCB (also named as insulated metal substrate, or IMS) is widely used in high-power LED assemblies to distribute and dissipate heat [1, 2]. It offers good thermal performance at a reasonable cost. In general, MCPCB has four key components: core, dielectric layer, copper traces, and solder mask. Figures 4.3 and 4.4 show the schematic diagram and cross-section of an MCPCB, respectively.

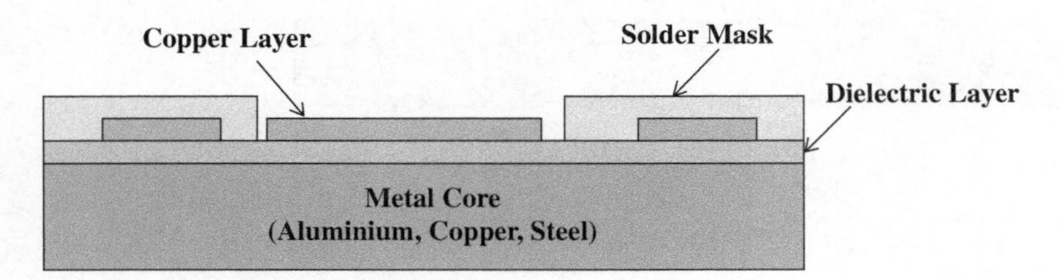

Figure 4.3 MCPCB schematic diagram.

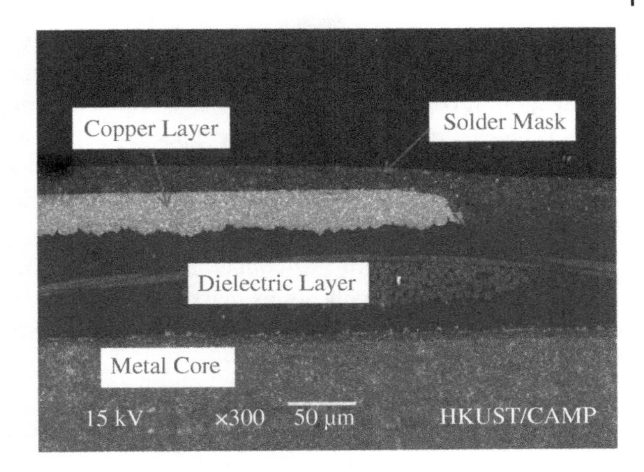

Figure 4.4 MCPCB cross-section.

Table 4.1 PCB core material properties.

Materials	Density (g/cm³)	CTE (10^{-6}/°C)	Thermal conductivity(W/(m · °C))
FR4	1.85	~12–14 (in-plane) ~70 (through-plane)	0.8–1.1 (in-plane) [3, 4] 0.29–0.34 (through-plane) [3, 4]
Copper	8.96	16.6	401
Aluminum	2.7	23.1	237
Steel	7.8–8.0	13.0	~16

The core of an MCPCB is made of metal, which provides mechanical integrity and serves as a heat spreader. Copper, steel, or aluminum is commonly used as the core of an MCPCB. Compared with FR4, a metal core has a much higher thermal conductivity (Table 4.1), which can effectively dissipate the heat generated from the LED components. Among those metals, copper has the highest thermal conductivity. However, copper is expensive and heavy, which substantially increases the manufacturing, setup, and shipping costs of the LED assemblies. Aluminum is lighter and has a good thermal performance with a reasonable cost. Therefore, aluminum is widely used as the core in most LED applications.

The thickness of the aluminum core is typically 1–3 mm. A thicker MCPCB has a better mechanical rigidity, but it is heavier and more expensive. Law et al. evaluated the effect of the aluminum core thickness on the overall thermal performance [5]. Their study included three types of MCPCB with different aluminum core thicknesses (0.9, 1.6, and 2.1 mm). High-power LEDs (Cree Xlamp MX6) were mounted onto the MCPCBs, and the junction temperature of the LEDs and overall thermal resistance were measured. The results showed that samples with a thicker aluminum core had a lower junction temperature and overall thermal resistance. However, it was not very effective to reduce the junction temperature of the LED by increasing the aluminum core thickness. According to their study, the junction temperature and the thermal resistance were reduced only by around 20% if the aluminum core thickness was increased from 0.9 to 2.1 mm. The increase in core thickness leads to more material consumption and hence increases the cost and weight of the LED assemblies.

The overall circuit designs of most LED assemblies are usually simple. Single-layer MCPCBs are normally sufficient to connect all components. A prepreg is used in MCPCB fabrication. It consists of a dielectric layer and a copper layer. The dielectric layer electrically isolates the metal core of the copper layer. The circuit layout is patterned onto the copper layer by a regular photolithographic process. The copper layer is usually 0.5 to 6 oz thick (1 oz = 35 μm thick). A thicker copper layer can spread the heat more effectively. However, it is difficult to achieve a fine line width/gap if a thick copper layer is used. Currently, MCPCBs with 1 oz copper, 5 mil line width, and 5 mil line gap are commonly available in the LED applications market. The photo-imaged prepreg is finally laminated onto the metal core by hot press. Since there is only one layer, it is not necessary to precisely align the prepreg with the metal core.

A layer of solder mask is applied to cover the copper layer. The solder mask opening patterns are also defined by a regular photolithographic process. The solder mask layer protects the copper layer from the environment. During the reflow process, the molten solder will wet on the copper layer. The solder mask layer defines the bond pad location (i.e. component location). It confines the solder joint geometry and prevents the solder wets along the copper traces. Different solder mask colors are available (green, red, blue, white, etc.). In order not to affect the final optical performance of the LED assemblies, white solder masks are used.

MCPCBs are not exclusive to LED applications. They are also widely use in other high-power and heavy-duty applications, such as power converters, telecommunications, high-voltage regulators, power supplies, etc. [6]. In these applications, the circuit is more complicated, which requires double-layer or multilayer MCPCBs. Multilayer MCPCBs consist of two or more copper layers on the same side of the metal core. This requires additional alignment and a hot press process to laminate the multilayers together with the metal core. There are normally no through holes on a multilayer MCPCB and hence it only supports surface mount type components.

There is also double-side MCPCB available on the market. Unlike multilayer MCPCBs, photo-imaged prepregs are laminated on both sides of the metal core. The electrical interconnects between the copper layers on both sides are made by PTHs. The through holes pass through the metal core, which is electrically conductive. An extra insulation layer is required to be deposited on the through hole sidewall; otherwise, all through holes will be electrically shorted together. Finally, copper is plated on the insulation layer to complete the electrical connections. Double-side MCPCBs are usually more expensive as they require additional alignment, insulation, and plating processes. Figure 4.5 shows a cross-section view of a double-side MCPCB with a PTH.

As the metal core is electrically conductive, a dielectric layer is used to provide insulation between the metal core and the copper traces. The dielectric layer is usually 35–125 μm thick. The breakdown voltage of the dielectric layer is proportional to the thickness. A thicker dielectric layer has a higher breakdown voltage [7]. However, a thick dielectric layer also has a higher thermal resistance which leads to a poorer thermal performance. The dielectric layer is either an organic layer (FR4 epoxy, polyimide, Teflon, etc.) or an epoxy with an inorganic filler (SiO_2, Al_2O_3, BN, etc.).

In general, the thermal conductivity of the dielectric layer is usually low. The dielectric layer of the MCPCB is always the bottleneck of the overall heat dissipation path. The thermal property of the dielectric layer will directly affect the junction temperature of the LED

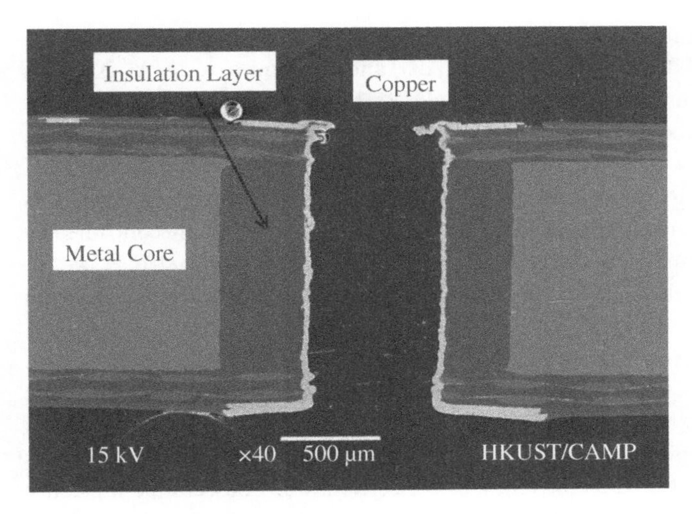

Figure 4.5 MCPCB with PTH.

chip [8]. Kim et al. studied the effect of the dielectric layer thickness and thermal conductivity on the junction temperature [9]. In general, a thin and thermal conductive dielectric layer yields a lower thermal resistance and thus the LED will have a lower junction temperature.

The overall thermal conductivity of the dielectric layer can be improved by adding high thermal conductivity fillers to the epoxy [10–12]. The fillers can improve thermal performance while maintaining a high breakdown voltage for electrical insulation. The thermal conductivity of different filler materials is listed in Table 4.2 [13].

There are different issues when selecting the filler materials. Diamond fillers can be added to the dielectric layer to improve thermal conductivity [14, 15], but this will substantially increase the cost. Silicon carbide (SiC) has a high thermal conductivity but is very hard,

Table 4.2 Dielectric layer filler material properties [13].

Filler material	CTE (10^{-6}/°C)	Thermal conductivity (W/(m · °C))
SiO_2	0.5–14	1–10
Al_2O_3	7.1	18–37
Si_3N_4	2.3	30–100
SiC	3.7	65–270
AlN	3.3	230
BeO	6.8	240
BN	3.7	300
Diamond	2.3	2000

Source: Lee, G., Lin, J. The introduction of high heat dissipation material. *Proc. 4th International Microsystems, Packaging, Assembly and Circuits Technology Conference (IMAPCT 2009)*, Taipei, 21–23 October 2009, pp. 240–243.

which causes difficulties in the manufacturing process. Also, the electric resistivity of SiC is relatively low compared with aluminum nitride (AlN) and aluminum oxide (Al_2O_3). This may lead to a low breakdown voltage. Beryllium oxide (BeO) has good thermal conductivity value but is toxic. In general, silicon dioxide (SiO_2) and Al_2O_3 are widely used in the market.

Lee et al. studied the effect of the filler content on the thermal conductivity of the dielectric layer. In their study, a higher thermal conductivity was achieved when more filler was added to the dielectric layer [13]. The packing density and the filler content are controlled by the filler size. A higher packing density can be achieved by reducing the filler size. However, a high packing density will also lead to a lower adhesion strength, poorer workability, and a higher production cost.

A ceramic dielectric layer can be used to improve the overall thermal performance. Ceramic material, such as alumina, has a good thermal conductivity and high dielectric strength. Cho et al. proposed an MCPCB with an alumina dielectric layer by aerosol deposition [16]. In their study, the thermal resistance of the ceramic dielectric layer was significantly lower than that of the conventional dielectric layer. The ceramic dielectric layer also has a very low leakage behavior at 100 V bias. However, the material costs and process costs of the ceramic dielectric layer are relatively high.

The thermal conductivity of the dielectric layer is still low, even when different types of filler are applied. Directly attaching the LED component to the metal core of the MCPCB is the most effective way to dissipate the heat. Heat is directly transferred to the high thermal conductivity metal core without passing through the dielectric layer. However, metal is electrically conductive, and so this approach may lead to an electrical short circuit. As a result, this method is not applicable for all LED components.

In some high-power LED components, a thermal pad is used exclusively for heat dissipation. The thermal pad is usually located directly under the LED chip, which has the shortest thermal dissipation path. The Cree XLamp is one of the examples that have this thermal pad design. The corresponding layout is shown in Figure 4.6 [17]. The thermal dissipation path is electrically isolated from the anode and the cathode of the component. If only the thermal pad of the LED component is directly attached to the metal core of the MCPCB, it will not lead to an electrical short circuit. This can substantially improve the overall thermal performance of the assemblies.

Several MCPCB designs allow the thermal pad of the high-power LED to be directly attached to the metal core. In general, a via is formed on the dielectric layer and is filled with a high thermal conductivity material. The thermal pad of the LED is then mounted onto the filled via. Alignment between the vias and the copper traces, via drilling and via filling, are the major challenges in fabricating this type of MCPCB.

Wang et al. proposed a SuperMCPCB structure [18]. In their design the thermal pad of the LED component was mounted on a copper post extruded from the metal core. Mashkov et al. suggested soldering the thermal pad of the LED component to an FR4 PCB with a copper pin [19]. Wu et al. proposed a similar structure. They drilled a hole through the dielectric layer and the metal core. The hole was then filled with copper by electroplating [20]. Juntunen et al. utilized a ceramic dielectric layer with a through via. The via was filled with screen printed silver [21]. They also tried to use FR4 as the dielectric layer. There were copper PTHs on the FR4 laminate which connected the LED thermal pads and the metal core [22]. All results showed that there were significant reductions in the overall thermal

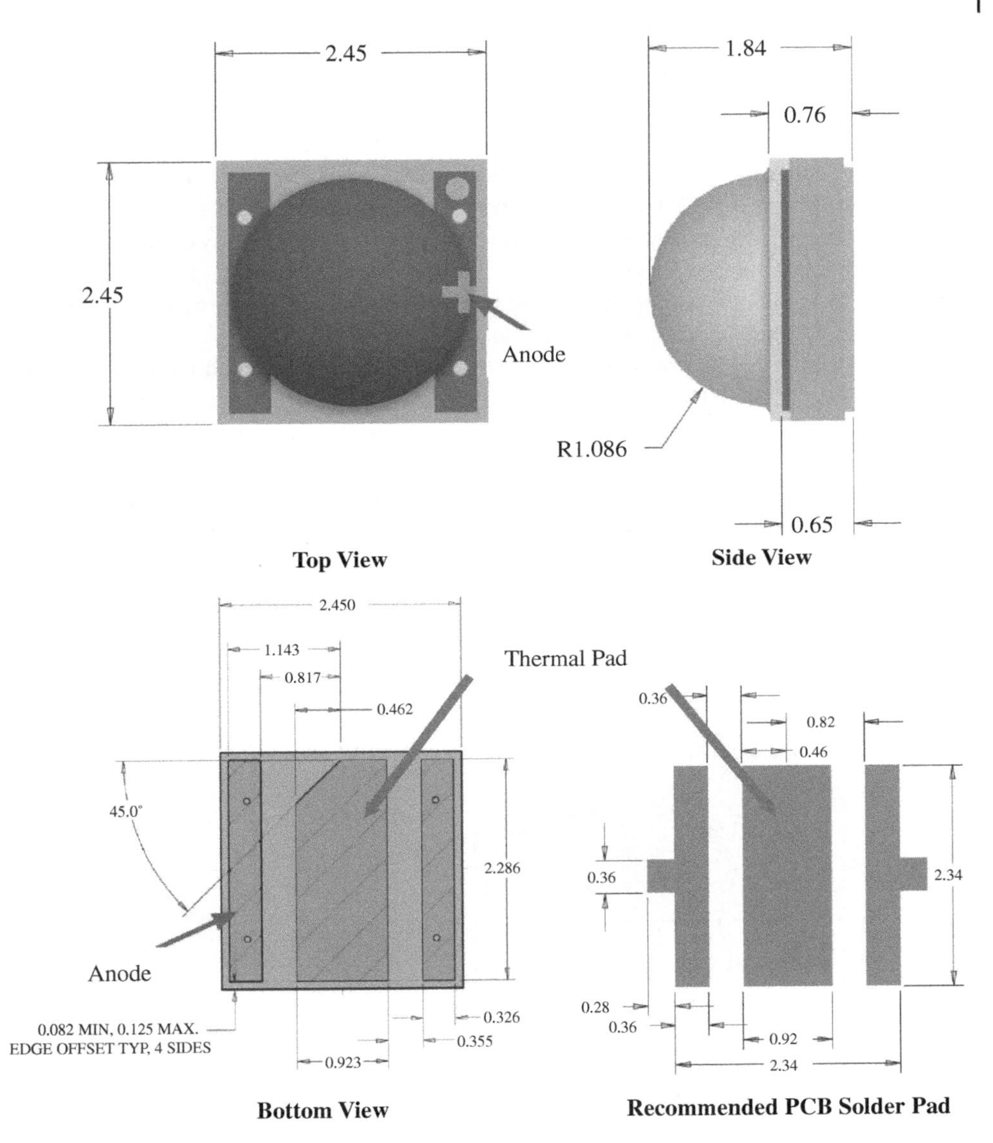

Figure 4.6 Cree XLamp bond pad layout [17]. Source: Modified from Cree Prdouct Falmily Data Sheet, Cree XLamp® XT-E® LEDs, Product Data Sheet. Cree Inc. Durham, NC, USA. 2011.

resistance of the MCPCBs. The heat was transferred to the metal core directly from the thermal pad without passing through the low thermal conductivity dielectric layer.

4.2.2 Printed Circuit Board with Thermal Vias

Although regular FR4 PCBs usually have poor thermal performance, which is not suitable for high-power LED applications, they have several advantages over MCPCBs. An FR4 PCB is cheaper than an MCPCB, which lowers the overall cost of the LED assembly. It is easier to fabricate fine pitch features on FR4 PCB. Readers can find 3 mil line gap/width FR4 PCBs widely available on the market. As mentioned in Section 4.2.1, an extra insulation layer on the side wall of the through hole is required to make a double-side MCPCB to prevent an electrical short circuit. This tremendously increases the cost of double-side MCPCBs. Also, an FR4 PCB is lighter than an MCPCB. For these reasons, people are trying to improve the thermal performance of FR4 so that it can be used in high-power LED applications.

The thermal performance of an FR4 PCB can be improved significantly by adding thermal vias. The thermal vias are plated with copper, which has a high thermal conductivity. Heat generated from the LED component is then transferred through the thermal vias to the heat sink. Figure 4.7 shows the schematic diagram.

The fabrication process of thermal vias is similar to that performed for PTHs. In conventional FR4 PCBs, PTHs are only used to electrically connect the circuits in different layers. The PTHs are normally plated with copper. Although copper has a very high thermal conductivity, which can transfer heat effectively, PTHs are not optimized for heat dissipation. It is because only a conformal copper layer covers the side wall of the through hole, as shown in Figure 4.8. The thickness of that copper layer is usually 0.5–1 oz (17–35 μm). This thin layer does not offer sufficient heat dissipation. There are some numerical studies

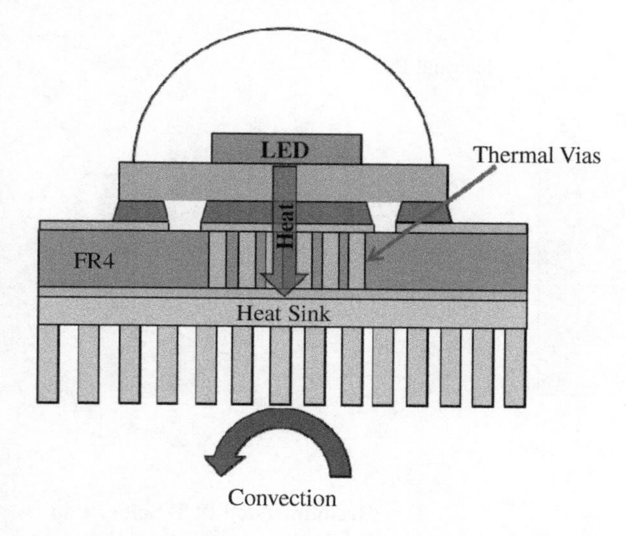

Figure 4.7 FR4 PCB with thermal vias.

Figure 4.8 PTH with conformal copper layer.

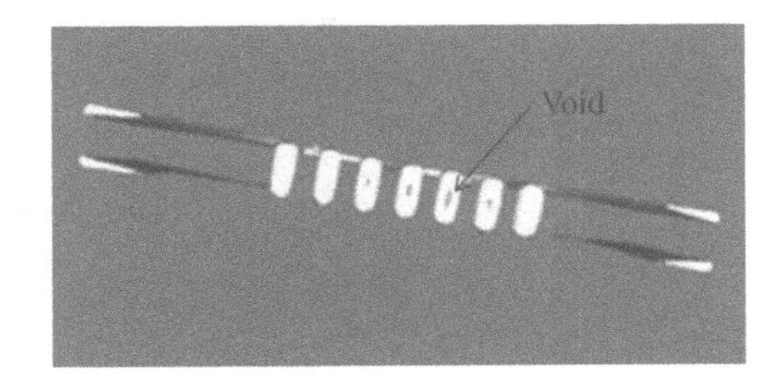

Figure 4.9 Void trapped in thermal vias.

showing that if the through hole is full-filled by copper the overall thermal resistance of the PCB is reduced and hence suitable for high-power LED applications [23, 24].

There are two challenges when implementing the thermal vias on an FR4 PCB for LED assemblies. First, it is very difficult to completely fill the through hole with copper. The plating process has to be well controlled; otherwise, a void will be formed in the copper layer. Figure 4.9 shows an X-ray inspection of thermal vias with voids. The void may affect the overall thermal performance and reliability. If the plating parameters are not optimized, the vias may be over-plated (or under-plated). This will affect the flatness of the top and bottom surfaces of the PCBs, and hence may introduce problems to the subsequent assembly process. Figure 4.10 shows some common defects of thermal via copper filling. Second, not all LED components are designed with an electrically isolated thermal pad. If the electrodes of the LED component are directly connected to the thermal vias, it will increase the risk of electrostatic discharge (ESD) failure.

MacDermid Enthone have developed effective copper-filled thermal vias [25]. Figure 4.11 shows the cross-section inspection of a thermal via plated by their technology. It is void free and flat on both the top and the bottom surface.

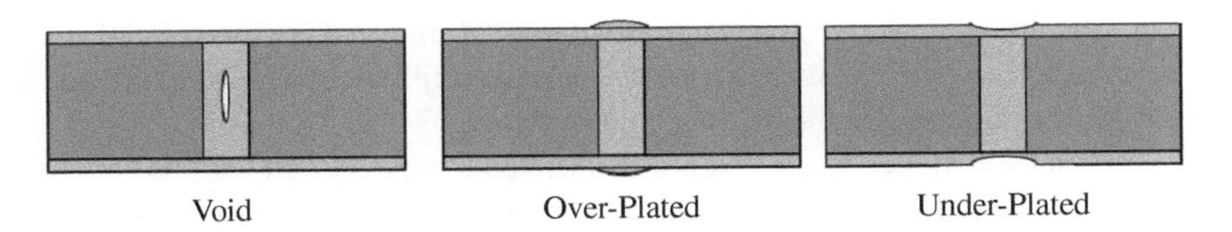

Figure 4.10 Thermal vias filling defects.

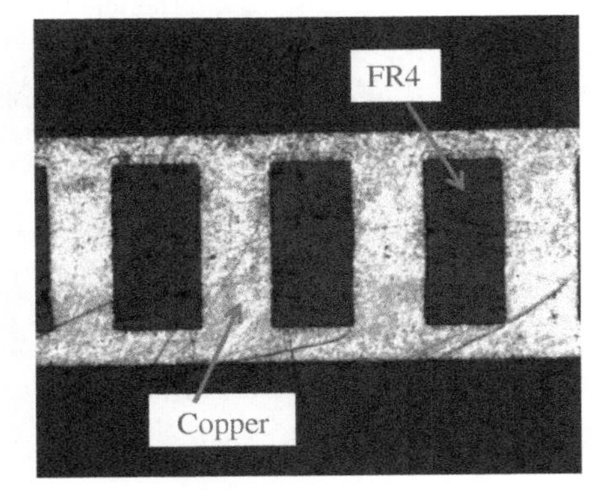

Figure 4.11 Full-filled thermal via by LuMac Copper Systems.

Table 4.3 Experiment test matrix (MCPCB and FR4 with thermal vias thermal performance comparison).

PCB type	Via diameter	Filling mode
MCPCB	Nil	
	0.25 mm	Conformal
FR4 with thermal vias	0.35 mm	Conformal
	0.35 mm	Full-filled

A study was carried out to compare the thermal performance of an FR4 PCB with thermal vias and MCPCB with 100 μm thick dielectric layer. There were 14 vias fabricated on each sample. Table 4.3 shows their test matrix. The effects of via diameter and the filling mode (conformal or full-filled) were studied. The PCB used in the study is shown in Figure 4.12.

The white light CREE XLamp XP-E LED, which has an electrical isolated thermal pad design, was used as the test vehicle. It was mounted on the sample by a regular surface mount soldering process. The cross-section inspection is shown in Figure 4.13. The junction temperature of each sample was measured and recorded by a Mentor Graphic T3Ser system. During measurement, the bottom side of the sample was attached to a thermostat which was maintained at 25 °C. (The measurement method is discussed in detail in Chapter 5.) Figure 4.14 shows the junction temperature results.

The experimental results showed that the LED component attached to the MCPCB had the highest junction temperature. This was because the thermal conductivity of the MCPCB dielectric layer was very low. Heat could not be effectively transferred from the LED component to the metal core through the dielectric layer. On the other hand, the thermal pad of the component was directly mounted on the thermal vias. Heat was dissipated effectively through the thermal vias. The effects of the via diameter and filling mode are significant. A larger full-filled thermal via had the best thermal performance.

The overall thermal resistance of the sample was calculated by Eq. (4.1) and is shown in Figure 4.15. In the calculation, only the power that contributed to heat generation was considered. It was equal to the difference between the total electric input power to the LED

Figure 4.12 FR4 PCBs with thermal vias.

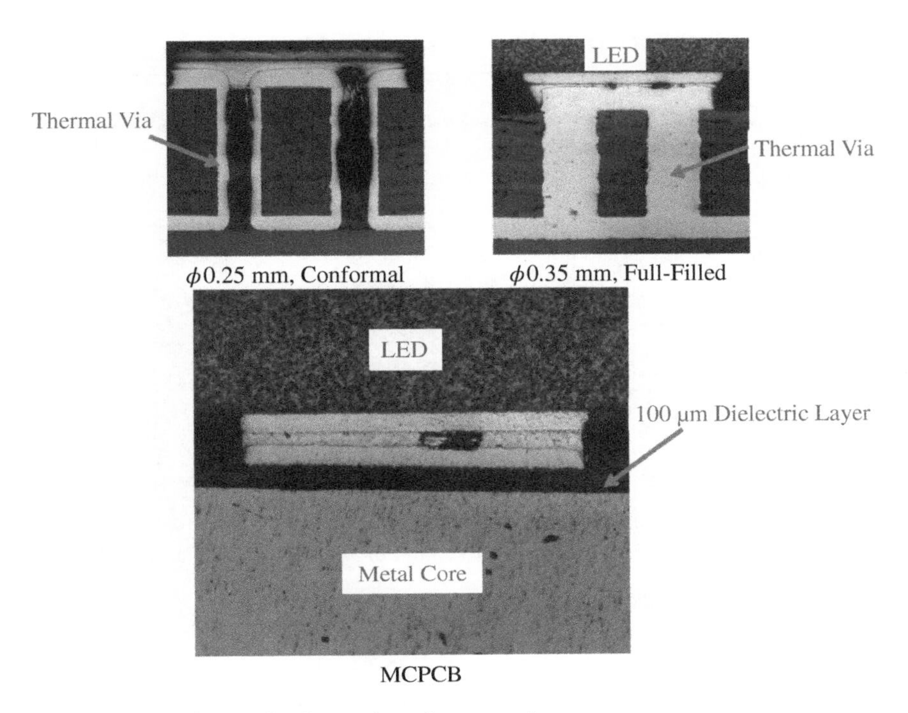

Figure 4.13 Cross-section inspection of test samples.

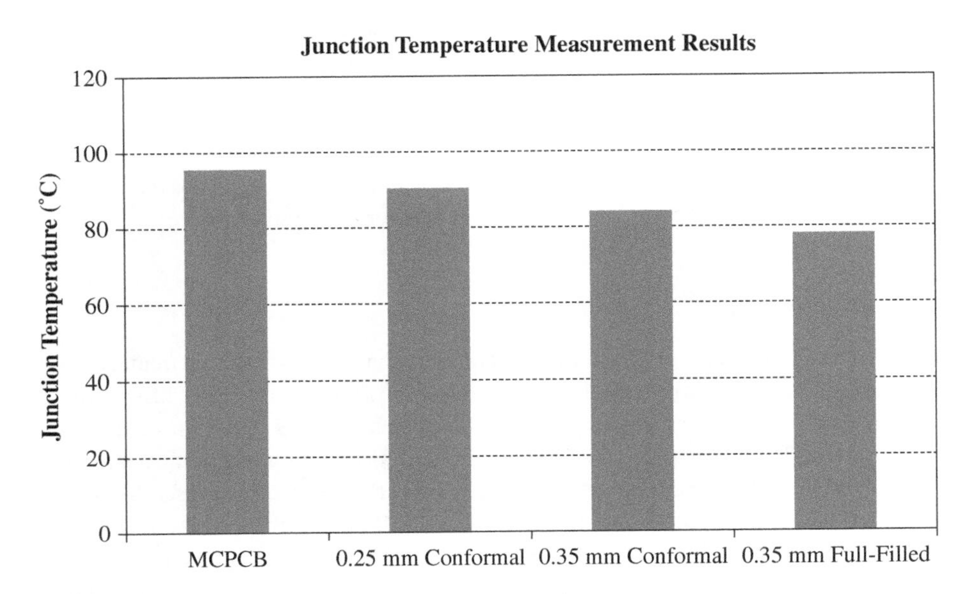

Figure 4.14 Junction temperature measurement results.

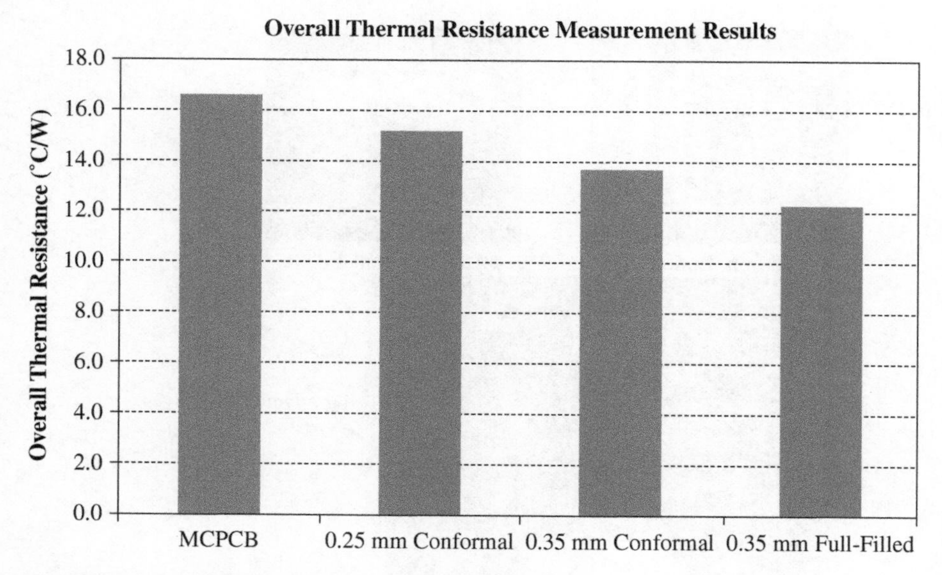

Figure 4.15 Overall thermal resistance measurement results.

and the optical output power from the LED. The thermal resistance that corresponded to the PCB was also calculated and is shown in Figure 4.16. The results indicated that regular MCPCBs had the worst thermal performance. Conversely, the FR4 PCB with full-filled thermal vias had a very low thermal resistance. Experimental results showed that the thermal performance of the FR4 PCB could be significantly improved by adopting full-filled thermal vias.

$$R_{\text{overall}} = \frac{T_{\text{j}} - 25}{P_{\text{electric}} - P_{\text{optical}}} \tag{4.1}$$

where R_{overall} is overall thermal resistance, T_{j} is junction temperature, P_{electric} is total electric input power to the LED, and P_{optical} is optical output power from the LED.

4.2.3 Wave Soldering

As mentioned in Chapter 2, the PTH component does not have a good thermal management design. It cannot be used with medium- and high-power applications. PTH LED components are normally used in low-cost and low-power applications, as shown in Figure 4.17. After being assembled to the board, the PTH component has a strong mechanical strength, which is why PTH sockets and connectors can still be found in some LED assemblies.

After inserting the components into the PTH of the PCB, solder material is applied from the back side of the PCB. The molten solder wets along the component lead and the side wall of the through hole. The solder is then solidified and a solder joint is formed. The solder can be applied manually. For large volume manufacturing, a wave soldering process is used.

Most wave soldering machines have similar components and working principles. Generally, there are three key zones in a typical waver soldering process. They are fluxing,

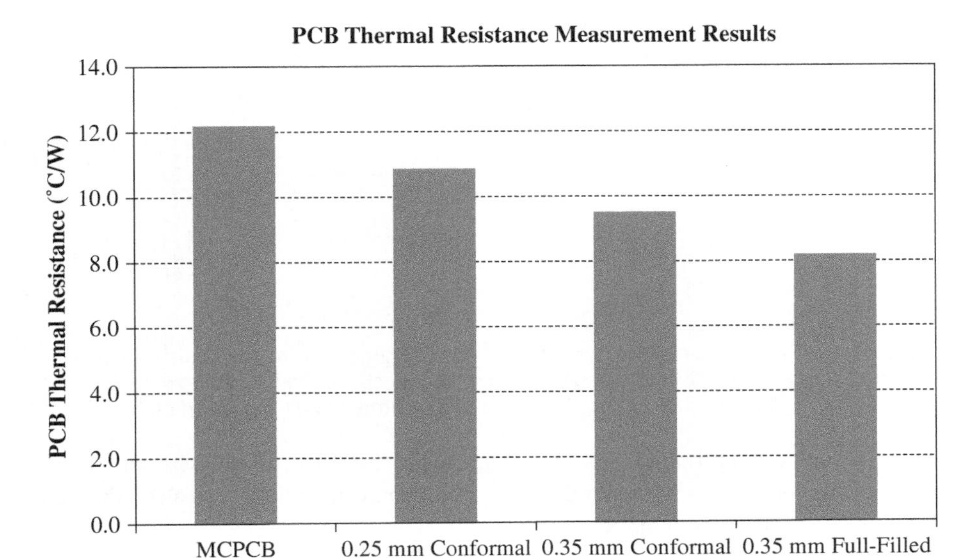

Figure 4.16 PCB thermal resistance measurement results.

Figure 4.17 PTH LED assembly.

preheating, and soldering. There is a conveyor which transfers the PCB through these zones. Figure 4.18 shows the process flow of wave soldering.

The PCB is first transferred to the fluxing zone. Liquid flux is sprayed onto the back side of the PCB by a fluxer. The flux used in wave soldering is usually mixed with solvent to reduce viscosity, making it easier to get into the PTH. The major function of the flux is to remove oxide on the component lead and PTH sidewall. This provides a clean surface for

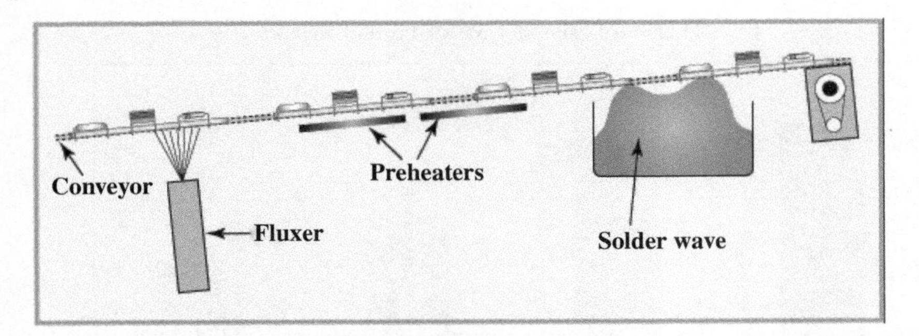

Figure 4.18 Typical wave soldering process [26]. Source: PCBgogo. How to distinguish between wave soldering and reflow soldering? June 2019. https://www.pcbgogo.com/Blog/How_To_Distinguish_Between_Wave_soldering_And_Reflow_Soldering_html [www.pcbgogo.com].

the subsequent soldering process. After the soldering process, flux residues will remain on the surface of the board. Depending on the flux types, it may be necessary to clean the flux residues with hot water or an organic solvent.

Next, the PCB will reach the preheating zone. The PCB is heated up to above 100 °C. The flux is activated and the solvent in the flux is evaporated at this high temperature. Preheating the PCB is an important step to prevent thermal shock. If the PCB is not preheated and is suddenly exposed to the high temperature of molten solder (normally >230 °C), the huge temperature difference will cause serious damage to the PCB. The preheating step can reduce the degree of board warpage after the wave soldering process. During the wave soldering process, the bottom side of the board is exposed to the hot molten solder. The temperature at the bottom part of the PCB is higher than the top part. The temperature difference will introduce different degrees of thermal expansion. This is the major cause of board warpage. Preheating can reduce the temperature gradient between the bottom and the top of the PCB. If the temperature difference between the molten solder and the PCB is large, the molten solder will solidify quickly. The solder will solidify before it fully fills the PTH. This will lead to incomplete PTH filling and will affect reliability.

After preheating, the PCB will go to the soldering zone. There is a tank of molten solder, and a pump generates standing waves in the molten solder. The bottom surface of the PCB will have direct contact with the molten solder, as shown in Figure 4.19. The molten solder wets along the component lead and the side wall of the PTH. After the PCB exits the soldering zone, the temperature decreases gradually to room temperature and the molten solder solidifies. A PTH solder joint is finally formed. If a flux which requires residue cleaning is used, the PCB should be rinsed with hot water or organic solvent.

4.2.4 Surface Mount Reflow

SMDs are widely used in LED assemblies and conventional electronic devices. The board level assembly processes of SMD components are different from those for PTH components. In general, surface mount reflow is used and it involves three key steps: solder paste printing, pick-n-place, and reflow.

Solder material is applied onto the PCB (or MCPCB) by solder paste printing. Solder paste is a mixture of tiny solder particles (Figure 4.20) and flux. Normally, solder paste consists

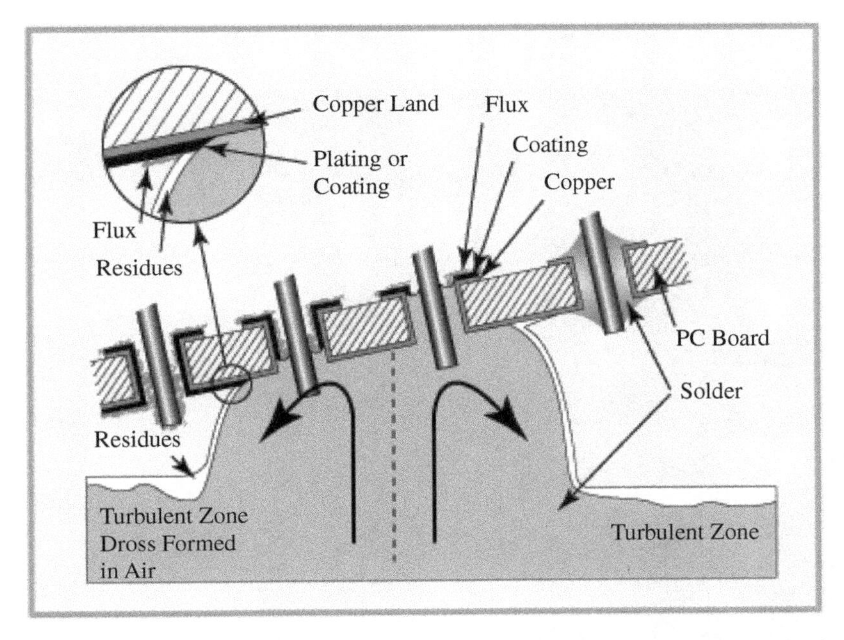

Figure 4.19 Wave soldering [26]. Source: Based on PCBgogo. How to distinguish between wave soldering and reflow soldering? June 2019. https://www.pcbgogo.com/Blog/How_To_Distinguish_Between_Wave_soldering_And_Reflow_Soldering_.html.

Figure 4.20 Solder particles in solder paste.

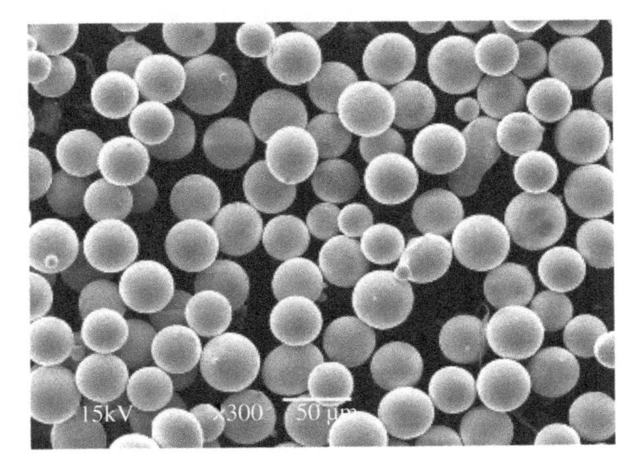

of around 20% of flux. A stencil is used during the solder paste printing process. The stencil is usually made by a thin stainless steel sheet (usually 100–150 μm thick). The openings on the stencil are matched with the bond pad locations on the PCB. In the printing process, we have to make sure there is close contact between the stencil and PCB after the alignment process. Solder paste is then printed on the bond pads of the PCB. Figure 4.21 shows a typical solder paste printing result. The amount of solder being transferred to the PCB bond pad is controlled by the stencil opening area, stencil thickness, and the solder content in the paste.

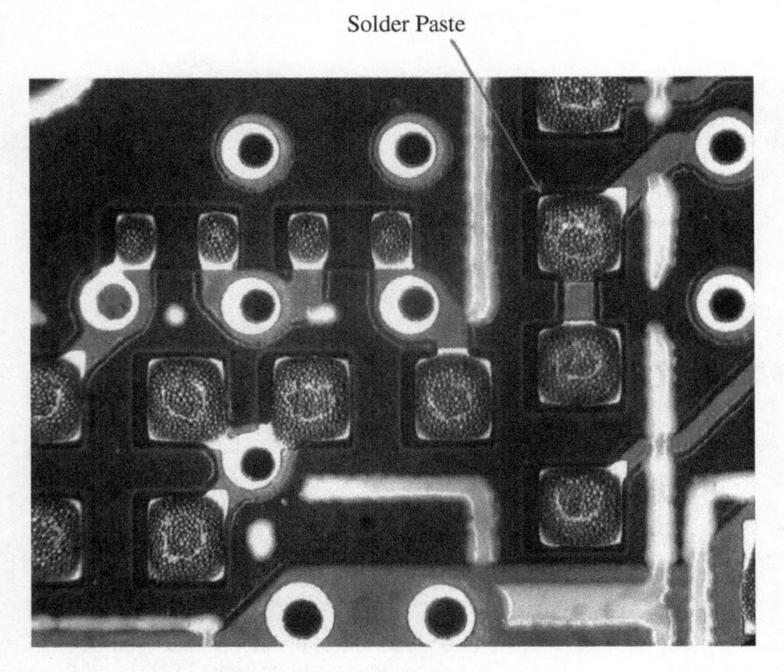

Figure 4.21 Solder paste printing results.

Several factors affect the quality of solder paste printing, such as the viscosity and tackiness of the paste, printing speed, etc. The solder particle size is also a critical parameter. If the bond pad is small and the pitch is fine, smaller solder particles are required. JEDEC has published a standard which categorizes the solder pastes into six types based on particle size (Table 4.4) [27]. In most applications, type 3 solder is normally used. If the solder paste is too fine, the solder particles will be oxidized easily.

After solder paste printing, the components are placed onto the board by a pick-n-place machine. During the reflow process, the surface tension of the molten solder will pull the

Table 4.4 Solder paste type (according to particle size) [27].

Type	None larger than (μm)	Less than 1% larger than (μm)	80% Minimum between (μm)	10% Maximum less than (μm)
1	160	150	150–75	20
2	80	75	75–45	20
3	50	45	45–25	20
4	40	38	38–20	20
5	30	25	25–15	15
6	20	15	15–5	5

slightly misaligned component back to its original position. This self-alignment capability can lower the pick-n-place accuracy requirement.

The board is then transferred to a reflow oven for solder reflow. A typical reflow oven has multiple heating zones. A conveyor will transfer the board through those heating zones. The board together with the components and solder will then be subjected to a temperature profile. The profile can be tuned by adjusting the conveyor speed and the temperature of each heating zone. In some LED assemblies, such as LED tubes, LED components are mounted on a long MCPCB. The length of the MCPCB may be longer than one heating zone. Normally, a reflow oven with at least 6–8 heating zones is required to achieve a uniform temperature distribution on the board.

A typical reflow temperature profile can be divided into different sections: preheat, ramp up, reflow, and cooling. The functions of the preheat section on the surface mount reflow process is the same as wave soldering. The board is first quickly heated to a high temperature. This is called the soak temperature and is usually above 150 °C. The board is maintained at soak temperature for a while. The soak temperature and dwell time are determined by the flux type. The flux in the solder paste is activated and the solvent is evaporated. This preheat section can also prevent damage due to thermal shock.

After that the board temperature will ramp up to the peak temperature, which is around 240–260 °C if SAC solder is used. The ramp rate determines how fast the temperature increases from the soak temperature to the peak temperature. A higher ramp rate can increase the throughput but will cause damage to the board and components. Ceramic components may crack if the temperature changes rapidly. Usually, the ramp rate is around 1–3 °C/s in order to maintain a high throughput.

The solder melts when the temperature is above its liquidus temperature. The time the board temperature is above the liquidus temperature is important and needs to be closely controlled. A shorter period can increase the overall throughput, but it is necessary to have sufficient time for the molten solder wets on the bond pad to form an intermetallic compound (IMC) with the bond pad. However, it will cause thermal damage to the board and the components if they are exposed to this high temperature for too long. A thicker IMC layer will also be formed and may affect the long-term solder joint's reliability. In general, the board is maintained above the liquidus temperature for 60–150 seconds.

Finally, the board will be cooled down to room temperature. The cooling rate from the peak temperature to the liquidus temperature will affect the grain size in the solder joint. A faster cooling rate will lead to a finer grain size which has a better mechanical reliability. However, if the cooling rate is too fast, it will cause thermal shock and damage the board.

As mentioned above, the temperature and the duration of each section are critical and will affect the quality and the reliability of the product. The solder paste and component vendors usually recommend a reflow profile for their products. For example, Cree recommends following the profile stated in the standard J-STD-020C [28] for their XLamp XB-D [17]. Figure 4.22 and Table 4.5 show the requirements listed in the standard. In practice, it is not possible for all components on the board to have the same temperature profile. Therefore, the standard recommends the temperature of all components should fall within the profile band.

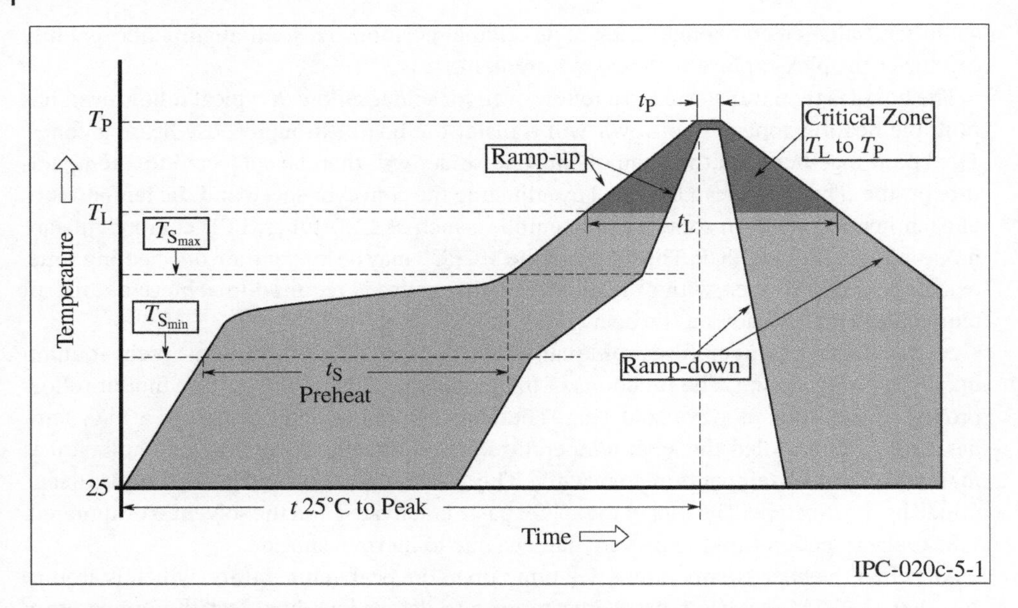

Figure 4.22 Typical reflow profile [28]. Source: Moisture/Reflow Sensitivity Classification for Nonhermetic Solid State Surface Mount Devices, IPC and JEDEC Solid Sate Technology Association, J-STD 020C, 2004.

Table 4.5 Reflow profile parameters (for Cree XLamp XB-D) [17].

Profile feature	Lead-based solder	Lead-free solder
Average ramp-up rate($T_{S_{max}}$ to T_P)	3 °C/s max.	3 °C/s max.
Min. preheat temp. ($T_{S_{min}}$)	100 °C	150 °C
Max. preheat temp. ($T_{S_{max}}$)	150 °C	200 °C
Preheat time (t_s)	60–120 s	60–180 s
T_L	183 °C	217 °C
Time maintained above T_L (t_L)	60–150 s	60–150 s
Peak temp. (T_P)	215 °C	260 °C
Time with 5 °C of actual peak temp. (t_P)	10–30 s	20–40 s
Ramp-down rate	6 °C/s max.	6 °C/s max.
Time 25 °C to T_P	6 min max.	8 min max.

Source: Modified from Cree Product Family Data Sheet, Cree XLamp® XT-E® LEDs, Product Data Sheet. Cree Inc. Durham, NC, USA. 2011.

4.3 Chip-on-Board Assemblies

Section 4.2 discusses the board level assembly process in detail. The LED chip is first packaged into a component and then assembled onto a board. Although the efficiency of LED chips has been enhanced substantially in recent years, the optical output of a single chip or

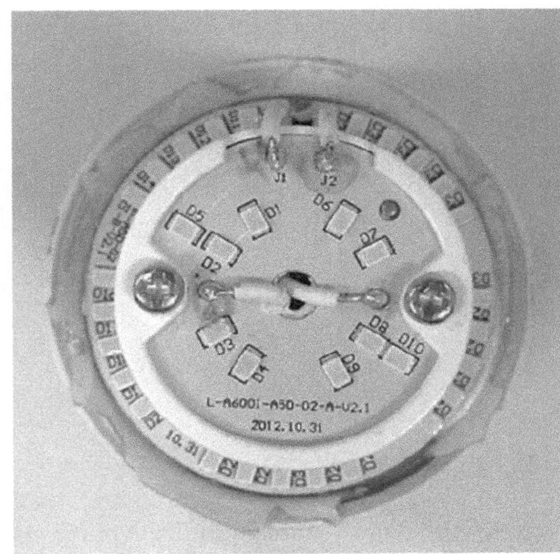

Figure 4.23 Board level assembly with multiple LED components.

component is still not sufficient for most applications. It is necessary to put multiple LED components on a board to achieve an acceptable optical performance. Figure 4.23 shows an example of an LED lightbulb with multiple surface mount LED components. Each LED chip has to be packaged, tested, and mounted on the board. LED packaging materials and processes are expensive and hence increase the overall cost [29, 30].

Board level assembly offers better design flexibility. The designer can choose the proper LED components for their applications. Each component is tested and the performance is well characterized. The overall performance – such as view angle, color temperature, angular distribution of light intensity, etc. – can be easily changed by modifying the board design and layout. However, if the applications are already known and specified, COB assemblies may offer a more cost-effective solution. The structure of an LED COB assembly is simple, as shown in Figure 4.24.

In addition to low cost, COB assemblies have better optical and thermal performance. Owing to the package size, LED components cannot be placed in a fine pitch in board level LED assemblies. Each discrete LED component on the board level assembly is considered a point light source. If a luminaire has multiple point light sources, this will lead to poor quality for general lighting applications. First, the illuminance uniformity is highly dependent on the LED component pitch. This is discussed in Section 4.4. In general, it is necessary to add a diffuser to improve illuminance uniformity. However, a diffuser will block part of the light and reduce the overall lumen output.

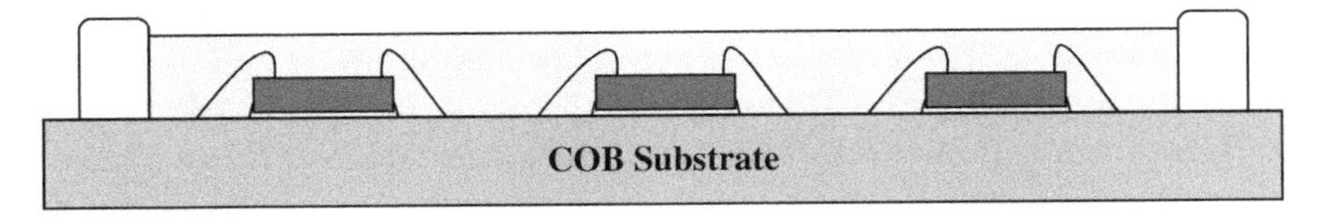

Figure 4.24 LED COB assembly structure.

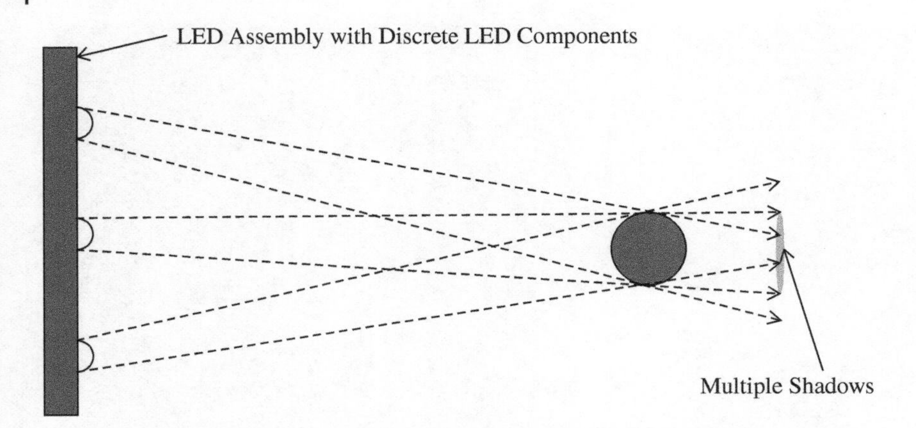

Figure 4.25 Multiple shadows from LED assembly with discrete LED components.

Multiple shadows is another problem of the board level LED assembly. Since the LED components are placed apart from each other, the light emitted from each LED shines at the object at a slightly different angle, as shown in Figure 4.25. As a result, there are multiple shadows at the back of the object. This may cause the user discomfort [31].

COB assembly allows mounting multiple chips on the substrate in a finer pitch. It is possible to have a large light-emitting surface by placing multiple medium-power LEDs in one COB assembly. Uniform light is emitted from this large surface. It is not necessary to add a diffuser to achieve a uniform illuminance. The large light-emitting surface can also eliminate the multiple shadow problem as there is only one light source. Therefore, COB assemblies are widely used in general lighting applications, for example downlights and spotlights. There are various COB sizes available on the market. For example, in the LUXEON COB product line, the diameter of the light-emitting surface ranges from 9 to 19 mm, as shown in Figure 4.26 [32]. The light-emitting surface of a COB assembly is not necessarily circular. There are also COB assemblies with square or rectangular light-emitting surfaces for different applications.

The good thermal performance is also an advantage of COB assemblies. In board level LED assemblies, the heat generated from the junction is dissipated to the environment through the chip, die attach material, package substrate, solder joint, and MCPCB. COB assembly has a shorter heat dissipation path. Heat is transferred to the environment through the chip, die attach material, and MCPCB. Figure 4.27 illustrates the difference. This short heat dissipation path has a lower overall thermal resistance. Therefore, COB assembly usually has a lower junction temperature and better long-term reliability when compared with regular board level LED assemblies [33, 34].

The COB fabrication process has three key steps: die mounting, interconnection, and encapsulation. Bare LED chips are first directly attached to the board. In most LED COB assemblies, multiple LED chips are usually used to obtain sufficient optical output. MCPCBs are commonly used in LED COB assemblies. The metal core in the MCPCB can spread the heat in a more effective way than that of FR4 substrate, and hence a lower junction temperature is obtained. However, as mentioned in Section 4.3, the dielectric layer has a high thermal resistance, which affects overall thermal performance [35, 36]. If the

Figure 4.26 LUXEON COB [32].

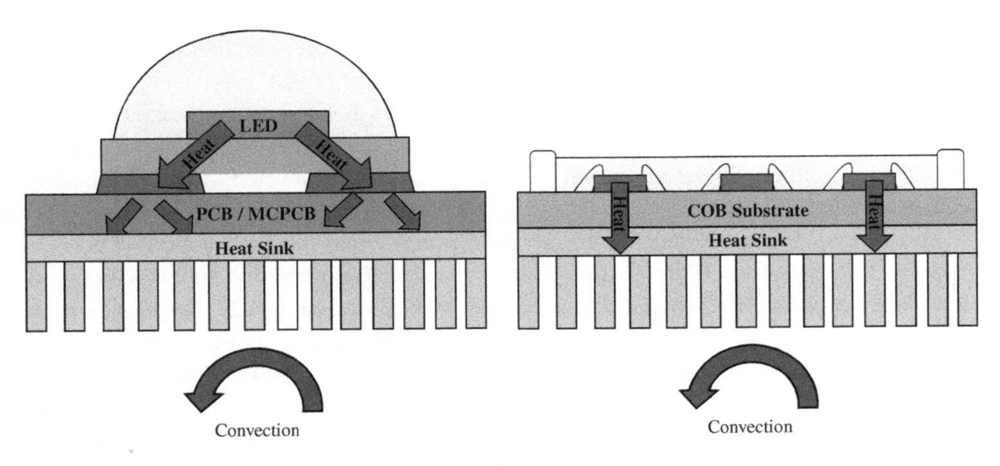

Figure 4.27 SMD vs. COB (heat dissipation path).

LED chip is directly attached to the metal core, it will have a higher risk of ESD failure. Ceramic substrate is used in some high-end COB assemblies. They have a high thermal conductivity and are not electrically conductive. It is not necessary to deposit a dielectric layer on top of the ceramic substrate. A copper layer can be directly deposited and patterned on the substrate [37]. Therefore, COB assemblies with ceramic substrates normally have a better thermal performance [38, 39]. Kim et al. compared the AlN and Al_2O_3 ceramic substrates [40]. Owing to the fact that AlN has a higher thermal conductivity, the thermal performance of AlN substrate was better in their experiment. COB with AlN substrate has a higher radiant flux than that of COB with Al_2O_3 substrate. As shown in Figure 4.28, the difference was significant when the assemblies were driven at a high current.

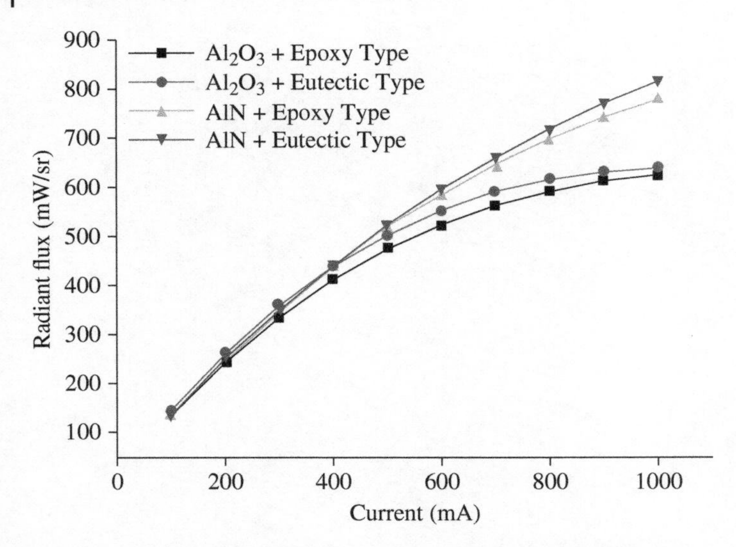

Figure 4.28 Al_2O_3 vs. AlN (radiant flux comparison) [40]. Source: Kim, J.S., Jeon, S.L., Le, D.W., et al. Thermal and optical properties of COB type LED module based on Al_2O_3 and AlN ceramic submounts. *Journal of Applied Sciences*, 2010, 10(24): 3388–3391.

After the die bonding process, electrical interconnects between the chips and the board are usually made by wire bonds. The circuity on the board links the LED chips together. The LED chips are normally connected in a series so that the electrical current across each LED is the same. However, if too many LEDs are connected in a series, it will lead to a very high driving voltage. Therefore, multiple series strings are connected in parallel in some COB assemblies.

Glob top dispensing and dam and fill are the common encapsulation methods in conventional IC COB assembly. Glob top dispensing is the simplest method to apply the encapsulant. The encapsulant is directly dispensed on top of the chip. Figure 4.29 shows an example of a COB assembly with glob top dispensing. A dome shape is formed by the surface tension of the encapsulant. The flow and the geometry of the encapsulant are difficult to control and are determined by the volume, viscosity, and contact angle of the material. The glob top geometry will directly affect optical performance, such as angular intensity and color temperature uniformity. Therefore, COB assembly normally does not adopt the glob top dispensing method.

In dam and fill dispensing, a dam is first prepared on the substrate. The dam is made by dispensing a high-viscosity epoxy on the board, as shown in Figure 4.30. In order to increase the throughput, the dam is formed on the substrate by molding in some COB assemblies. A low-viscosity encapsulant is then dispensed to fill the area inside the dam. In LED COB assemblies, phosphor slurry is used. The dam stops the flow of the low-viscosity encapsulant and confines the final shape. Figure 4.31 schematically shows the process flow. This method has a better control on the geometry of the encapsulant.

There are other encapsulation methods and configurations. Yuen et al. tried to improve optical performance by adopting remote phosphor in LED COB assemblies [42]. They first filled the dam with clear silicone. After curing, a remote phosphor layer was deposited on

Figure 4.29 Glob top dispensing [41]. Source: Source: NICHE-TECH. 2019. http://www.nichetechcorp.com/ (cited date 3 September 2020).

Figure 4.30 Dam on COB substrate. Source: Philips.

top of the clear silicone by dispersed dispensing. The structure is shown in Figure 4.32. Juntunen et al. compared the thermal performance of COB assemblies with different configurations [43]. Since the conversion of blue light to yellow light by phosphor will generate heat, this may increase the junction temperature if phosphor slurry is directly applied to the top of the LED chip. In their study, they measured the junction temperature of COB assemblies. Remote phosphor and conventional dispersed dispensing were included. In their remote phosphor setup, the clear silicone was covered by a phosphor plate, as shown in Figure 4.33. However, both studies showed that there were no significant improvements in optical and thermal performance.

The dam and fill method has two limitations. First, the curvature of the light-emitting surface is controlled by the surface tension of the slurry and is usually quite flat. A dome shape is more preferable because less light is trapped by total internal reflection. Second, all chips within the dam are covered by the same type of phosphor materials. In some applications, a few red LED chips are added to improve the color rendering index (CRI) of the

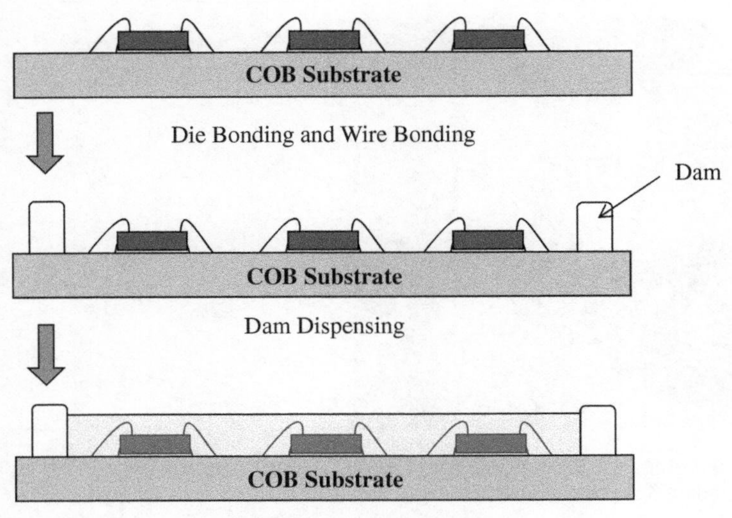

Figure 4.31 LED COB assembly fabrication process.

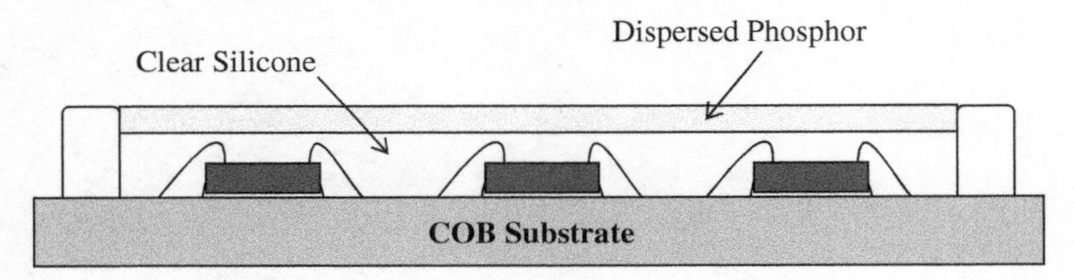

Figure 4.32 COB assembly with remote phosphor layer.

product. Most phosphors for LED application have high conversion efficiencies when they are excited by blue (or UV) light only. As a result, a portion of red light will be blocked or scattered if the chips are covered by the phosphor particles.

Owing to the aforementioned reasons, some LED COB assemblies are encapsulated by compression molding. The geometry of the encapsulant is determined by the mold cavity. A small dome can be formed on each LED. The selective molding method can be used to cover the red LED chips by clear silicone only, while the blue LED chips are covered by phosphor. Figure 4.34 shows some examples.

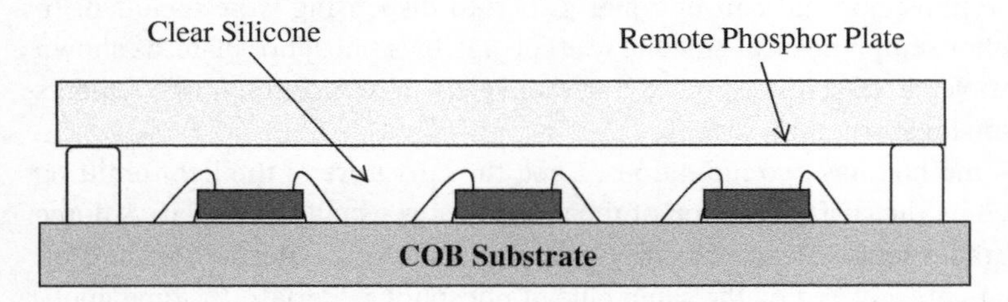

Figure 4.33 COB assembly with remote phosphor plate.

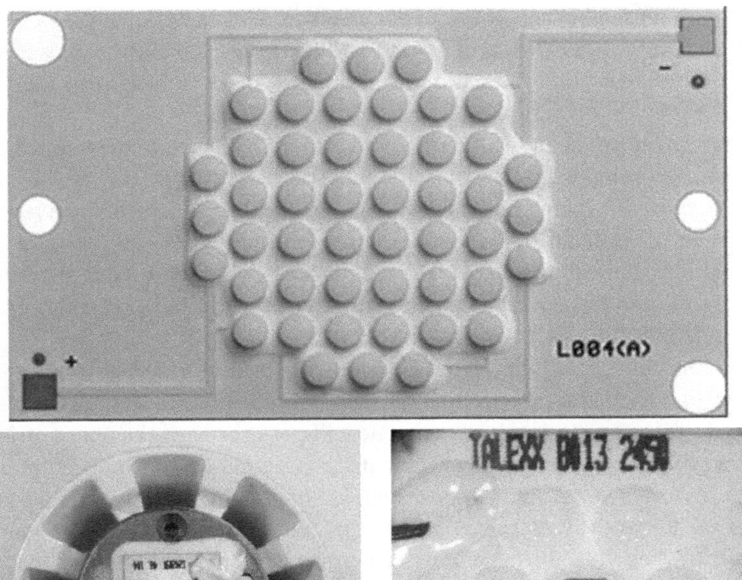

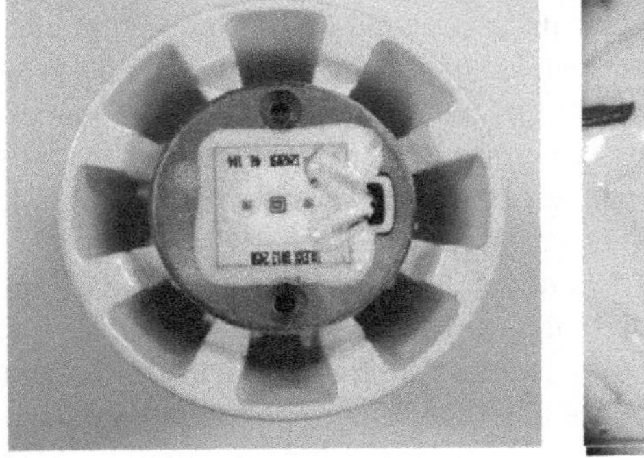

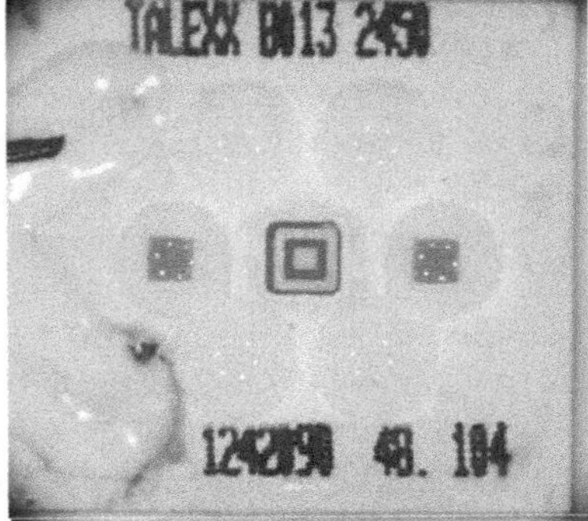

Figure 4.34 COB encapsulant by molding.

After the manufacturing process, the COB assembly is mounted on the heat sink. A thin layer of thermal interface material (usually thermal grease) is applied between the COB and the heat sink to reduce the contact thermal resistance. Wires are then soldered to the terminals on the COB to complete the electrical connection with other devices. Some COB assemblies can use solder-less connectors [30]. The connectors have mechanical contacts with the terminal on the COB. This does not involve any soldering process and hence simplifies the installation and replacement procedures.

4.4 LED Modules and Considerations

The optical output of a single LED component or chip is usually not sufficient for most applications. Therefore, an LED module normally has multiple LED packages. Similarly, there are multiple chips in one COB assembly to achieve enough lumen output. The number of LEDs, LED connection method, and the location of those LEDs are the key considerations in LED module design. These will affect the electrical, optical, and thermal performance, and also the cost of the whole module.

The LED connection method is one of the major electrical considerations. It will determine voltage and current requirements and hence provide a guideline to the driver design or selection. The circuit layout on the PCB or substrate also has to match with the LED connection method.

In general, the optical output of an LED component is directly related to the applied current. Because of this, the components on the LED module should be driven by the same constant current. No two LED components have identical electrical characteristics, owing to the variations in the manufacturing and packaging process. If the LED components are connected in parallel, the voltages across the components will be the same. However, the currents passing through those components are different. Therefore, LED components are normally connected in series to maintain a constant current.

Although connecting the LED components in series can ensure a constant and same current passing through each component, there are two disadvantages. If one LED has a problem which leads to an open circuit, no current will pass through the series strings. All LEDs will then go off at the same time. The catastrophic failure of the LED module may cause a safety hazard in some applications, such as street lighting. This can be solved by adding a Zener diode or an LED protector [44] in parallel with each LED component. When one LED in the string fails, the current will pass through the Zener diode or the protector. This maintains the current flow in the whole string. The amount of optical output decrease depends on the number of LED components connected in the string. This protective measure is also useful in preventing LED failure due to overvoltage and overcurrent caused by system switching, ESD, and nearby lightning strikes.

Second, the voltage across the whole LED string is related to the number of LEDs and the forward voltage of the LED component. Some applications, such as streetlights and spotlights, require many LED components to achieve a sufficient lumen output. A high voltage may cause safety hazards and electrical breakdown. In these situations, multiple serial strings are connected in parallel.

There are two major considerations when optimizing the optical performance of the LED module: the total lumen output and illuminance uniformity. The total lumen output can be simply estimated by the sum of the lumen output of each individual LED component. However, this only describes the overall brightness of the module. The light quality of the LED module cannot be solely evaluated by this index.

In most applications, the LED module is used to illuminate a target surface. The target surface may be a street, floor, wall, desk, etc. Illuminance (lux), which is defined by the light intensity per unit area, is used to describe the "brightness" on the target surface. The illuminance uniformity on the target surface is important. The spacing of the LED components on the LED module may affect the final illuminance uniformity. The spacing should be matched with the light pattern of the LED components. If it is not optimized, light spots or dark zones will result.

The illuminance uniformity can be improved by adding more LED components to the module. However, this will increase the weight and cost. In the board level assembly process, the throughput is mainly controlled by the pick-n-place process. More components on the module will increase the pick-n-place process and hence lengthen the cycle time. Adding a diffuser is another approach, and yet this will reduce the total lumen output.

The color temperature uniformity is an important parameter when the LED module is used as a display or backlight application. The color temperature variation is correlated to the angular color temperature uniformity of the LED component. It is necessary to select components which have uniform angular color temperature distribution if the color temperature uniformity is a critical parameter. Otherwise, yellow spots or yellow rings will be observed. In general, there are several simulation tools which can help to optimize the design of the LED module. The effect of the light pattern and the spacing of the LED components on the illumination, as well as color temperature uniformity, can be calculated.

Thermal management is always an important task when designing the LED module. A poor design will lead to a high junction temperature. Lumen drop and color shift are the immediate negative effects of a high junction temperature. In the long run, a high junction temperature will reduce the lifetime of the product.

The spacing of the LED component is a key parameter in the thermal management of LED modules. The heat generated from the LED component is transferred to the board. A portion of the board is heated up. If the LED components are placed too close to each other, the heating zone of each LED will intersect. This increases the overall thermal resistance of the board level assembly and hence increase the junction temperature [45]. COB assembly has a similar phenomenon. Wu et al. showed that the COB assembly with a smaller chip gap has a higher junction temperature. The increase in junction temperature reduces the luminous efficiency of the assembly [46].

The temperature uniformity of the LED module is another critical concern. Long et al. showed that temperature uniformity on a COB assembly played a crucial role in reliability [47]. It is necessary to achieve a more uniform temperature distribution on the board. The temperature uniformity may also affect the optical performance. Since the optical characteristics of an LED are highly dependent on the junction temperature, then, if the model has a nonuniform temperature distribution, the junction temperature of each component will be different. Normally, the component at the center of the LED module has the highest junction temperature [48]. The component with a higher junction temperature will have a lower lumen output. Subsequently, this will lead to illuminance nonuniformity. The temperature uniformity of the module can be calculated by the finite element method. There are also analytical solutions for some general configurations [49].

In many applications, the LED module must be installed in an enclosed housing to ensure it is both water- and dustproof. Heat transfer by natural convection is the dominant cooling scheme. Yung et al. showed that the size of the housing plays an important role in overall performance [50]. Figure 4.35 shows their simulation results. Also, the inclination of the module inside the housing will affect the temperature of the LED module, as shown in Figure 4.36.

In summary, the spacing of the LED components on the module is critical. It will affect the optical and thermal performance of the system. The spacing also determines the number of LEDs on the module and the lumen output required for each LED. The cost of the module is directly related to the types and number of LEDs selected. Chan et al. combined all considerations and proposed a guideline for spacing optimization [51]. According to them,

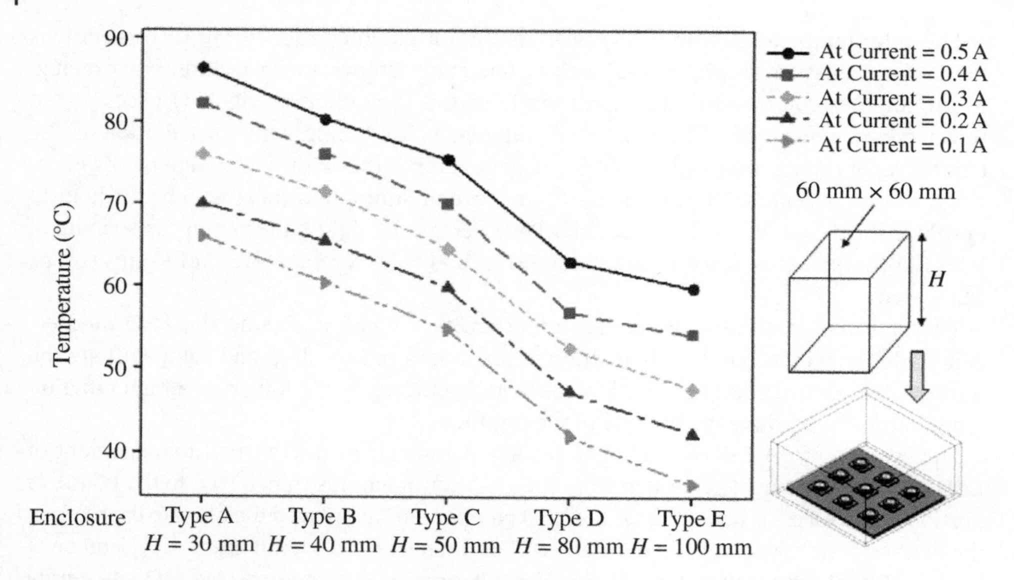

Figure 4.35 Effect of housing size on temperature. Source: Reprinted with permission from Elsevier [50].

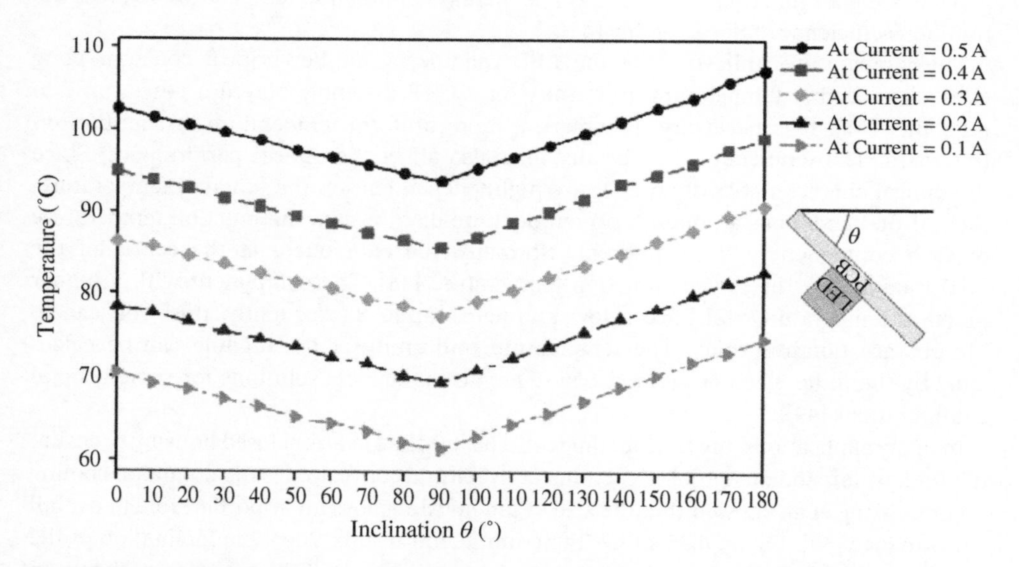

Figure 4.36 Effect of inclination on temperature. Source: Reprinted with permission from Elsevier [50].

if the LED component is driven at a higher power, fewer LEDs will be needed to achieve the required lumen output. This leads to a higher junction temperature and a worse thermal performance. This design has a lower cost, as fewer LEDs will be employed. Conversely, if more LEDs are used to lower the power consumption of each individual LED, this will result in a lower temperature but at a higher cost.

References

1 Bridgelux RS (2012). *Array Series Product Data Sheet DS25*. Livermore, CA: Bridgelux Inc.

2 PhlatLight (2011). *White LED Illumination Products, CBM-360 Series, Product Data Sheet*. Billerica, MA: Luminus Devices, INC.

3 Azar, K. and Graebner, J.E. (1996). Experimental determination of thermal conductivity of printed wiring boards. In: *Proceedings of the Twelfth Annual IEEE Semiconductor Thermal Measurement and Management Symposium (SEMI-THERM XII)*, Austin, TX (5–7 March 1996), 169–182. IEEE.

4 Sarvar, F., Poole, N.J., and Witting, P.A. (1990). PCB glass-fibre laminates: thermal conductivity measurements and their effect on simulation. *Journal of Electronic Materials* 19 (12): 1345–1350.

5 Law, S.B., Permal, A., and Devarajan, M. (2012). Effective heat dissipation of high power LEDs mounted on MCPCBs with different thickness of Aluminum substrates. In: *Proceedings of the 10th IEEE International Conference on Semiconductor Electronics (ICSE 2012)*, Kuala Lumpur, Malaysia (19–21 September 2012), 707–710. IEEE.

6 SFCIRCUITS. (2005). What is a metal core PCB (MCPCB)? http://www.sfcircuits.com/pcb-production-capabilities/metal-core-pcb (accessed 3 September 2020).

7 Lin, K.H., Xu, Z.H., and Lin, S.T. (2011). A study on microstructure and dielectric performances of alumina/copper composites by plasma spray coating. *Journal of Materials Engineering and Performance* 20 (2): 231–237.

8 Su, Y., Yang, Y., and Yang, W. (2012). A thermal performance assessment of panel type packaging (PTP) technology for high efficiency LED. In: *Proceedings of the 14th International Conference on Electronic Materials and Packaging (EMAP 2012)*, Hong Kong (13–16 December 2012), 1–5. IEEE.

9 Kim, I., Cho, S., Jung, D. et al. (2013). Thermal analysis of high power LEDs on the MCPCB. *Journal of Mechanical Science and Technology* 27 (5): 1493–1499.

10 Yoo, M., Cho, H., Kim, S. et al. (2012). Electric field assembled anisotropic dielectric layer for metal core printed circuit boards. *IEEE Electron Device Letters* 33 (11): 1607–1609.

11 Shew, S.C., Huang, H.J., and Shaw, H.J. (2013). Design and estimation of a MCPCB-flat plate heat pipe for LED array module. In: *Proceedings of the 2013 IEEE International Conference on Mechatronics and Automation*, Takamatsu, Japan (4–7 August 2013), 158–163. IEEE.

12 Yung, K.C., Liem, H., and Choy, H.S. (2013). Heat dissipation performance of a high-brightness LED package assembly using high-thermal conductivity filler. *Applied Optics* 52 (35): 8484–8493.

13 Lee, G. and Lin, J. (2009). The introduction of high heat dissipation material. In: *Proceedings of the 4th International Microsystems, Packaging, Assembly and Circuits Technology Conference (IMPACT 2009)*, Taipei (21–23 October 2009), 240–243. IEEE.

14 Horng, R.H., Chiang, C.C., Tsai, Y.L. et al. (2009). Thermal management design from Chip to package for high power InGaN/sapphire LED applications. *Electrochemical and Solid-State Letters* 12 (6): H222–H225.

15 GLXPCB. (2008) Structure of MCPCB. https://glxpcb.com/mcpcb/?lang=en (accessed 3 September 2020).

16 Cho, H.M. and Kim, H.J. (2008). Metal-core printed circuit board with alumina layer by aerosol deposition process. *IEEE Electron Device Letters* 29 (9): 991–993.

17 Cree (2011). *Product Family Data Sheet, Cree XLamp® XT-E® LEDs, Product Data Sheet*. Durham, NC: Cree Inc.

18 Wang, N., Hsu, A., Lim, A. et al. (2009). High brightness LED assembly using DPC substrate and SuperMCPCB. In: *Proceedings of the 4th International Microsystems, Packaging, Assembly and Circuits Technology Conference (IMPACT 2009)*, Taipei, Taiwan (21–23 October), 199–202. IEEE.

19 Mashkov, P., Pencheva, T., and Gyoch, B. (2011). PCB thermal performance for power LEDs. In: *Proceedings of the 34th International Spring Seminar on Electronics Technology (ISSE)*, Tratanska Lomnica (11–15 May), 33–38. IEEE.

20 Wu, J., Chen, C., Horng, R. et al. (2013). An efficient metal-Core printed circuit board with a copper-filled through (blind) hole for light-emitting didoes. *IEEE Electron Device Letters* 34 (1): 105–108.

21 Juntunen, E., Sitomaniemi, A., Tapanien, O. et al. (2012). Thermal performance comparison of thick-film insulated aluminum substrates with metal Core PCBs for high-power LED modules. *IEEE Transactions on Components, Packaging and Manufacturing Technology* 2 (12): 1957–1964.

22 Juntunen, E., Tapaninen, O., Sitomaniemi, A. et al. (2014). Copper-core MCPCB with thermal Vias for high power COB LED modules. *IEEE Transactions on Power Electronics* 29 (3): 1410–1417.

23 Bissuel, V., Daniel, O., Brizoux, M. et al. (2010). Impact of PCB via and micro-via structures on component thermal performances. In: *Proceedings of the 16th International Workshop on Thermal Investigations of ICs and Systems (THERMINIC)*, Barcelona, Spain (6–8 October 2010), 1–6. IEEE.

24 Yang, K.S., Wu, Y.L., Chen, I.Y. et al. (2009). An investigation of thermal spreading device with thermal via in high power LEDs. In: *Proceedings of the 4th International Microsystems, Packaging, Assembly and Circuits Technology Conference (IMPACT 2009)*, Taipei, Taiwan (21–23 October), 195–198. IEEE.

25 MacDermid Enthone. (2017). Copper filled thermal vias for heat management. https://electronics.macdermidenthone.com/news-and-events/copper-filled-thermal-vias-for-heat-management (accessed 3 September 2020).

26 PCBgogo. (2019). How to distinguish between wave soldering and reflow soldering. https://www.pcbgogo.com/Blog/How_To_Distinguish_Between_Wave_soldering_And_Reflow_Soldering_.html (accessed 3 September 2020).

27 Electronic Industries Alliance and IPC. (1995). Requirements for soldering pastes: J-STD 005. http://www.lg-advice.ro/J-STD-005.pdf (accessed 23 February 2021).

28 IPC and JEDEC Solid Sate Technology Association. (2004). Moisture/reflow sensitivity classification for nonhermetic solid state surface mount devices: J-STD 020C. http://ferroxcube.home.pl/envir/info/J-STD-020C%20Proposed%20Std%20Jan04.pdf (accessed 23 February 2021).

29 Keeping, S. (2014). *The Rise of Chip-on-Board LED Modules*. Digi-Key Corporation Article Library.

30 Weber, R. (2013). COB LEDs Simplify SSL Manufacturing, Drive Broader Deployment. *LEDs Magazine* July/Aug: 39–43.

31 Okamoto, K. (2011). Lighting Fair 2011 report: from an era of brightness to an era of light quality part 1. LED Next Stage Review 2011. Tokyo, Japan (9–12 March 2011).

32 LUXEON CoB. (2020). Core range. https://www.lumileds.com/products/cob-leds/luxeon-cob (accessed 3 September 2020).

33 Sim, J., Ashok, K., Ra, Y. et al. (2012). Characteristic enhancement of white LED lamp using low temperature co-fired ceramic-chip on board package. *Current Applied Physics* 12: 494–498.

34 Tsai, M.Y., Chen, C.H., and Kang, C.S. (2008). Thermal analyses and measurements of low-cost COP package for high-power LED. In: *Proceedings of the 58th Electronic Components and Technology Conference (ECTC 2008)*, Lake Buena Vista, FL (27–30 May 2008), 1812–1818. IEEE.

35 Labau, S., Picard, N., Gasse, A. et al. (2009). Chip on board package of light emitting diodes and thermal characterizations. In: *Proceedings of the 59th Electronic Components and Technology Conference (ECTC 2009)*, San Diego, CA (26–29 May 2009), 848–853. IEEE.

36 Yin, L., Yang, L., Yang, W. et al. (2010). Thermal design and analysis of Mulit-chip LED moudle with ceramic substrate. *Solid-State Electronics* 54: 1520–1524.

37 Ru, H., Wei, V., Jiang, T. et al. (2011). Direct plated copper technology for high brightness LED packaging. In: *Proceedings of the 6th International Microsystems, Packaging, Assembly and Circuits Technology Conference (IMPACT 2011)*, Taipei (19–21 October 2011), 311–314. IEEE.

38 Xie, Z., Li, C., Yu, B. et al. (2013). A novel COB structure with integrated multifunction. *Journal of Semiconductors* 34 (5): 055001-1–055001-4.

39 Ha, M. and Graham, S. (2012). Development of a thermal resistance model for chip-on-board packaging of high power LED arrays. *Microelectronics Reliability* 52: 836–944.

40 Kim, J.S., Jeon, S.L., Le, D.W. et al. (2010). Thermal and optical properties of COB type LED module based on Al_2O_3 and AlN ceramic submounts. *Journal of Applied Sciences* 10 (24): 3388–3391.

41 NICHE-TECH. (2019). Epoxy encapsulant. http://nichetech.com.hk/index.php/en/products/epoxyEncapsulant (accessed 23 September 2020).

42 Yuen, P.H., Shiung, H.H., and Devarajan, M. (2013). Influence of phosphor packaging configurations on the optical performance of chip on board phosphor converted warm white LEDs. In: *Proceedings of the 5th Asia Symposium on Quality Electronic Design (ASQED)*, Penang, Malaysia (26–28 August 2013), 329–333. IEEE.

43 Juntunen, E., Tapaninen, O., Sitomaniemi, A. et al. (2013). Effect of phosphor encapsulant on the thermal resistance of a high-power COB LED module. *IEEE Transactions on Components, Packaging and Manufacturing Technology* 3 (7): 1148–1154.

44 LITTELFUSE. (2020). LED lighting. https://www.littelfuse.com/industries/led-lighting.aspx (accessed 3 September 2020).

45 Petroski, J. (2004). Spacing of high-brightness LEDs on metal substrate PCB's for proper thermal performance. In: *Proceedings of the 9th Intersociety Conference on Thermal and Thermomechanical Phenomena in Electronic System (ITHERM' 04 Vol 2)*, Las Vegas, NV (1–4 June 2004), 507–514. IEEE.

46 Wu, H., Lin, K., and Lin, S. (2012). A study on the heat dissipation of high power multi-chip COB LEDs. *Microelectronics Journal* 43: 280–287.

47 Long, X., Liao, R., Zhou, J. et al. (2013). Thermal uniformity of packaging multiple light-emitting diodes embedded in aluminum-core printed circuit boards. *Microelectronics Reliability* 53: 544–553.

48 Yung, K.C., Liem, H., Choy, H.S. et al. (2014). Thermal investigation of a high brightness LED array package assembly for various placement algorithms. *Applied Thermal Engineering* 63: 105–118.

49 Cheng, T., Luo, X., Huang, S. et al. (2010). Thermal analysis and optimization of multiple LED packaging based on a general analytical solution. *International Journal of Thermal Sciences* 49: 196–201.

50 Yung, K.C., Liem, H., Choy, H.S. et al. (2010). Thermal performance of high brightness LED array package on PCB. *International Communications in Heat and Mass Transfer* 37: 1266–1272.

51 Chan, Y.S. and Lee, S.W.R. (2011). Spacing optimization of high power LED arrays for solid state lighting. *Journal of Semiconductors* 32 (1): 014005-1–014005-4.

5

Optical, Electrical, and Thermal Performance

5.1 Evaluation of Optical Performance

The essential function of an LED device is emitting light; therefore, it is important to properly characterize the light emitted. Light intrinsically is an electromagnetic wave which can be completely described in terms of its frequency and intensity in spatial distribution. Since in most of the application scenarios of LED devices humans are involved, human eyes are considered another important issue in evaluating an LED device because of their different sensitivity to photons of different wavelengths. Although the LED can be completely described by means of its spectrum, directly using frequency and intensity are usually inconvenient. Therefore, other quantities are derived from the LED light spectrum for different application scenarios. There quantities are introduced in this chapter. Different specific characterization equipment have also been developed for evaluating these quantities so that the performance of an LED device can be appropriately assessed.

5.1.1 Basic Concepts of Radiometric and Photometric

Light is essentially electromagnetic radiation, hence radiometric concepts can be introduced here for quantitatively describing light emitting from a light source. The most frequently used quantities are listed in Table 5.1. If an adjective "spectral" is added before these quantities, it means the quantities with respect to the specified wavelength.

Considering a radiant source emitting light in a two-dimensional space (i.e. imaging a light source on paper), the angular or directional power intensity should be quantified by power per angle. By expanding such concept to a three-dimensional space, the solid angle is therefore introduced. In the SI unit system, steradian (sr) is the unit for the solid angle. Similar to the radian (rad), which describes the plane angle subtended by a circular arc as the length of the arc divided by the radius of the arc, a steradian can be defined as the solid angle subtended at the center of a unit sphere by a unit area on its surface. For an area on a sphere, the solid angle (Ω) subtended to it is

$$\Omega = \frac{A}{r^2}(\text{sr}) \tag{5.1}$$

where A is the area and r is the radius of the sphere. The area of a sphere is $4\pi r^2$ which means a sphere has a solid angle at the center equals 4π.

From LED to Solid State Lighting: Principles, Materials, Packaging, Characterization, and Applications, First Edition.
Shi-Wei Ricky Lee, Jeffery C. C. Lo, Mian Tao, and Huaiyu Ye.

Table 5.1 Radiometric quantities (SI).

Quantity (symbol)	Unit (symbol)	Notes
Radiant energy (Q_e)	Joule (J)	Energy of electromagnetic radiation
Radiant flux (Φ_e)	Watt (W)	Total radiant energy emitted, reflected, transmitted or received, per unit time and sometimes referred to as radiant power
Radiant intensity ($I_{e,\Omega}$)	Watt per steradian (W/sr)	Radiant flux emitted, reflected, transmitted, or received, per unit solid angle
Radiance ($L_{e,\Omega}$)	Watt per steradian per square meter ($W/(sr \cdot m^2)$)	Radiant flux emitted, reflected, transmitted, or received by a *surface*, per unit solid angle per unit projected area
Irradiance flux density (E_e)	Watt per square meter (W/m^2)	Radiant flux *received* by a *surface* per unit area
Radiosity (J_e)	Watt per square meter (W/m^2)	Radiant flux *leaving* (emitted, reflected, and transmitted by) a *surface* per unit area. This is sometimes also confusingly called intensity

There are three different types of cells, which are called cone cells, following their shape sensing light in the normal human eye. They have their own sensitivity to different wavelengths of light. These three types of cells are known as "L," "M," and "S," representing long, medium, and short wavelengths. Combining the three different responses of the cone cells to light, brightness and color are generated. However, quantitatively depicting the human psychological perception of light is very difficult and the results vary with different measurement methodologies. The most widely accepted standards were created by the International Commission on Illumination (CIE). Two major standards, CIE 1931 and CIE 1978, are frequently used in the LED industry.

In CIE 1931 color space, the luminosity function was established to quantify brightness of light. The profile of the luminosity function is given in Figure 5.1 where "photopic" implies a well-lit environment and "scotopic" means a dark environment. The physical meaning of the luminosity function is the normalized sensitivity of the human eye to light at different wavelengths. The luminous flux, which depicts the brightness of the light, can be defined as

$$\Phi_V = 683.002 \cdot \int_0^\infty \bar{y}(\lambda)\Phi_{e,\lambda}(\lambda)d\lambda \tag{5.2}$$

where

Φ_V is the luminous flux, in lumens;
$\Phi_{e,\lambda}$ is the spectral radiant flux, in watts per nanometer;
$\Phi_{e,\lambda}$ is also known as the luminosity function, dimensionless;
λ is the wavelength.

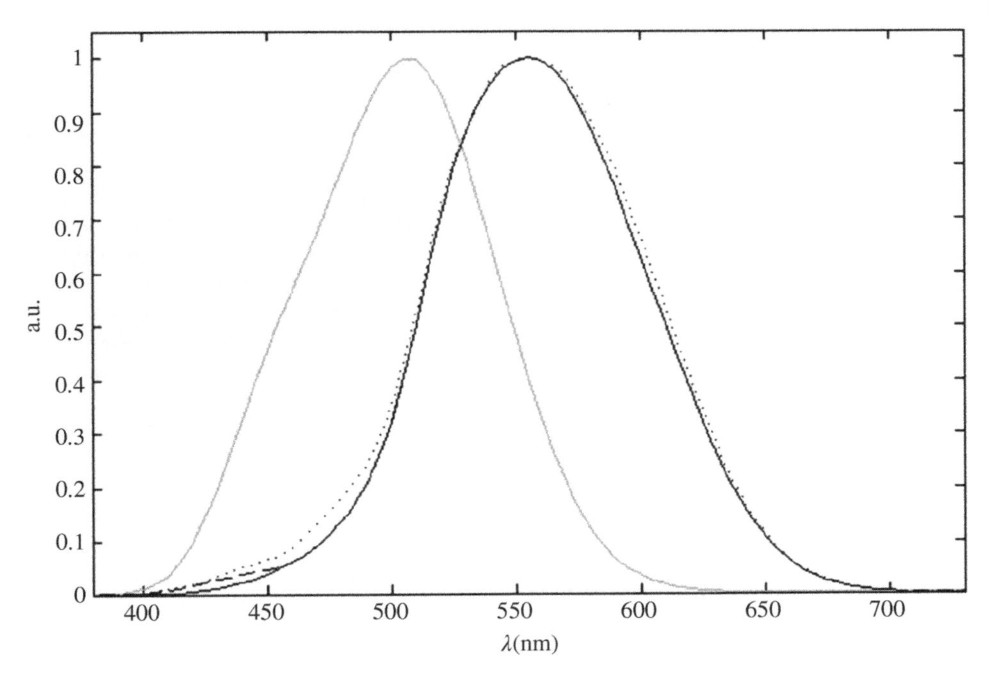

Figure 5.1 Photopic (black) and scotopic (gray) luminosity functions. The solid plot is the CIE 1931 standard, the dashed plot Judd-Vos 1978 modified data, and the dotted plot is Sharpe, Stockman, Jagla and Jägle 2005 data. Source: Modified from Sharpe, L.T., Stockman, A., Jagla, W., and Jägle, H. (2005). A luminous efficiency function, $V^*(\lambda)$, for daylight adaptation. *Journal of Vision* 5(11), 3. https://doi.org/10.1167/5.11.3.

For angular intensity, the lumen per solid angle, lm/sr, is usually converted to candela, which is luminous power per solid angle.

Meanwhile, in a dark environment, another cell in the human eye is activated to sense light: the rod cell. Correspondingly, the rod cell has its luminosity function similar to the cone cell, called the scotopic luminosity function. Since there is only one cell as the photoreceptor in a dark environment, color becomes dimmer in a dark environment.

Similar to the radiometric quantities, another set of quantities that are used for describing brightness is listed in Table 5.2.

The light response of a cone cell can be measured by shining light of different wavelengths on the particular cone cell [1]. The response with respect to the wavelength is known as the color matching function. However, directly using the original human eye color matching function to characterize brightness or color is not convenient; therefore, CIE 1931 color space standardized the color matching function, $\bar{x}(\lambda)$, $\bar{y}(\lambda)$, and $\bar{z}(\lambda)$, as shown in Figure 5.2 [2]. Integrating the color matching functions with respect to wavelength gives us the tristimulus values, which can be further used to determine the color of the light. Assuming there is a beam of light with a spectral radiance, $L_{e,\Omega,\lambda}$, its tristimulus values can be calculated according to

$$X = \int_{\lambda} L_{e,\Omega,\lambda}(\lambda)\bar{x}(\lambda)\mathrm{d}\lambda$$

Table 5.2 Photometric quantities (SI).

Quantity (symbol)	Unit (symbol)	Notes
Luminous flux (Φ_e)	lm	Energy of electromagnetic radiation
Luminous intensity (I_v)	cd, which is lumen per solid angle	Total radiant energy emitted, reflected, transmitted, or received, per unit time and sometimes referred to as radiant power
Luminance (L_v)	cd/m^2	Luminous flux per unit solid angle per unit projected source area. The candela per square meter is sometimes called the nit
Illuminance (E_v)	Lux, which is lumen per square meter	Luminous flux incident on a surface

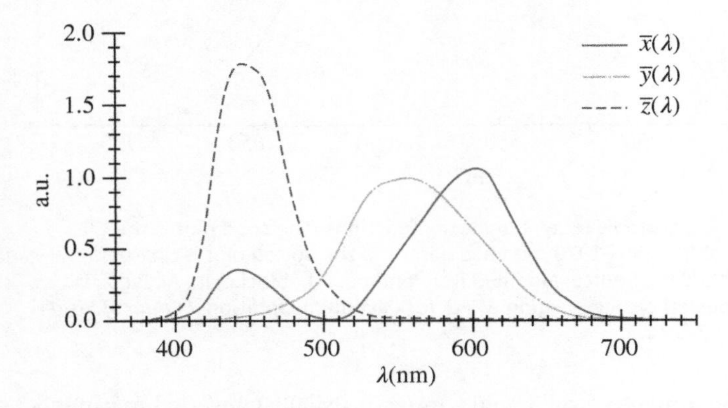

Figure 5.2 The CIE 1931 color matching function.

$$Y = \int_{\lambda} L_{e,\Omega,\lambda}(\lambda)\bar{y}(\lambda)\mathrm{d}\lambda$$

$$Z = \int_{\lambda} L_{e,\Omega,\lambda}(\lambda)\bar{z}(\lambda)\mathrm{d}\lambda \tag{5.3}$$

Because humans tend to perceive the green light brighter than the other two colors, the tristimulus value, Y, is set to be luminance which is used to describe the brightness of the light. It should be noticed that the Y, which is originated from the response of the M cone cell, is slightly deviated from the real light response, as illustrated in Figure 5.3.

In order to practically exhibit the color of the tristimulus value regardless of brightness, the tristimulus values of a given light can be normalized by the equations in Eq. (5.4).

$$\begin{cases} x = \dfrac{X}{X+Y+Z} \\ y = \dfrac{Y}{X+Y+Z} \\ z = \dfrac{Z}{X+Y+Z} \end{cases} \tag{5.4}$$

where x and y are called color coordinates in the CIE 1931 color space. Inversely, if a light whose color (x,y) and brightness (Y) is determined, its tristimulus value can be

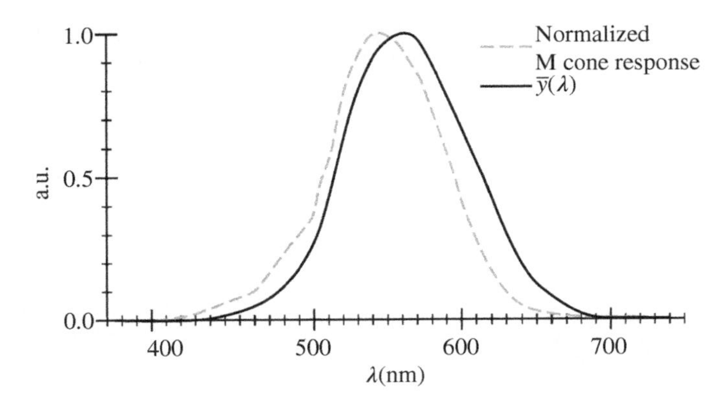

Figure 5.3 Comparison between the response of M cone cell and the *y* tristimulus function.

obtained by

$$\begin{cases} X = \dfrac{x}{y}Y \\ Z = \dfrac{1-x-y}{y}Y \end{cases} \tag{5.5}$$

For a common display, a pixel consists of three different lighting units which are red (R), green (G), and blue (B). Different colors are displayed by adjusting their relative light intensity. Here are the procedures to calculate the R, G, and B value for a color with the known coordinate (x, y). For the *R*, *G*, and *B* pixels, their spectrum is known and consequently the tristimulus can be calculated. Note that there are three spectrums and for every spectrum there are three tristimulus values. Therefore, there are totally nine tristimulus values that can be organized in a matrix form so that the tristimulus value of the combination of different *R*, *G*, and *B* values can be easily calculated by the given equation

$$\begin{bmatrix} X \\ Y \\ Z \end{bmatrix} = \begin{bmatrix} X_R & X_G & X_B \\ Y_R & Y_G & Y_B \\ Z_R & Z_G & Z_B \end{bmatrix} \begin{bmatrix} R \\ G \\ B \end{bmatrix} \tag{5.6}$$

where *R*, *G*, and *B* are the light intensity of the three pixels; *X*, *Y*, and *Z* denote the three different tristimulus value; and the subscript R, G, and B indicate from which light source the tristimulus value is generated. Solving the equations and substituting *X*, *Y*, and *Z* by Eq. (5.5), it can be shown that

$$\begin{bmatrix} R \\ G \\ B \end{bmatrix} = \begin{bmatrix} X_R & X_G & X_B \\ Y_R & Y_G & Y_B \\ Z_R & Z_G & Z_B \end{bmatrix} \begin{bmatrix} x/y \\ 1 \\ (1-x-y)/y \end{bmatrix} Y \tag{5.7}$$

The equation entails that giving a luminance and a color coordinate, the required combination of *R*, *G*, and *B* can be calculated. Using such a concept, a color map can be drawn with a fixed luminance, as shown in Figure 5.4, which is called a chromaticity diagram. It should be noted that this chromaticity diagram is originated from the CIE 1931 color

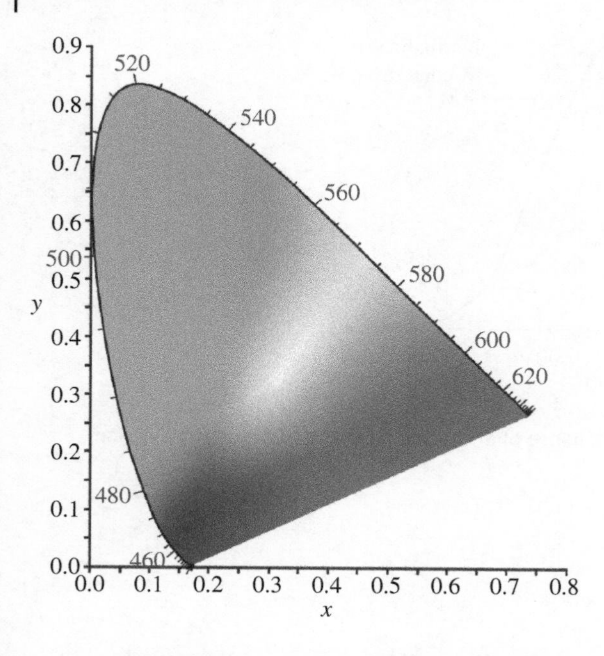

Figure 5.4 The CIE 1931 chromaticity diagram.

matching function which is intentionally adjusted. Different strategies of constructing the color matching function will result in different chromaticity diagrams. The CIE 1976 diagram is another chromaticity diagram which is frequently used.

Given that there are two lights with respect to two points on the chromaticity diagram, $P_1(x_1,y_1)$ and $P_2(x_2,y_2)$, the color coordinate of the mix light $P_{mix}(x_{mix},y_{mix})$ becomes

$$\begin{cases} x_{mix} = \dfrac{X_1 + X_2}{X_1 + Y_1 + Z_1 + X_2 + Y_2 + Z_2} \\ y_{mix} = \dfrac{Y_1 + Y_2}{X_1 + Y_1 + Z_1 + X_2 + Y_2 + Z_2} \end{cases} \tag{5.8}$$

Substituting the tristimulus value by the color coordinates and luminance (i.e. Y_1 and Y_2), the equations come to

$$\begin{cases} x_{mix} = \dfrac{x_1 + \dfrac{Y_2}{Y_1}\dfrac{y_1}{y_2}x_2}{1 + \dfrac{Y_2}{Y_1}\dfrac{y_1}{y_2}} \\ y_{mix} = \dfrac{y_1 + \dfrac{Y_2}{Y_1}y_1}{1 + \dfrac{Y_2}{Y_1}\dfrac{y_1}{y_2}} \end{cases} \tag{5.9}$$

Let $\lambda = \frac{Y_2}{Y_1}\frac{y_1}{y_2}$, the equations become

$$\begin{cases} x_{mix} = \dfrac{x_1 + \lambda x_2}{1 + \lambda} \\ y_{mix} = \dfrac{y_1 + \lambda y_2}{1 + \lambda} \end{cases} \tag{5.10}$$

It can be seen that P_{mix} is collinear with the line P_1P_2, and because λ is positive, P_{mix} should locate between P_1 and P_2. This means two light sources can generate a mix light with its color coordinate on the graph's line of the two sources on the chromaticity diagram while three light sources can generate lights in the triangle area.

A monochromic light is a light with a very narrow spectrum. Using the prescribed method, monochromic lights of different wavelengths within the human visible range (380–780 nm) can form a curve on a chromaticity diagram, which is called the spectral locus. Previous discussion has proved that any possible light should be a mix of monochromic lights which indicates that all possible colors should be located inside the spectral locus.

In order to quantify the capability of a normal human to discriminate different colors, Dr. David L. MacAdam developed a concept called MacAdam ellipse with a chromaticity diagram [3]. It is also known as standard deviation color matching (SDCM). The MacAdam ellipses are described in steps. If there are light sources whose color coordinates falls within 1 SDCM or 1-step MacAdam ellipse, 68.26% of the normal population cannot distinguish the color difference among these light sources. A 3 SDCM suggests that 99.44% of the normal population cannot differentiate the color within the ellipse. Some MacAdam ellipses of CIE 1931 are illustrated in Figure 5.5. It can be noticed that the sizes of these ellipses vary

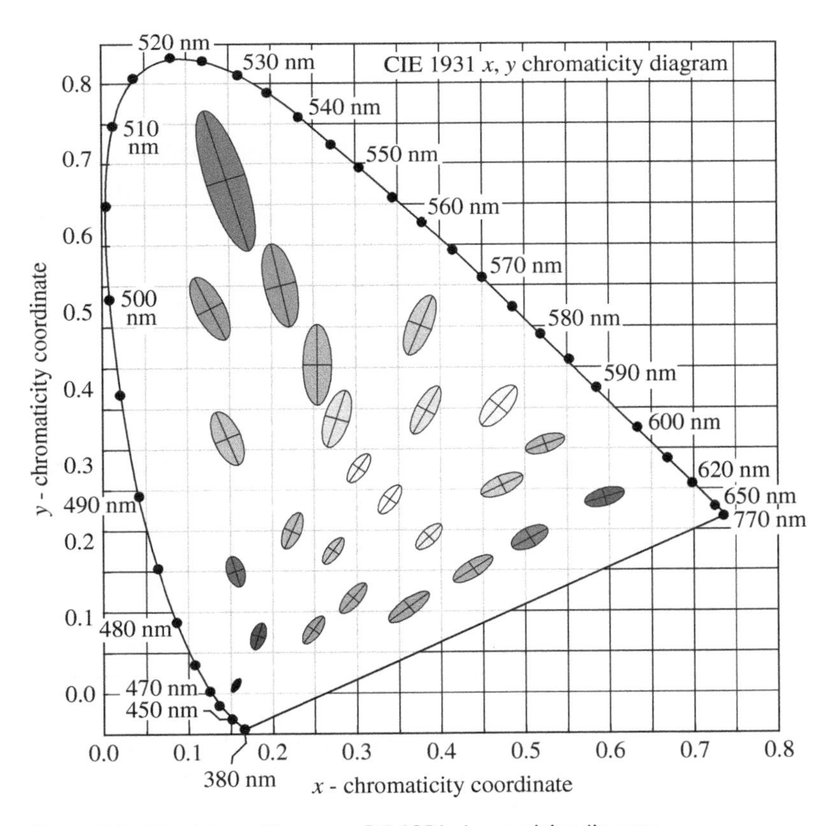

Figure 5.5 MacAdam ellipses on CIE 1931 chromaticity diagram.

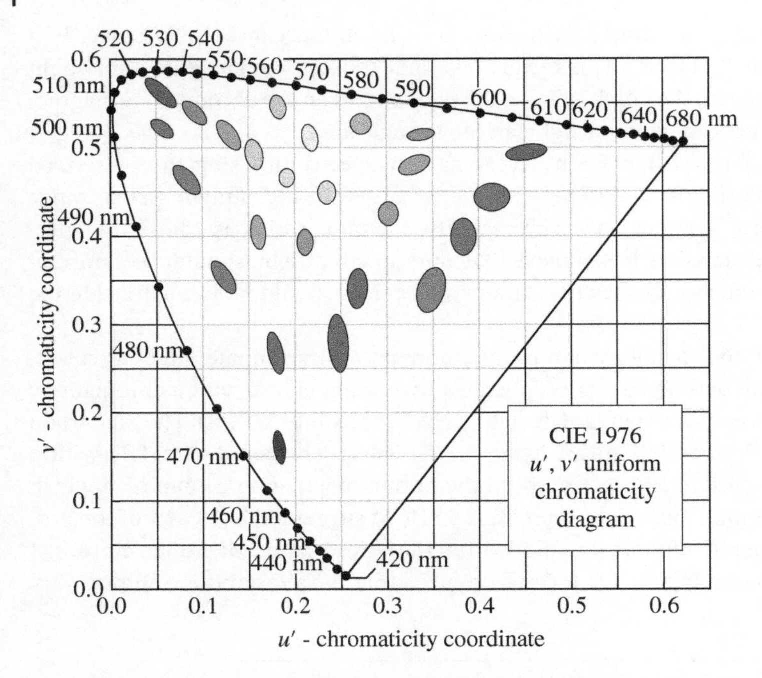

Figure 5.6 MacAdam ellipses on CIE 1976 chromaticity diagram.

significantly among different color. To addressing this drawback of CIE 1931 chromaticity diagram, another chromaticity diagram, CIE 1976, was developed as given in Figure 5.6 [4]. It can be seen that the MacAdam ellipse of CIE 1976 is much more uniform compared to CIE 1931. The color coordinates in CIE 1976, (u', v') can be transfer from CIE 1931 by the following equations

$$\begin{cases} u' = \dfrac{4x}{-2x + 12y + 3\lambda} \\ v' = \dfrac{9y}{-2x + 12y + 3\lambda} \end{cases} \tag{5.11}$$

According to Planck's law, the spectrum of the black body radiation of different temperatures can be obtained as

$$P_{\lambda}(T, \lambda) = \frac{2\pi hc^2}{\lambda^5} \times \frac{1}{\exp\left(\dfrac{hc}{\lambda k_B T}\right) - 1} \tag{5.12}$$

where h is the Planck constant, c is the speed of light in a vacuum, k is the Boltzmann constant, λ is the wavelength of the electromagnetic radiation, and T is the absolute temperature of the body.

Using the aforementioned method, a curve can be drawn in the chromaticity diagram according to the spectrum obtained from Planck's law. Such curves are often referred to as the black body radiation curve or Planckian locus, as illustrated in Figure 5.7. Any point on this curve represents a black body temperature. A given color coordinate on a chromaticity

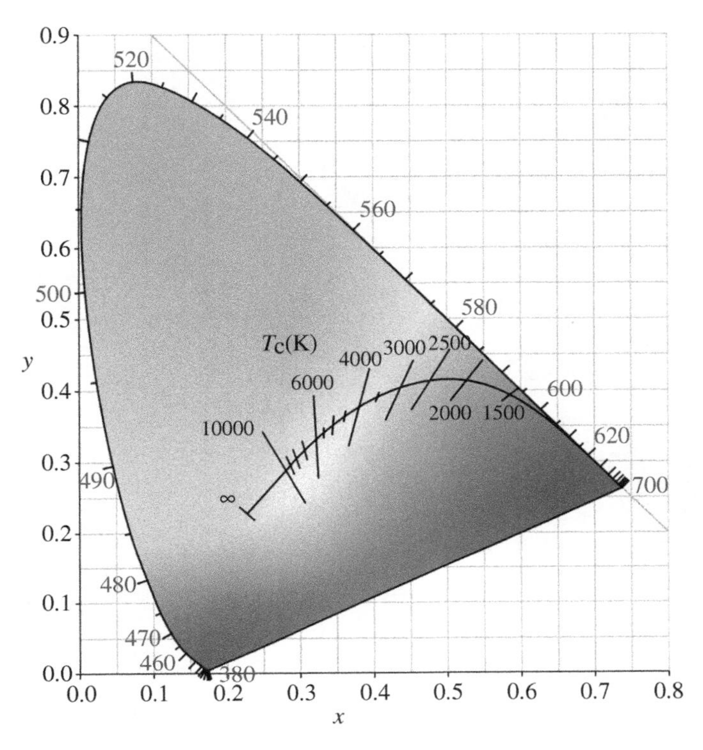

Figure 5.7 The Planckian locus on CIE 1931 chromaticity diagram [5]. Source: Wikipedia, Planckian locus. 2019. https://en.wikipedia.org/wiki/Planckian_ locus. Public Domain.

diagram can always find the closest point on the black body radiation curve and the corresponding temperature is called the color temperature of the given color ordinate. It can easily be noticed that if the given color coordinate is far away from the black body radiation curve there could be more than one color temperature. So it is meaningless to discuss this kind of color temperature. Usually, a color temperature can be assigned to a color coordinate only if the color coordinate falls within a 7 MacAdam ellipse of the black body radiation curve.

The color of an object relies greatly on illuminating light; however, not all light can properly present the true color of an object, as illustrated in Figure 5.8. The objects under the monochromatic light source from a high-pressure sodium lamp exhibit a significantly different appearance compared to the scene under a white LED light source. Color presentation is an important parameter of a light source. For quantitative assessment of the color presentation capability, the color rendering index (CRI) was developed. CRI is an index to quantify the ability of a light source to render the true color of different objects [6, 7]. The best light source for illumination should be sunlight. As a result, the black body radiation in 5000 K is the perfect light source which has a CRI of 100.

The value of the CRI is often called the international standard color rendering index value (CIE RA) on commercial LED products. Numerically, daylight as a perfect light source has the highest possible CIE RA: 100. Only the light source identical to the daylight spectrum would be ranked 100 RA.

Figure 5.8 Comparison of the color presented by high pressure sodium lamp (on the left) and LED (on the right) on a cement road surface.

Here is a brief explanation of how to calculate RA. The light source should first be confirmed to be close enough to the Planckian locus because the CRI is intended for a white light source. It is meaningfulness to discuss the CRI of a light source that is not recognized as white light. The light source under test and the standardized light (5000 K black body radiation) are shined on eight different standard samples of different colors, as listed in Table 5.3 as specified in CIE 1995. The color coordinate difference between the test light source and the standardized light source can be calculated. The RA is the arithmetic mean of these differences.

Table 5.3 The original test color samples (TCS) for evaluating the color-rendering index of a light source.

Name	Appearance under daylight	Swatch
TCS01	Light grayish red	
TCS02	Dark grayish yellow	
TCS03	Strong yellowish green	
TCS04	Moderate yellowish green	
TCS05	Light bluish green	
TCS06	Light blue	
TCS07	Light violet	
TCS08	Light reddish purple	

5.1.2 Irradiance Measurement Calibration

The irradiance, which is radiant flux received per unit area, is the most fundamental physical quantity in light characterization. Almost all the optical characteristics of an LED under test can be derived from the spectral irradiance generated from the light source of a specific direction and distance. Once the irradiance can be precisely measured, other parameters – such as radiant flux, radiant intensity, lumen, etc. – can be easily calculated. Counting the number of photons received is a straightforward approach for irradiance measurement. There are manifold devices that can sense photons, for example the photomultiper [8] or photodiode [9]. These types of devices can transform the light signal into an electrical signal; nevertheless, the relationship between the input light intensity and output current or voltage is usually known. Consequently, calibration is required.

A standardized light source with a known radiation including the intensity, the angular distribution, and the spectrum is an essential prerequisite but is not easy to achieve.

One example of a calibration light source is the tungsten ribbon lamp which is essentially an incandescent lamp, as shown Figure 5.9 [10–12]. The light of the tungsten ribbon lamp is generated by heating up the tungsten ribbon by applying electrical power. The light is essentially thermal radiation and the radiation emitted is regarded as blackbody radiation. The light emits from the plane surface of the tungsten ribbon and consequentially the radiation emits in a Lambertian pattern. When the tungsten ribbon lamp is being driven under a

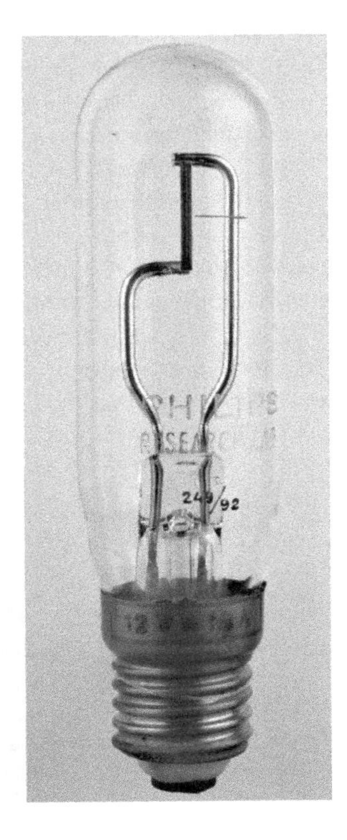

Figure 5.9 Philips research lab. 12 V 16A tungsten ribbon calibration lamp [10]. Source: J. Voogd, Tungsten Ribbon Lamps for Optical Measurements. *Philips Technical Review* 5(3), March 1940.

certain current, its resistance can be monitored. The electrical resistivity of tungsten is temperature dependent. As a result, the temperature can be calculated from a precalibrated temperature-resistance curve. Once the temperature is achieved, the radiation intensity and spectrum can be easily calculated according to Planck's law.

A lamp with a known radiation spectrum and pattern is regarded as a standard calibration lamp. It can be utilized to calibrate light measurement devices. A spectrometer is a typical light measurement device which usually consists of a silicon photo diode combining with a prism or a diffraction grating. According to the photoelectric effect [13], the silicon photo diode can convert an incident light signal into a voltage or current signal which is linearly proportional to the intake photon number. The prism or the diffraction grating splits a beam of light from a standard calibration lamp into a light spectrum. Placing the photo diode in a different location or rotating the prism or grating the light at different wavelengths can be converted into a voltage or current signal which is further extracted by additional circuitry. Consequentially, the electrical signal can be correlated to the known radiation and therefore the spectrometer is well calibrated and can be used to measure spectral irradiance flux density.

5.1.3 Common Measurement Equipment

A spectrophotometer like what is shown in (Figure 5.10) is the basic component of a light characterization equipment. The spectral irradiance flux density is the most essential quantity in light characterization. All other radiometric quantities are originated from spectral irradiance flux density. Providing that the distance and angle between the spectrophotometer and the light source is known, the radiometric quantities can be calculated.

Usually, an optical fiber is attached to guide the incident light to the spectrometer. One end of the glass fiber creates a light receiving surface and the spectrophotometer measures the spectral irradiance flux density ($W/(m^2 \cdot nm)$) shining on that surface. An extra light receiving apparatus, a fiber collimating lens, may add to the glass fiber to guarantee that the incident light can properly transmit into the glass fiber. The incident light from the fiber is then separated spectrally by a diffraction grating or a prism. A silicon photo diode detector converts the light signal to an electrical signal and mechanics allow the optical path to be changed so that the light spectrum can sweep across the photodetector. As a result,

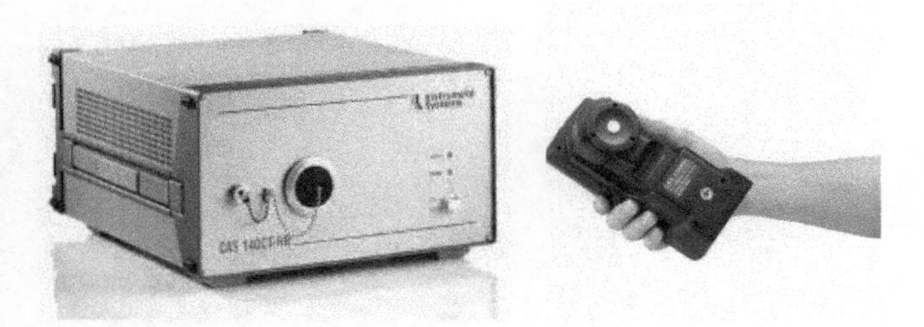

Figure 5.10 Philips' commercial desktop spectrometer (Instruments System® CAS 140CT-HR) and a handheld spectrometer (Konica Minolta® CL-500A). Source: PHILIPS A commercial desktop spectrometer (Instruments System® CAS 140CT-HR) and a hand held spectrometer (Konica Minolta® CL-500A).

light of different wavelengths can be captured by the photodetector. The incident light may raise the detector's temperature and can increase signal noise; therefore, cooling devices are used to maintain the detector temperature when precise measurement is required. The small electrical signal captured by the detector will further be amplified and converted to digits by peripheral circuitry and then transmitted to a host machine.

There always exist different systematic errors, such as attenuation from the optical system, temperature shift of the detector, or the nonlinear response of the detector. Careful calibration by a standard lamp should be performed beforehand so as to eliminate any potential errors. Given that there is a standard lamp which is considered a black body with a known temperature, the spectral irradiance flux density can be precisely predicted based on Planck's law. By correlating the electric signal captured by the spectrometer and the calculated data, the device can be precisely and accurately calibrated.

In many application scenarios, a light source can be regarded as a point light source as long as the receiving plane has a distance to the light source more than 20 times that of the features size of the light source. For example, providing that there is a light tube of 1 m, if the light receiving plane is over 20 m away from the light tube, the light tube is regarded as a point source. In such cases, spectral radiant intensity of the light source is an issue. Radiant intensity is the angular distribution of radiation power. This quantity is commonly measured by a specific device called a goniophotometer.

Since the light source here should be regarded as a point source, the special lactation of the detector of the spectrometer can be described by the distance between the detector and the source and two rotational freedoms. A goniophotometer can rotate the light source, and the detector is placed at a fixed location. By rotating the light source in all directions, the detector equivalently travels to all locations on a sphere whose center is aligned with the light source. The radius of the sphere is equal to the distance between the light source and the detector. The distance fulfilling the point light source requirement could be unacceptably large for practical usage. Sometimes a specially designed optical path is implemented so as to reduce the total space occupied by the goniophotometer, as illustrated Figure 5.11. A typical measurement result obtained by a goniophotometer is illustrated in Figure 5.12.

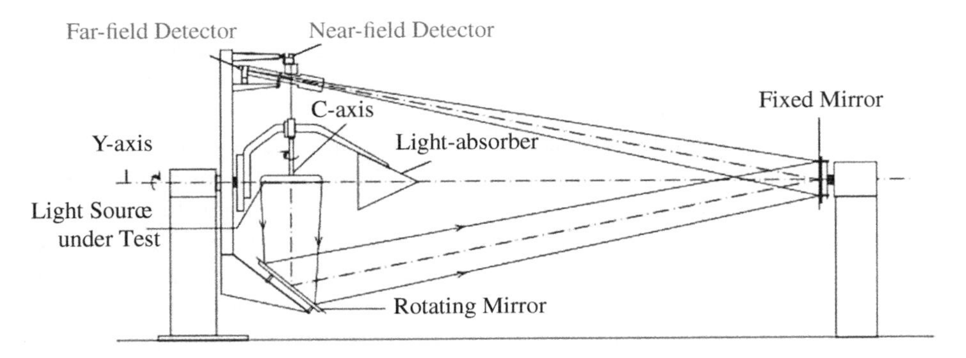

Figure 5.11 The light path from the light source under test to the photodetector is particularly arranged to reduce the room required for the equipment [14]. Source: Everfine. http://www.everfine.cn/productsinfo.php?cid=8&id=21 (cited date 3 September 2020).

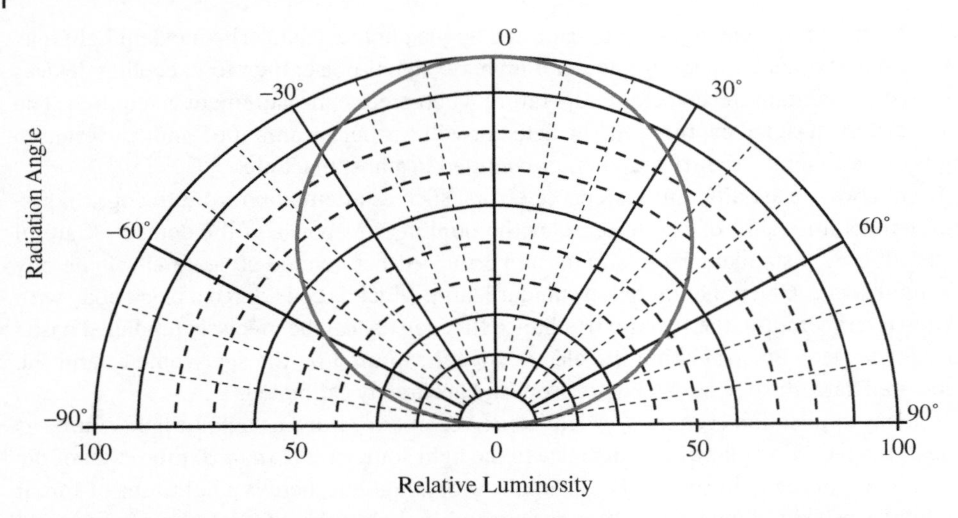

Figure 5.12 Typical polar light pattern of a 5050 LED component [15]. Source: Bridgelux SMD 5050 3w 18V Product Data Sheet DS62 Rev B. Bridgelux Inc. Livermore, CA, USA. 2016.

The angular light distribution of an LED device is often known as a light profile. The illuminating performance of an LED device is often evaluated by its light profile. For this purpose, the light source can be regarded as a point source because in most of the application scenarios the surface that is expected to be illuminated is far away from the light source. There is another type of goniophotometer which no longer considers the light source as a point source. The light source is regarded as a plane surface. The photo diode detector becomes a charge-coupled device (CCD). The CCD works with a lens so that the light source plane can form an image on the CCD. Every pixel on the CCD is an independent light sensor to measure the radiance of the corresponding location on the source plane. This type of goniophotometer is commonly called a source imaging goniometer. Using this method to construct the light distribution of a light source allows us to reproduce the true light pattern, which will benefit light tracing simulations for secondary optical design [16].

The radiant flux is the most important parameter of an LED device. Though the radiant flux can be obtained from a goniophotometer, it would be too time-wasting and costly. Another equipment, which is usually called "integrating sphere," is developed for fast measuring the spectral radiant flux only. An integrating sphere is a hollow sphere typically made of cast aluminum. Its inner wall is coated with a layer of highly reflective material which is barium sulfate in most commercial products. Light inside the integrating sphere is reflected several times so that a uniform illuminance can be obtained on the inner surface of the integrating sphere. At least two small windows are needed to be opened on the integrating sphere of which one is for the light source under test and the other is for the spectrophotometer. Commonly, there is an obstacle in front of the detector to avoid light directly shining on the detector. In reality, there is always light consumed inside the integrating sphere by the reflector layer or the fixture and the illuminance might be not even on the inner surface of the integrating sphere, hence the integrating sphere should be calibrated by a standard lamp beforehand. For different configurations, calibrations must

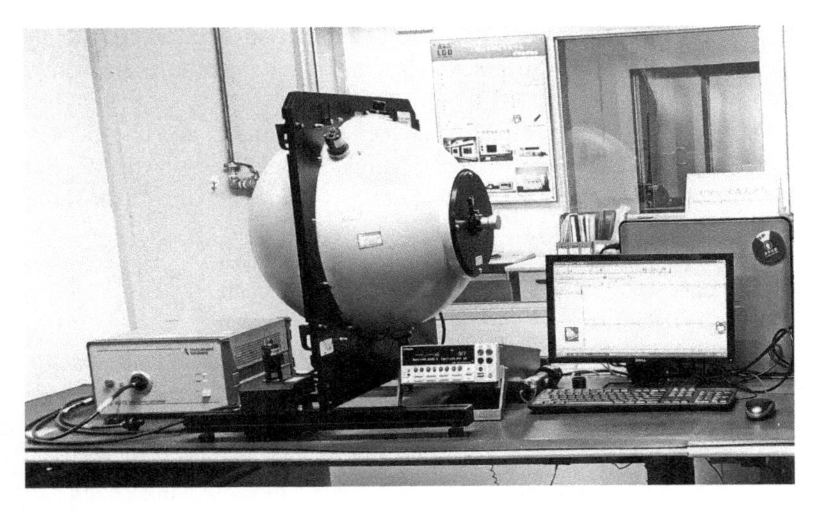

Figure 5.13 Typical integrating sphere system including a spectrometer (Instrument System CAS-140CT).

be performed individually, since the slightest change of the measurement system may affect the accuracy of the measurement result.

Common LED devices only emit light toward one side and thus can be installed on the wall of the integrating sphere. Such a configuration is known as the 2π configuration for the possible light-emitting solid angle is equal to 2π. However, for other devices, such as lightbulbs and light tubes, which emit light in all directions, it should be hung at the center of the integrating sphere. Accordingly, such configurations are called 4π configurations. Practically, the 4π configuration is more difficult to implement because of the mechanical and electrical interconnection and light source heat dissipation.

A spectrophotometer is commonly connected to an integrating sphere by a glass fiber. The entrance surface of the glass fiber is aligned with the integrating sphere surface. The spectrophotometer can measure the illuminance. With the equal illuminance assumption and a known diameter of the integrating sphere, the total spectral radiation output from the light source can be discovered. A typical integrating sphere system is shown in Figure 5.13.

5.2 Power Supply and Efficiency

The LED for its unique electrical and thermal characteristics should be properly driven so as to acquire the best performance. Different types of LED configuration require different types of driving circuitry. Meanwhile, LED efficiency is also critical to the application of LEDs. In this section, the intrinsic electrical characteristics are introduced and then some typical driving circuitry briefly described, before concluding with a description of LED efficiency.

5.2.1 Electrical Characteristics of LED

Although the electrical characteristics of LED are rather simple compared to other electronic devices, they should be thoroughly understood for better application. The LED is essentially a diode with only two terminals. Its electrical characteristics can be completely described by four parameters, which are current, voltage, junction temperature, and sometimes time.

There are several papers discussing the relation between the V_f and T_j and also the forward current, I_f [17–22]. An LED obeys the Shockley diode model in which the forward current, I_f, is a function of the forward voltage, V_f, and the T_j. The mathematical expression is given by

$$I_f = I_{sat} \exp\left(\frac{qV_f}{nk_B T_j}\right) \tag{5.13}$$

where I_{sat} is the reverse bias saturation current (or scale current), n is the ideality factor, also known as the quality factor or sometimes emission coefficient, k_B is the Boltzmann constant, $1.3806488 \times 10^{-23}$ J/K, T_j is the absolute temperature of the p–n junction, and q is the magnitude of charge of an electron (the elementary charge, $1.602176565 \times 10^{-19}$ C).

I_{sat} is a temperature-dependent parameter and can be described by employing a band gap energy, E_g, as

$$I_{sat} = C_{sat} \exp\left(-\frac{E_g}{k_B T_j}\right) \tag{5.14}$$

where C_{sat} is a quasiconstant factor. The Varshni formula yields

$$E_g = E_{g,0K} - \frac{\alpha T_j^2}{T_j + \beta} \tag{5.15}$$

where $E_{g,0K}$ is the band gap energy at 0 K and α and β are two positive fitting parameters. As the junction temperature under consideration ranges from the room temperature, 293 K, to about 413 K, Eq. (5.15) can be expressed in linear form approximately as

$$E_g = E_{g,300K} - \alpha'(T_j - 300) \tag{5.16}$$

where α' is another positive fitting parameter. By combining Eqs. (5.13)–(5.16), finally V_f can be obtained as

$$V_f(I_f, T_j) = \frac{nk_B}{q} T_j \ln\left(\frac{I}{C_{sat}}\right) + \frac{E_{g,300K} - \alpha'(T_j - 300)}{q} \tag{5.17}$$

To calculate the total voltage across the LED, the serial resistance in an LED circuit has to be taken into account. It can be assumed that an LED chip with a size of 1 mm^2 is driven under 1 mA forward current. Even if its specified contact resistance is $0.01\,\Omega \cdot \text{cm}^2$ and other serial resistance in the circuit, such as resistances of wires and connector contacts, can reach 1 Ω, its corresponding voltage drop is only 2 mV. For this reason, this voltage drop can be ignored, especially when the forward current is small. Transforming Eq. (5.17) into a simpler mathematical form and substituting the constant k_B and q into it, finally, results in

$$V_f(I_f, T_j) = 0.86173 \times 10^{-4} n T_j \ln\left(\frac{1}{C_{sat}}\right) + V_{f,0} - AT_j \tag{5.18}$$

And since the exact values of the parameters, n, C_{sat}, A, and $V_{f,0}$ have to be calibrated, we use a more simplified form, which is

$$V_f(I_f, T_j) = V_N T_j \ln(I) + V_{f,0} - V_A T_j \tag{5.19}$$

The sensitivity of V_f to T_j is commonly called the K-factor. Since the expression relates forward current, I_f, and T_j to V_f, the K-factor can be easily calculated by differentiating V_f with respect to T_j as

$$K = \frac{\partial V_f}{\partial T_{jnct}} = V_N \ln(I) - V_A \tag{5.20}$$

From this equation, one of the most important characteristics of the K-factor can be observed that its value will increase with current as the following equation

$$\frac{\partial K}{\partial I} = \frac{V_N}{I} \approx \frac{0.86173 \times 10^{-4} n}{I} \tag{5.21}$$

Usually, a diode has a diode capacitance and in some special applications that require dynamics or transient response of LED time is involved. The diode capacitance is in nano farad level, and under a normal operating conditions, whose forward current is about 1 A and forward voltage 3.3 V, the time constant is about 10 ns and the corresponding bandwidth is around 10 MHz. Consequently, it would not affect pulse width modulation (PWM) controlling, which is about 10 kHz level but has to be seriously considered when the LED is used for optical communication, which requires the operating frequency to be as high as possible [23, 24].

5.2.2 Power Supply for LEDs

From previous discussion, it can be seen that an LED has a negative voltage-temperature coefficient under a certain current. This entails that if an LED is driven under a constant voltage of which the circuit is the simplest, when its junction temperature is rising, it will draw in more current and boost the temperature even higher. Such a positive feedback may induce a thermal runaway problem as the junction temperature keeps rising. For this reason, the LED should be driven in constant current mode.

Sometimes, LED chips have to be connected in parallel for a higher current and higher light output. However, LEDs in parallel very likely have different electrical characteristics. Consequently, the current tends to flow into the one with a lower forward voltage and so potentially damage the LED. One way to avoid this is to connect the LEDs serially first and then connect them in parallel. In such a configuration, the overall forward voltage is the sum of the individual ones and so the slight difference of the forward voltage between different chips can be eliminated. As a result, the driving current can be evenly distributed to all series of LEDs.

After years of LED development, engineers no longer need to design a circuit using discrete components. There are many LED products available on the commercial market containing constant current driving IC (integrated circuit). These LED products can be easily used with a few additional peripheral components.

Most of LEDs for general lighting use wall-plug alternating current (AC) power whose voltage ranges from 110 to 240 V with a frequency of 50 or 60 Hz. The voltage is far too high for directly driving LEDs, and a transformer and AC/DC converter must be added to the LED device. However, the transformer and the peripheral AC/DC converting circuit will substantially increase the size and weight of the LED device. Because of this drawback, high-voltage LED modules are gaining increasing interest recently. One of the low-cost solutions for high-voltage LED devices is connecting a few dozen (wall-plug power voltage divided by the forward voltage) LEDs in series. Another strategy is to connect the discrete LEDs serially as the wafer fabrication stage and integrate them into a high-voltage LED chip. Because of the highly nonlinear relation between driving voltage and light output, if LEDs are directly driven by 220 V AC, voltage overshoot becomes inevitable. Providing that an LED chip has a normal operating voltage of 3.3 V and a driving AC voltage in a sinusoidal waveform, the maximum driving voltage should be set to 3.35 or 3.4 V, which is slightly higher than the normal operating voltage. Otherwise, during a cycle of the input AC power, in only a very short time the LED is truly turned on and (most of the time) the LEDs will remain dark. Directly driving a series of LEDs also brings a light flickering problem. AC power frequency (i.e. 50 or 60 Hz) is still too low for human eyes and can cause health problems over the long term. To counter these disadvantages, LEDs are arranged in rectifier configuration so that in half of the power period half of the LEDs are on. As a result, the overall on/off frequency is doubled.

Considering the electrical characteristic of LED, PWM is the best choice for dimming. The power source delivers a square wave current. The light output of the LED can be adjusted by changing the duty cycle. One major problem that comes with PWM control is that when the frequency of PWM is not high enough it can become harmful to the human eye, especially over the long term.

5.2.3 Power Efficiency

In an LED chip, electrons are driven under the electrical field and some of them recombine in the junction region of the chip and emit photons. The quantum efficiency, η, of an LED device is usually defined as

$$\eta_{\text{EQE}} \equiv \frac{\text{photons/sec}}{\text{eletrons/sec}} = \eta_{\text{IQE}}\eta_{\text{extract}} \tag{5.22}$$

where the η_{IQE} is internal quantum efficiency (IQE) and η_{EQE} is the external quantum efficiency. The IQE depends on the current injection efficiency and recombination efficiency as

$$\eta_{\text{IQE}} = \eta_{\text{injection}}\eta_{\text{recombination}} \tag{5.23}$$

where the $\eta_{\text{injection}}$ is current injection efficiency and $\eta_{\text{recombination}}$ is recombination efficiency. The main mechanism of $\eta_{\text{injection}}$ is current leaking from the quantum well. The $\eta_{\text{injection}}$ may significantly drop under high injection current density [25]. For those electrons that recombine with the holes, some of them emit photons – this is called radiative recombination – while the remaining are nonradiative recombination. In radiative recombination, the energy carried by electrons is converted into light, but in nonradiative recombination the energy is converted to heat. The Auger recombination is the main mechanism of non-radiative recombination. In fact, the major efficiency droop under a high-injection current is due to a low $\eta_{injection}$. $\eta_{injection}$ is current injection efficiency $\eta_{extract}$ depicts the proportion of photons that can escape from the LED chip. It largely depends on the light pattern of the LED chip and the refractive index of the surrounding medium.

The quantum efficiency is one of the most important mechanisms regarding LED efficiency. However, since it does not involve the energy amount carried by an electron or a photon, it still cannot represent the power efficiency of an LED device. From an application point of view, wall-plug efficiency, η_{wpe}, is more important. The η_{wpe} is defined as

$$\eta_{wpe} \equiv \frac{P_{opt}}{P_{electric}} \tag{5.24}$$

where the P_{opt} is total light output and the $P_{electric}$ is electrical power. Usually, the general light is the application that concerns the wall-pug efficiency the most. In general lighting scenarios, the current injection could be over 500 mA/mm^2. In a high injection current, not only the quantum efficiency but also the ohmic loss should be considered. The electrical resistance of ohmic contact and the current spreading layer may contribute a large proportion of the total joule heat. Furthermore, it is impossible to completely extract the light produced in the multiple quantum well (MQW) because of the total internal refraction effect. Hence, the LED packaging can also substantially affect the overall power efficiency.

Power supply efficiency contributes to overall efficiency, though most of the commercial power supply does exhibit a 90% efficiency. Moreover, the power supply efficiency is closely related to the quality of components, such as the transformer and the capacitor. When an LED device product is designed, the cost of the power supply and the benefit from the saved energy should be considered.

5.3 Consideration of LED Thermal Performance

The thermal performance of an LED generally involves two aspects. One is thermal characterization and the other is thermal management. The main objective of the thermal management of an LED focuses on keeping the device, and especially the junction, at an acceptable temperature. To achieve this, it is essential that you understand bottlenecks in the thermal path, and so now we shall discuss the thermal characteristics of LED devices.

5.3.1 Thermal Characterization Methods for LEDs

The analysis above gives a linear relation between the T_j and the forward voltage, V_f, providing the test vehicle is subjected to a constant forward current. If the T_j of the LED is intended to be estimated by measuring the corresponding forward voltage, a calibration should be performed beforehand. This idea is reported in several papers and is widely accepted [26–29].

The typical procedures for the calibration are shown in Figure 5.14. One of the important assumptions of this method is that the T_j is equal to the environment temperature.

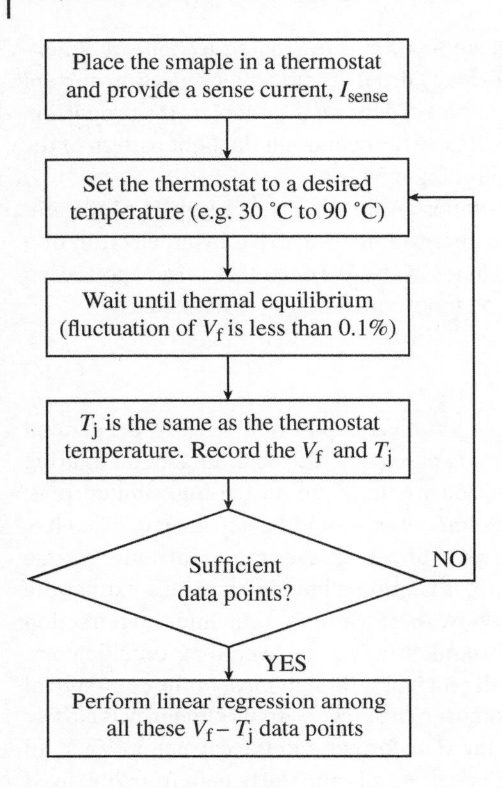

Figure 5.14 Typical calibration procedures for the *K*-factor.

But once the electrical power is sent to the test vehicle, the T_j immediately starts to rise. This self-heating problem can only be eliminated by giving the I_f in a very short time and measuring the V_f instantly. The reasonable period of the source measurement has to be within 50 μs. Therefore, the high-performance source and meter are required.

The main inconvenience of this method is that the calibration has to be carried out for all the I_f concerned, and for most of the time the T_j of different power levels is of interest. Another drawback is that for this method only the history of the T_j rising can be achieved.

For the measurement method mentioned above, the driving current is used for both heating and sensing and there are many inconveniences using the driving current for sensing. To address those drawbacks, the current to sense the T_j is separated from the driving current. The *K*-factor calibration is performed under a small current which cannot generate significant self-heating. Such a current is named after the sense current, I_{sense}, which is usually a few milliamperes. Then, the test vehicle is heated up by a driving current, I_{drive}, which typically ranges from a few hundred milliamperes to a few amperes. In the test, after the test vehicle is sufficiently heated, the current is switched from the I_{drive} to the I_{sense}. By measuring the V_f right after the current switching, the corresponding T_j can be calculated from the voltage change with the *K*-factor.

The practical T_j measurement is illustrated in Figure 5.15. Above all, a calibration should be conducted beforehand so as to get the *K*-factor. To begin with, the test vehicle is heated up by I_{drive} until the fluctuation of the forward voltage is less than 0.1%, indicating that the thermal equilibrium state is reached. Subsequently, the current is switched from I_{drive} to

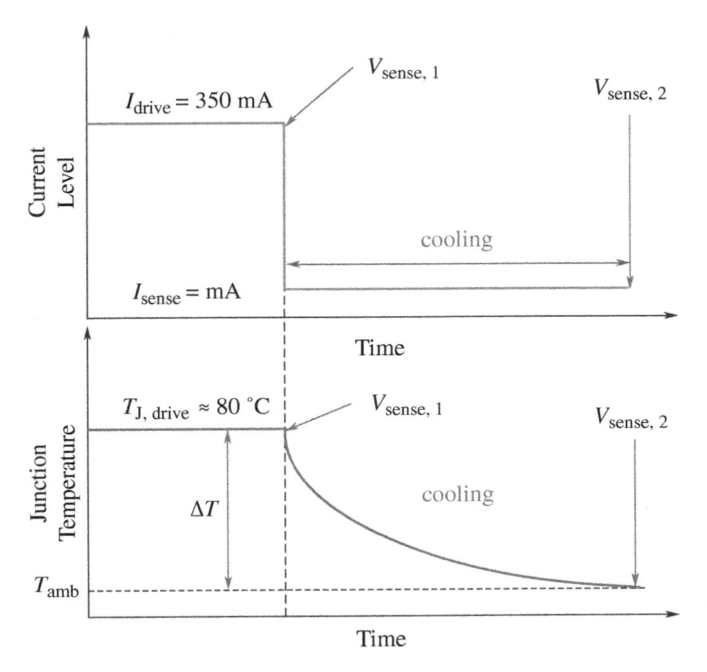

Figure 5.15 Junction temperature measurement procedures by the forward voltage method.

I_{sense} and the corresponding forward voltage, $V_{sense,1}$ is recorded. Because of the thermal equilibrium, the delay in current setup and voltage measurement would not bring significant T_j drop and the measurement would still be accurate. After the current switching, the test vehicle starts to cool down. The cooling transient is considered finished by monitoring the forward voltage fluctuation as well, and the forward voltage at the end of cooling is also recorded as $V_{sense,2}$. Finally, the T_j drop during cooling can be calculated by

$$\Delta T_j = \frac{V_{sense,1} - V_{sense,2}}{K} \tag{5.25}$$

And the T_j for the specific driving current can be obtained by adding in the ambient temperature

$$T_j = \Delta T_j + T_{amb} \tag{5.26}$$

where T_{amb} is the ambient temperature. Here the definition of the ambient temperature can be diverse depending on the experiment setup. But it is always the temperature when the test vehicle is completely cooled down.

In addition, this method has another important advantage in that it can easily record the history of the T_j change versus time in the cooling period, which enables transient thermal analysis to investigate the thermal characteristic of the test vehicle in detail, which will also allow the interpretation of thermal resistance data.

One of the most important thermal analysis methods is the thermal electrical analogy method. The governing equation of thermal conduction is Fourier's law

$$\vec{q} = -k\nabla T \tag{5.27}$$

where $\vec{q}$ is heat flux density, k is the thermal conductivity, and T is the temperature. The thermal capacitance, C_θ, is defined as

$$C_\theta = \frac{qt}{\Delta T} \tag{5.28}$$

where q is heat flow, t is time, and ΔT is temperature change. On the other hand, the governing equation of electrical conduction is the Maxwell equation

$$\vec{J} = \sigma\vec{E} = -\sigma\nabla V \tag{5.29}$$

where $\vec{J}$ is current density, σ is electrical conductivity, and V is voltage. The electrical capacitance is defined as

$$C_{\text{electrical}} \equiv \frac{It}{\Delta V} \tag{5.30}$$

where I is current, t is time, and ΔV is voltage difference. It can be noticed that the two governing equations have the same mathematical form and definition of capacitance. The voltage is equivalent to temperature while current is equivalent to heat flow. Consequently, the knowledge developed in the electrical network can be immediately applied to the thermal network.

In an LED device, the different parts often have a large area compared to the thickness. For example, the high-power LED chip is commonly 1 mm by 1 mm; however, the thickness is only about 0.1 mm. Consequently, the heat flow path in an LED device can be considered a one-dimensional (1D) path. Previous power efficiency analysis revealed that most of the joule heat is generated on the junction of the LED chip. The heat is then dissipated through different layers (i.e. chip substrate, chip bonding layer, chip carrier, insulator, and so on). Though there are various types of LED devices, in most cases they can meet the 1D heat conduction assumption.

The temperature difference of a structure (e.g. the LED chip) along with the heat flow, ΔT, can be calculated by

$$\Delta T = q\frac{T}{kA} \tag{5.31}$$

where q is the heat flow that passes through the structure, T is the thickness, A is the area, and k is the thermal conductivity. The thermal resistance can be accordingly defined as

$$R_\theta = \frac{T}{kA} \tag{5.32}$$

Other than thermal resistance, different parts also have their thermal capacitance. These thermal resistor and capacitor pairs compose a network, as given in Figure 5.16. In a thermal steady state, the capacitor is regarded as open in the circuit. In the aforementioned thermal electrical analogy, heat flow is equivalent to current. Therefore, a current source is used to mimic the heat generated in the junction of the LED chip. The heat flow passes through different thermal resistors belonging to different parts of an LED device and finally come to a "dissipation" resistor. This resistor is often set to simulate convection behavior, which is the most common way to dissipate the heat to the environment. The heat convection transfer can be depicted by the equation

$$q = hA(T_{\text{dissipation}} - T_{\text{amb}}) \tag{5.33}$$

where h is the convection coefficient, which is mainly determined by the air flow across the surface; A is the surface area; $T_{dissipation}$ here is the nominal surface temperature of the heat sink; and T_{amb} is the ambient temperature. Since voltage and temperature are equivalent in the circuit, the environmental temperature is considered as the reference temperature while, from the electrical point of view, the ground is taken as the reference for the voltage. For this reason, the end of the dissipation resistor should be shorted to the ground. But in order to conserve the ladder structure, the shorting wire is replaced by a capacitor of infinite capacitance. When the transient state of the thermal network is investigated, thermal capacitors should be connected to the network. Consider the scenario in which the LED is turned on and its temperature continues to rise. Once thermal equilibrium is achieved, a thermal capacitor in the network should be fully charged. The heat stored in the capacitor can then be calculated from the difference between the node temperature and the reference temperature. Accordingly, from an electrical point of view, one node of the capacitor should be grounded. Similar to the definition of thermal resistance, the thermal capacitance can be evaluated by

$$C_{\theta} = C_{p}AT$$

where C_p is the specific heat and A and T are the area and thickness of which the product is the part volume.

The network ladder given in Figure 5.16 has been studied for decades. Because the governing equations of both electrical conduction and thermal conduction have the same mathematical form, conclusions from the electrical network analysis can be directly applied to thermal analysis. Usually, resistances of different parts in an LED device is what we consider since it can help us detect the bottleneck in the thermal path. Previous network analysis has shown that, if a step heat power is inputted into a resistor–capacitor ladder

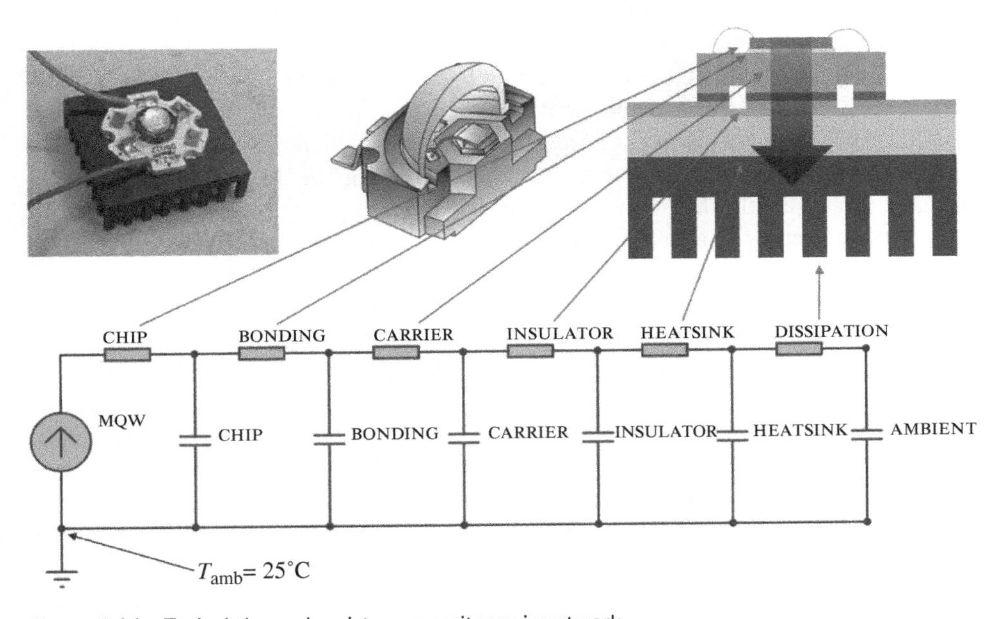

Figure 5.16 Typical thermal resistor–capacitor pair network.

similar to Figure 5.16, the resistances and capacitances in the ladder can be extracted from the change of the source temperature along with time. In the illustration diagram, there are only six resistor–capacitor pairs, but in reality a practical model often consists of thousands or even millions of these pairs. A part of a structure in an LED device thus would be represented by multiple pairs.

In most cases, the temperature voltage calibration is so accurate that the error can be lower than 0.1 K. The major error of this characterization method mainly comes from two parts. One is inaccurate measurement of the total light output which will lead to an over- or underestimation of the thermal power. The other one comes from parasitic electrical resistance. The wires to the LED device, the traces on the printed circuit board (PCB), or the gold wire bonding in the device can all contribute to parasitic resistance. The forward voltage that is consumed by the parasitic resistance is not able to separate and as a result the thermal power is overestimated.

Given that there are N pairs of thermal resistance, $R_1, R_2, \ldots, R_n, \ldots, R_N$ and capacitance, $C_1, C_2, \ldots, C_n, \ldots, C_N$. After all the exact R_n and C_n in the thermal model are extracted from the junction temperature data, a cumulative resistance, $R_{\text{Cum},n}$, can be defined as

$$R_{\text{Cum},n} = \sum_{i=1}^{n} R_i \tag{5.34}$$

and similarly, a cumulative capacitance, $C_{\text{Cum},n}$, can be defined as well. A diagram can be plotted wherein the X axis is the cumulative resistance and the Y axis is the cumulative capacitance. Because the diagram of such function can represent the thermal structure of a system, the function is called integral structure function. Meanwhile, a differential structure function can be obtained by differentiating the integral structure function. In the integral structure function, the cumulative capacitance keeps rising along with the cumulative resistance. The vertical line at the end entails an infinite capacitance which is the environment and therefore the corresponding cumulative thermal resistance is the thermal resistance between the junction and the environment. The thermal conductivity of different materials can vary (e.g. the thermal conductivity of epoxy is only 0.1 W/(m · K) and copper is over 300 W/(m · K)). Meanwhile, the specific capacitance different is much smaller than the structure capacitance. A flat line in the integral structure function represents a thermal insulator. Consequently, the thermal resistance of different parts in an LED device can be to be distinguished by this method. This method which utilizes the dynamic response of the thermal ladder network is often referred as transient thermal analysis.

Although some researchers claim that the thermal resistances of different parts can be measured by means of transient thermal analysis, it lacks numerical simulation or experimental test validation. In fact, there is still no standardized method to measure the thermal resistance of arbitrary parts inside a system. The JEDEC 51 is a set of standards about the thermal characterization. Among them, standard JEDEC 51-1 stipulates a method to measure the resistance from the junction or the heat source to a particular thermal interface. The principle of the method can be described as alternating the thermal interface material or geometry so that different resistances of the thermal interface can be created. Therefore, the difference in the thermal resistance would be shown in the structure function. Consequently, the location of the thermal interface along the cumulative thermal resistance can be determined and the thermal resistance from the junction to that thermal interface can be known. One typical example is the thermal resistance from the junction to the bottom of a metal core printed circuit board (MCPCB) which is frequently used for high-power LEDs. The MCPCB is usually attached to a heatsink with thermal grease. By measuring the different structure functions of different thermal greases, the thermal resistance can be obtained.

The forward voltage method for T_j measurement is usually implemented by the Mentor Graphic T3Ster® system, shown in Figure 5.17, which can provide up to 2 A of driving current and down to less than 50 mA sense current with a maximum 12.5 V forward voltage. The current switching can be finished within 10 μs.

The electrical connection diagram is shown in Figure 5.18. The I_{drive} (I_E^+) and I_{sense} are provided by two different modules. Both modules are connected to the test vehicle by two coaxial cables in parallel. The I_{sense} is sent to the test vehicle all the time and the current switching is realized by only shutting off the I_{drive}. The I_{sense} module has a very high dynamic response to guarantee fast switching. There is a digital voltage meter (DVM)

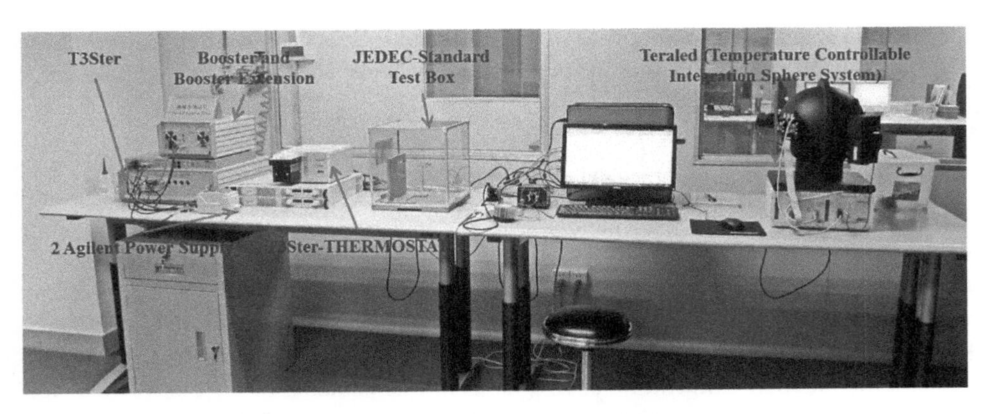

Figure 5.17 Mentor Graphic T3Ster® for the measurement of the forward voltage junction temperature.

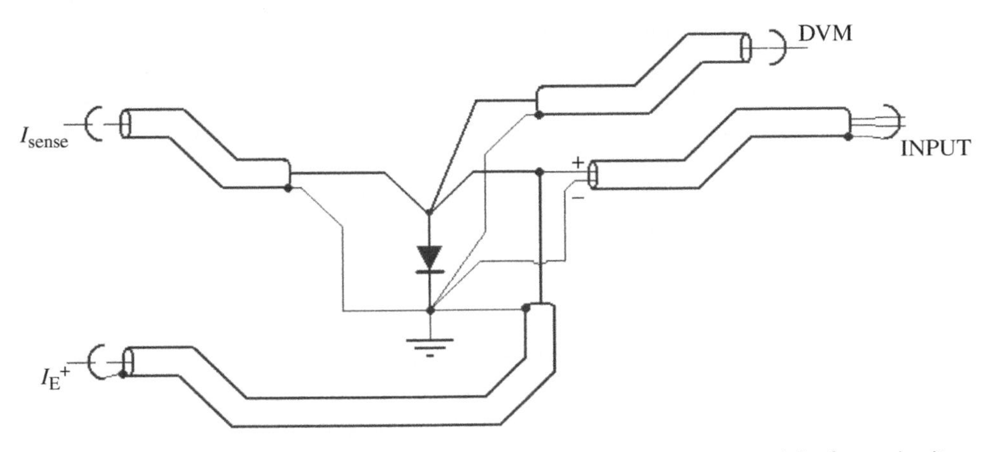

Figure 5.18 Electrical configuration for the driving current, sense current, and the forward voltage measurement.

to monitor the overall forward voltage level. However, with solely using this meter, it is impossible to fulfill the requirement of the T_j measurement. The K-factor is usually around 1.5 mV/K, nevertheless the forward voltage of an LED is around 3 V. As a result, the voltage measurement module needs to distinguish about 0.05% voltage change for the sake of realizing 1 °C temperature resolution. For most common analog digital (AD) converters, the required resolution is too high. In fact, it is unwise to measure the full range of the forward voltage since only the voltage change is of interest. Thus, a voltage signal is sensed by another measurement module (through the INPUT channel in the figure). In that measurement module, an offset voltage is provided so that the input voltage is subtracted to about 200 mV. The voltage is then converted to digitals by a 12-bit 1 MHz AD converter.

Furthermore, a thermostat provides a controllable temperature ranging from 0 to 100 °C for the K-factor calibration. During the calibration process, the test vehicle is attached to the thermostat with some thermal grease as a thermal interface material to reduce the thermal resistance. In the calibration, every 5 °C a voltage is recorded. To ensure the thermal equilibrium state, the system will first wait till the temperature of the thermostat reaches the target temperature, and afterward wait another 60 seconds to make sure the test vehicle is sufficiently heated. The calibration quality can be checked by examining the linearity of the voltage temperature diagram.

5.3.2 Thermal Management Methods

The objective of thermal management is to maintain a proper junction temperature while keeping the cost of thermal management as low as possible. Different application scenarios have different heat dissipation requirements. For example, a flashlight LED may not require very good heat dissipation though its peak power could be very high, while for a streetlamp LED, because of much stricter reliability requirements compared to other LED devices, the junction temperature should be as low as possible. Better thermal management always requires higher design and manufacturing costs. LED devices are often very cost sensitive, but at the same time the reliability of an LED device is closely related to its thermal performance. As a result, optimization should be achieved among cost, reliability, and thermal performance. Finding the bottleneck in the thermal path of an LED system is most critical to the thermal management. Only the efforts on the bottleneck can effectively enhance the thermal performance.

Referring to Figure 5.16, the thermal interface in an LED device is the major heat barrier in an LED device. The chip bonding layer, chip carrier, insulating layer on the PCB, and the thermal interface between LED package and heatsink are the four major thermal resistance sources in the entire LED device. The thermal resistance, R_θ, of a thermal interface can be estimated by

$$R_\theta = \frac{T}{kA} \tag{5.35}$$

where k is the thermal conductivity of the thermal interface material, T is the thickness, and A is the heat flow entrance area.

Wire bonding type LED chips are used in most commercial LED products. For this type of LED chip, it is usually bonded to a chip carrier by thermal conductive adhesive. Other advanced thermal conductivity adhesive has also been studied for decades. Epoxy filled with silver nano particles is the most ordinary adhesive. The silver particles aggregating together provide a heat path. Generally, the higher content of silver particles can enhance the thermal conductivity of the material; however, this can significantly increase the viscosity of the adhesive at the same time. The stickiness of the adhesive often results in a thicker chip bonding layer and consequently the R_θ can only be reduced slightly. It would be more effective to use soldering or a sintering method for enhancing the thermal performance. However, the soldering method requires special metallization on the back side of the LED chip. The reflow process in assembly can also increase the manufacturing cost. The sintering method requires special metallization as well, while it needs the chip, and the carrier can sustain high temperature and pressure during the sintering process.

Flip-chip type LEDs are now attracting more attention. Since the active layer of the LED chip is directly bonded to the chip carrier by the soldering method, it can provide a much better thermal performance compared to the wire bonding type LED. Sometimes, the chip is soldered onto the chip carrier by a solder ball grid array. In such cases, instead of the average thermal resistance of the bonding layer, more attention should be paid to the nonuniform junction temperature in the chip [30].

Lead frame type LEDs have been used for decades, but this type of packaging is no longer suitable for high-power applications. The high heat flow density results in significant CTE (coefficient of thermal expansion) mismatch between the copper lead frame and the molding compound. Ceramic substrate as the chip carrier is the mainstream for LED component assembly. Aluminum nitride or aluminum oxide are the two common materials for the ceramic substrate. A sheet of ceramic plate is fabricated by sintering aluminum nitride (AlN) or aluminum oxide (Al_2O_3) particles. A metal layer is deposited on the substrate by sputtering and plating. Silicon substrate is sometimes used as well. For these ceramic substrates, the thermal conductivity can easily exceed 100 W/(m · K) and the ceramic can provide excellent electrical insulation. However, the mechanical strength is still seen by many as responsible for the low fracture toughness of ceramics.

PCB, an indispensable part in an LED system, is the main source of overall thermal resistance (contributing, as it does, more than 30%). In most cases, the MCPCB is applied. An MCPCB has a metal core which is often aluminum or copper. The copper circuit layer is laminated on the metal core by a layer of polymer which is usually 0.1 mm thick epoxy resin. The thermal conductivity of polymer is usually lower than 1 W/(m · K). Considering the electrical insulating may require the substrate to sustain 2000 V, the polymer layer cannot be too thin. As a result, the thick polymer layer creates a large heat barrier in the entire LED system. For example, if a 3030 LED component which generates 0.7 W heat is surface mounted on a MCPCB with 0.1 mm thick epoxy insulation layer (0.2 W/(m · K)), the temperature difference is

$$\Delta T = P_\theta R_\theta = P_\theta \frac{T}{kA} = 0.7\,\mathrm{W} \times \frac{0.1\,\mathrm{mm}}{0.2\,\mathrm{W/(m\cdot K)} \times 9\,\mathrm{mm}^2} = 39\,\mathrm{K} \tag{5.36}$$

Larger LED packages, even though they occupy a larger area, can essentially reduce the thermal resistance. A 5050 LED with the same heat power only results in 14 K temperature on the same MCPCB.

The MCPCB or other thermal conductive substrate with a printed circuit is usually fastened to a heatsink. The gap between the substrate and the heatsink is commonly filled with thermal grease. Silicone oil (polydimethylsiloxane) is the most common matrix of thermal grease in the LED industry. Fillers such as alumina, boron nitride, and zinc oxide are mixed with the silicone oil to form a paste.

The thermal grease is in paste form and requires a dispensing machine. A preformed pad as a thermal interface is preferred in manufacturing but the thermal performance of most thermal pad products cannot fulfill the requirements of an LED. Phase change material is another type of common material used for the thermal interface.

Thermal adhesive, which is usually an epoxy-resin-based thermal interface material, is sometimes applied. The main problem of this type of thermal interface material is that after the assembly process the adhesive should be fully cured and consequently the device or the system becomes unreworkable.

Comparing the heat flow density inside the LED device, the heat flow density across the thermal grease is much lower. As a result, instead of the thermal dissipation performance, the manufacturing cost should have a higher priority in thermal management design.

References

1. Schacter, D.L., Gilbert, D.T., and Wegner, D.M. (2011). *Psychology*, 2e. New York: Worth.
2. CIE (1931). *Commission Internationale de l'Eclairage Proceedings*. Cambridge: Cambridge University Press.
3. MacAdam, D.L. Visual sensitivities to color differences in daylight. *JOSA* 32 (5): 247–274.
4. Schubert, E.F. (2006). *Light-Emitting Diodes*, 2e. Cambridge: Cambridge University Press.
5. Wikipedia, (2019). Planckian locus. https://en.wikipedia.org/wiki/Planckian_locus (accessed 3 September 2020).
6. CIE. (1995). Method of Measuring and Specifying Colour Rendering Properties of Light Sources: TR 13.3/1995. Vienna: Commission Internationale de l'Eclairage.
7. Nickerson, D. and Jerome, C.W. (1965). Color rendering of light sources: CIE method of specification and its application. *Illuminating Engineering, IESNA* 60 (4): 262–271.
8. Chynoweth, A.G. and McKay, K.G. (1956). Photon emission from avalanche breakdown in silicon. *Physical Review* 102 (2): 369.
9. Graeme, J.G. and Jerald, G.G. (1996). *Photodiode Amplifiers: Op Amp Solutions*. New York: McGraw-Hill.
10. Voogd, J. (1940). Tungsten ribbon lamps for optical measurements. *Philips Technical Review* 5 (3): 82–87.
11. Buckley, J.L. (1971). Use of a tungsten filament lamp as a calibration standard in the vacuum ultraviolet. *Applied Optics* 10 (5): 1114–1118.
12. Rutgers, G.A.W. and Jan, C.D.V. (1954). Relation between brightness, temperature, true temperature and colour temperature of tungsten. Luminance of tungsten. *Physica* 20 (7–12): 715–720.
13. Geist, J., Gladden, W.K., and Edward, F.Z. (1982). Physics of photon-flux measurements with silicon photodiodes. *JOSA* 72 (8): 1068–1075.
14. Everfine. (2020). GO-R5000 Full-space Rapid Goniophotometer. http://www.everfine.cn/productsinfo.php?cid=8&id=21 (accessed 3 September 2020).

15 Bridgelux Inc (2016). *Bridgelux SMD 5050 3w 18V Product Data Sheet DS62 Rev B.* Livermore, CA: Bridgelux.

16 Jenkins, D.R. and Mönch, H. (2000). P-81: Source imaging goniometer method of light source characterization for accurate projection system design. *SID Symposium Digest of Technical Papers* 31: 862–865.

17 Xi, Y. and Schubert, E.F. (2004). Junction-temperature measurement in GaN ultraviolet light-emitting diodes using diode forward voltage method. *Applied Physics Letters* 85 (12): 2163–2165.

18 Chen, N.C., Wang, Y.N., Tseng, C.Y. et al. (2006). Determination of junction temperature in AlGaInP/GaAs light emitting diodes by self-excited photoluminescence signal. *Applied Physics Letters* 89 (10): 101114.

19 Chen, N.C., Yang, Y.K., Lien, W.C. et al. (2007). Forward current-voltage characteristics of an AlGaInP light-emitting diode. *Journal of Applied Physics* 102 (4): 043706.

20 Keppens, A., Ryckaert, W.R., Deconinck, G. et al. (2008). High power light-emitting diode junction temperature determination from current-voltage characteristics. *Journal of Applied Physics* 104 (9): 093104.

21 Laubsch, A., Sabathil, M., Baur, J. et al. (2010). High-power and high-efficiency InGaN-based light emitters. *IEEE Transactions on Electron Devices* 57 (1): 79–87.

22 Wen, J., Wen, Y., Li, P. et al. (2011). Current-voltage characteristics of light-emitting diodes under optical and electrical excitation. *Journal of Semiconductors* 32 (8): 084004.

23 Yang, W., Zhang, S., McKendry, J.J.D. et al. (2014). Size-dependent capacitance study on InGaN-based micro-light-emitting diodes. *Journal of Applied Physics* 116: 044512.

24 Xia, Y., Williams, E., Park, Y. et al. (2004). Discrete steps in the capacitance-voltage characteristics of GaInN/GaN light emitting diode structures. *MRS Proceedings* 831: 233–238.

25 Zhao, H., Liu, G., Zhang, J. et al. (2013). Analysis of internal quantum efficiency and current injection efficiency in III-nitride light-emitting diodes. *Journal of Display Technology* 9 (4): 212–225.

26 Chhajed, S., Xi, Y., Gessmann, T. et al. (2005). Junction temperature in light-emitting diodes assessed by different methods. In: *Proc. SPIE 5739, Light-Emitting Diodes: Research, Manufacturing, and Applications IX*, 16–24. SPIE.

27 Siegal, B. (2006). Practical considerations in high power LED junction temperature measurement. In: *International Electronic Manufacturing Technology 2006*, Putrajaya, Malaysia, 62–66. IEMT.

28 Sa, E.M., Antunes, F.L.M., and Perin, A.J. (2007). Junction temperature estimation for high power light-emitting diodes. In: *IEEE International Symposium on Industrial Electronics*, 3030–3035. ISIE.

29 Roscam, A.F.D. and Pertijs, M.A.P. (2011). Light-emitting diode junction-temperature sensing using differential voltage/current measurements. In: *2011 IEEE Sensors Proc.*, 861–864. IEEE.

30 Tao, M., Lee, S.W.R., Yuen, M.M.F. et al. (2012). Effect of die attach adhesive defects on the junction temperature uniformity of LED chips. In: *2012 35th IEEE/CPMT International Electronics Manufacturing Technology Conference (IEMT)*, Ipoh, Malaysia, 1–7. IEEE.

6

Reliability Engineering for LED Packaging

Reliability engineering is technology to improve the ability of a light-emitting diode (LED) system maintaining its specific function during its designed life. It includes product design, analysis, and testing. System reliability should not only be achieved passively by testing and examination but also be ensured by proper design, manufacturing, and management. Figure 6.1 shows the three major steps of reliability engineering. First of all, a good design is the key factor to achieve product reliability. The manufacturing process must fulfill the design requirement so as not to lose any reliability. Design and manufacturing should be governed by correct management. Afterward, reliability tests should be performed and, hence, the lifespan of the product estimated. Failure analysis should also be conducted to understand the failure mechanisms. If the product failed the reliability assessment, this reliability engineering cycle should be repeated until the specification of the product is fulfilled.

For an LED product, its life should be ascertained according to the customer's needs. After the product is designed and manufactured, reliability tests should be conducted. Reliability tests mimic real application scenarios in order to obtain the product's lifespan. However the common lifespan of an LED product could be several years, which is too long for commercialization. To minimize the test time, reliability tests are often executed with accelerated conditions. The correct analysis method should be used for a precise life estimation. For samples that fail their reliability tests, failure analysis ought to be performed to understand the reason for the failure to improve product performance.

6.1 Concept of Reliability and Test Methods

6.1.1 Reliability of Electronic Components or Systems

Reliability is the ability of a system or component to function under stated conditions for a specified period of time [1]. Here, to function implies that the system or component should work normally as expected. Once the performance of a system or a component cannot fulfill its specification, failure occurs, no matter if it is recoverable or not. Even for the same system or component, different specifications can be assigned which will lead to different standards of failure. Therefore, reliability highly depends on the definition of failure.

From LED to Solid State Lighting: Principles, Materials, Packaging, Characterization, and Applications, First Edition.
Shi-Wei Ricky Lee, Jeffery C. C. Lo, Mian Tao, and Huaiyu Ye.

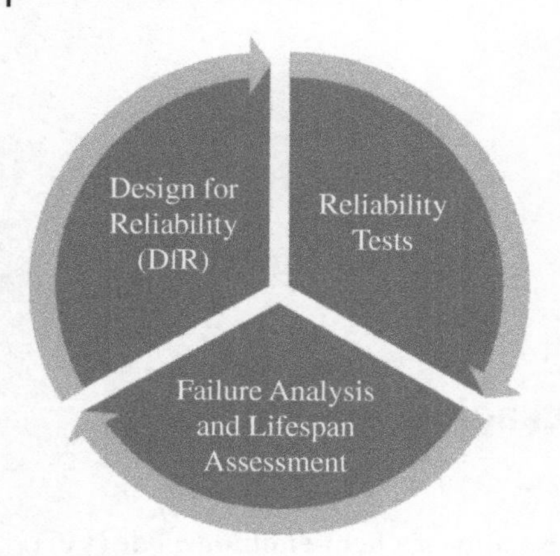

Figure 6.1 Three major steps of reliability engineering.

The main task of an LED component or device is to properly deliver light. The intensity or color of the light is often of most interest and used as the criterion for reliability. Commonly, if light output drops by more than 70%, the device is considered to have failed. Another failure criterion is the color shifting of an LED product. Since the failure criterion is defined by the deviation of the LED's original specifications, this type of failure is known as degradation failure. Meanwhile, catastrophic failures, which indicate that the LED system or the component does not emit light anymore, are usually not allowed in LED products. Usually, reliability estimation in the LED industry does not focus on this type of failure.

The number of LED components or device products that are supplied to customers can be huge. The number a product (e.g. LED 2835, 3030, or 5050) could reach billions a day for a manufacture line. It is impractical to guarantee the reliability of all the products and would not be regarded as a reliability problem as long as the number of failed products is within an acceptable range. For such a point of view, reliability tests should be conducted on enough samples so that the reliability results have statistical meaning.

Reliability tests and qualification tests are two very similar tests but they have a different purpose. Although the processes of these tests are almost the same, the objectives are very different. The reliability test requires a certain percentage of the total sample population that has failed to estimate the product's life, while the qualification test needs there to be no failure detected in the sample being tested. In a qualification test, a type of product with a sample size of 22–32 pieces is tested within a limited time or period. The moment a failure occurs, the test is stopped. If there is no failure, the product is considered to have passed and be ready for delivery. On the other hand, in a reliability test, samples should be tested for as long as possible. The failure time of different samples is recorded. The data will be further analyzed so that the product life can be extrapolated. Reliability tests are often conducted at the product design stage, while qualification tests are performed at the manufacturing stage, especially just before delivering the product to the customer.

6.1.2 Common Failure Mechanisms and Reliability Tests

A failure mechanism must explain what the root cause of a failure during a reliability test is and how this cause leads to the failure. A failure mechanism can be very complicated and failure analysis is the process whereby the failure mechanism is divined. Failure analysis is akin to how a doctor diagnoses a patient. The symptom discovered initially might not be the true root cause of the failure. Dedicated study and analysis must be carried out so as to find the true failure mechanism, thus allowing the reliability of the product to be improved. Following decades of development in the LED industry, people discovered that there are some particular failures that repeatedly occurred in LED products. These failures largely determined the reliability of the LED products. According to the potential failure mechanism, corresponding reliability test methods were developed to evaluate a product's lifespan.

Most of the failure mechanisms of both degradation failure and catastrophic failure are due to their being thermally overstressed. Causes of thermal overstress can be complicated. Improperly driving an LED device (for example overly high driving current, excessive overshoot in flashlight LED and high-voltage AC LED, imbalanced current distribution in a parallel LED circuit) can lead to a detrimentally high temperature. On the other hand, poor heat dissipation (such as inappropriate thermal path design, unexpected delamination or breaking of the thermal interface, insufficient convection heat dissipation) can also be harmful to LED products. In addition to these, local thermal overstress occurs from time to time in commercial LED products. Poor fabrication or degradation of the current spreading layer in an LED chip may induce a current crowding problem and consequently thermal overstress. An insufficient number of bond wires or too small a diameter of bond wires can also generate a high temperature in wire bonds. One cause of thermal overstress is, however, often neglected: the heat generated from phosphor. Phosphor is frequently used to convert the light from an LED chip to other colors. There is inevitably energy lost in this light conversion process. Though the amount of heat generated in phosphor is tiny, the temperature of the phosphor can be very high since the phosphor is encapsulated in silicone, which is a thermal insulator [2]. Figure 6.2 illustrates a study about a phosphor heating problem on remote phosphor coating LEDs. The T_{max} is the maximum temperature measured by infrared thermography, while the T_j is the junction temperature obtained by means of forward voltage drop. For the white LEDs, the T_{max} is substantially higher than the junction temperature, T_j, which suggests that phosphor itself can generate a significant amount of heat and raise its temperature. On the other hand, for blue LEDs, because there is no phosphor converting light energy to heat, the T_{max} is very close to T_j.

There are two aspects of detrimental effects from high temperature: overly deformed packaging structure and degradation of material. An LED device inevitably includes three parts: the LED chip, the chip carrier, and the encapsulant. The coefficient of thermal expansion and the mechanical modulus of the materials that constructs these three parts are very different. Consequently, when the structure of an LED is subjected to overheating, it will be become deformed and damage itself or other structures. For example, over-deformed silicone can create high level force and gold wire bonding that is enwrapped inside the silicone. The high force may stretch and break the gold wire or peel off the second wire bond, which is the weakest in the whole wire bond structure. The LED chip may be cracked due to the excessive stress built up from the coefficient of thermal expansion (CTE) mismatch between the chip substrate and the chip carrier. Meanwhile, overheating can induce substantial high stress in the interface between different parts. The deformed silicone may delaminate from the chip carrier and leave a crack in the interface. The chip carrier is often coated with silver as a light reflector. Without protection from silicone, contamination such as moisture and

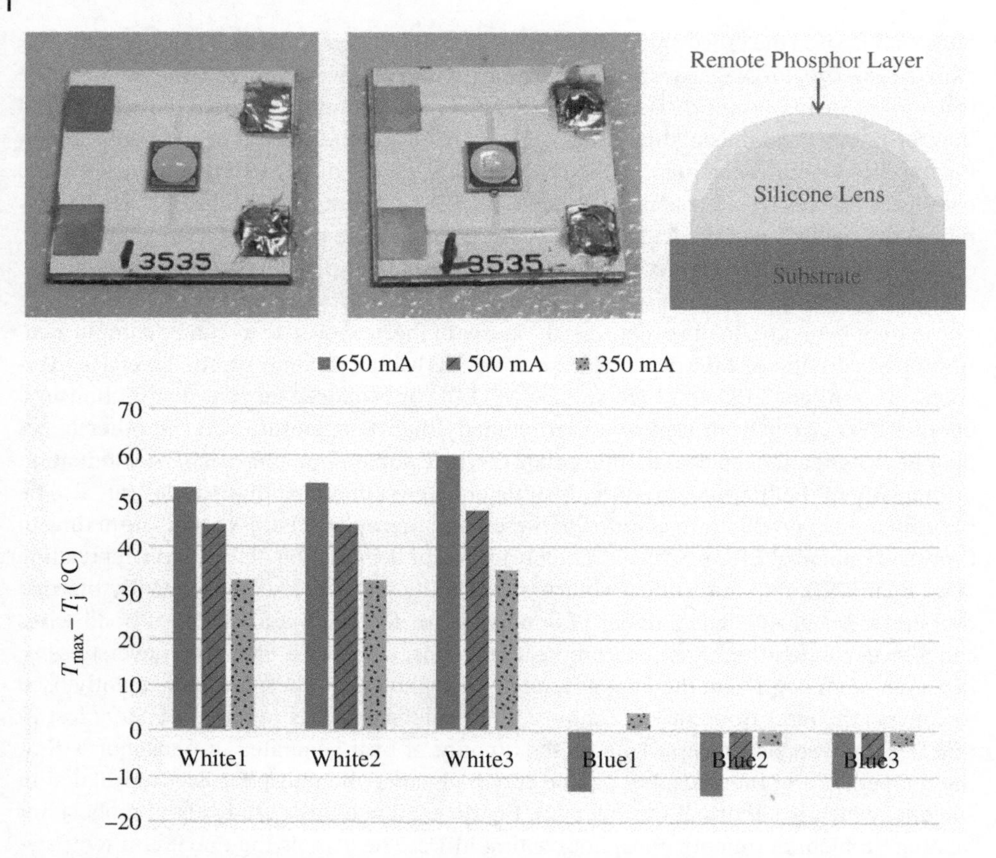

Figure 6.2 Comparison of max temperature on LED surface and junction temperature on white and blue LEDs.

sulfide may permeate along the delamination crack and tarnish the silver surface. If the contamination permeates deeper, it may attack the LED chip and cause catastrophic failure.

Degradation of the material is largely driven by temperature. A typical example is the yellowing of silicone or epoxy resin. The resin may be carbonized under high temperatures with exposure to chloride. In addition, high temperatures can also degrade the multiple quantum well (MQW) and ohmic contact of the LED chip. Meanwhile, high current density itself, other than the joule heat effect, can damage the chip as well [3, 4].

Most failure mechanisms relate to thermal and/or electrical overstress. Based on such knowledge, several reliability test methods have been developed. The most fundamental test method is the stress operating test. The LED samples are placed in a temperature oven where the temperature is higher than that experienced during normal operating conditions. They are being driven under a current that is higher than the rated level. The LED samples can be components or devices. Proper heat dissipation management should be applied to them so that the junction temperature remains at a normal level when at room temperature.

It is very easy to monitor the forward voltage of individual LEDs using a data logger similar to other electronic devices. However, it is much more important to capture light output because of the special failure criterion for LEDs. The light output, or more precisely the lumen output, is not that easy to measure. The simplest method is to take out the samples from the oven at a certain interval time and measure them with an integrating sphere. The drawback is that time and labor costs can be unacceptably high and at the same time the test may not be very reliable because it is impossible to record many data points, which substantially increases the error in data analysis.

It is very necessary to conduct on-site monitoring, and it will be noticed that only the relative change of the light output is needed instead of the absolute output. On the other hand, the light measurement unit is a fragile piece of equipment and should be isolated from elements of the harsh environment of the reliability test, such as high temperature. It is impossible to directly place the light measurement unit inside the chamber.

With all of these considerations, a special reliability system has been developed, as illustrated in Figure 6.3 [5]. Generally, it is a temperature chamber that contains the LED samples under test. There is an optical window opened on the chamber wall and a lux meter is attached to the window. The inner chamber and the fixtures are made of stainless steel, with a polished surface which can effectively reflect light. Stainless steel is considered inert and stable during the reliability testing period and therefore the reflectance will not change. The light from LED samples in the chamber is multiply reflected inside the chamber and finally reaches the lux meter. When the optical output of a particular sample is desired, the other samples are turned off so that the lux meter can only receive the light from this sample. By comparing the lux measured to the initial value, the relative light output change can be calculated. By successively turning on the different samples and measuring the light output respectively, the relative changes of the light output of all the samples can be obtained.

A temperature cycle is another test method often introduced for the purpose of evaluating the fatigue resistance of an LED device. A periodic temperature change will induce repeating deformation in the LED device, owing to thermal expansion. Different parts of

Figure 6.3 The real-time on-site light output monitoring system for LED reliability test. A lux meter is placed at the center of the oven wall.

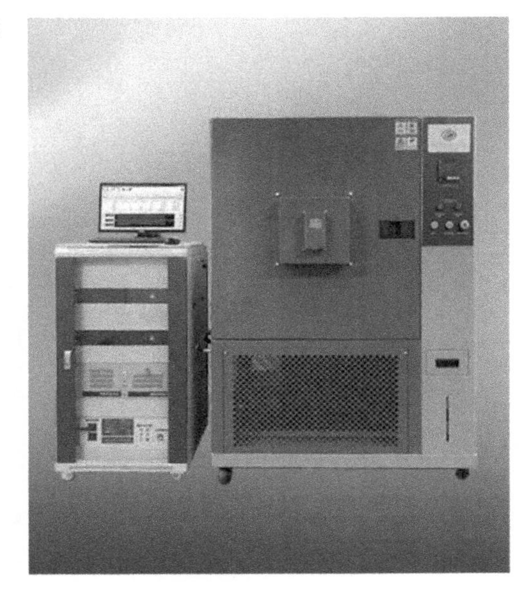

the LED device are therefore subjected to a cyclic stress. Fatigue failure is expected to occur in such circumstances. The samples under test are placed in a chamber. Air flow of high or low temperature is blown through the samples. A thermal couple is usually installed next to the tested samples to capture the in-situ temperature for better feedback control. The temperature ramp rate is a critical parameter for the thermal cycling test, and some standards specify a temperature ramp rate as a guideline for testing. Most of the materials in an LED system are softened in high temperatures, which means that the stress could be substantially higher in the low-temperature stages during a thermal cycling test. Consequently, fatigue failures tend to occur in the low temperature. Fatigue failure is often in a form of material cracking or delamination between different layers. Such failures may recover when the temperature rises from low to normal levels. For this reason, in-situ monitoring must be implemented for the temperature cycle tests. Otherwise, the cycle-to-failure will be significantly underestimated.

Silicone is the most common material encapsulating LED chips. In high temperatures and humid environments, moisture can easily permeate the silicone encapsulant, especially when LED devices are in an elevated temperature. In an LED package, a silver surface is often used as reflector. Silver is vulnerable to moisture. Besides, moisture is also very likely to penetrate the interface between different parts. The moisture may tarnish the metal reflector and reduce the light output. If the moisture reaches the LED chip, it may cause corrosion to the wire bonding pad or the electrodes and sometimes create a short circuit. Furthermore, if an LED component absorbs a significant amount of moisture during manufacturing, when it is going through a reflow process, the moisture may rapidly evaporate and create a so-called popcorn failure. Hence, it is important to evaluate the moisture resistance capabilities of an LED device. A typical high temperature/moisture oven is shown in Figure 6.4. The samples under test are stored in an environment with an elevated moisture

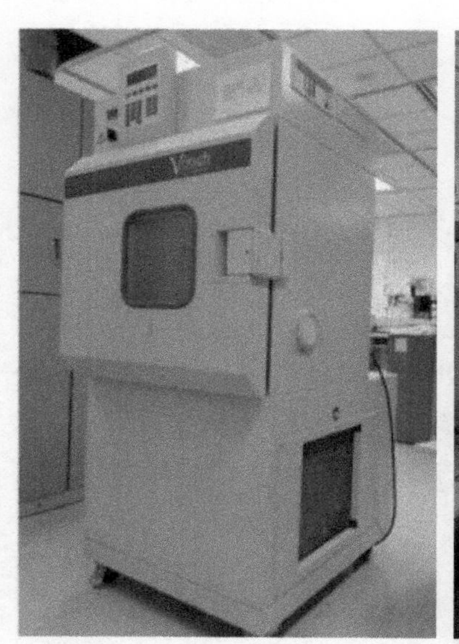

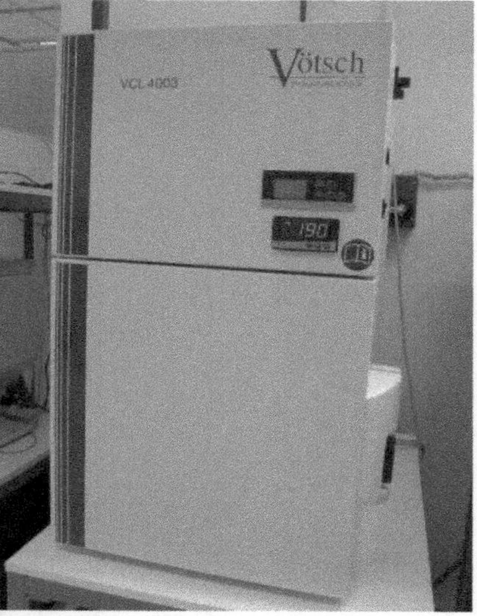

Figure 6.4 Typical thermal cycling chamber (left) and humidity chamber (right).

and temperature which are usually 85% of relative humidity and 85 °C. To examine moisture resistance, weight gain after the tests is commonly used as an index. Or sometimes, a reflow process is carried out on the samples to complete a high temperature/moisture test. Popcorn failure will then be checked by visual inspection.

6.2 Failure Analysis and Life Assessment

6.2.1 Methodology for Failure Analysis

Failures will occur after LED products are used for a certain time or are subjected to accelerated reliability tests. Understanding the root cause of the failure is one essential step for improving product reliability. However, finding the root cause of the failure is, in most cases, very difficult. It is similar to diagnosing a disease from the symptoms of a patient. Identifying the failure location in an LED system is only like knowing the symptom. It needs further diagnosis to reveal the disease hidden behind it. Unluckily, what can be directly observed is often not the true cause of the failure. For example, if open-circuit failure is found in an LED, microscopic inspection may find that the gold wire bonding was broken in the second-bond location. It might be due to poor bonding quality; however, more often it is induced by the thermal deformation of the silicone around the gold wire.

In failure analysis, the failure should first be located. For catastrophic failure, there is usually only one failure location. But for wear-out failure, which implies that light output decreases below a certain level, there might be several failure locations. The chip could be degraded in high currents and at the same time the encapsulating silicone could be yellowish. Different aging effects from different mechanisms should be taken into account.

In principle, nondestructive inspection approaches should be conducted first and then destructive methods used. Optical and electrical characterizations are the simplest approaches for diagnosing the problem. They can immediately give a lot information of potential failure locations. X-ray, including three-dimensional (3D) X-ray, inspection is another prominent nondestructive inspection method. It can discover some failures, like wire bond breaking, chip cracking, or voids in the soldering layer. Figure 6.5 illustrates the use of 3D X-ray imaging to detect voids in the thermal via of a printed circuit board (PCB). However, X-ray inspection is not capable of detecting delamination. To detect delamination, confocal scanning acoustic microscopy (CSAM) is often introduced.

Usually, the nondestructive method should be able to reveal failure locations. Destructive methods, such as cross-section inspection, chemical decapsulation, ion-milling, dye-and-pry, etc., are usually employed only when further confirmation of the failure location is required or when failure locations cannot be found by nondestructive methods. However, the destructive methods are very risky since they will damage the sample and eliminate other information. In many cases, the actual root cause of a failure may be lost by the destructive method and so destructive methods should be regarded as a last resort.

After the failure is located, the reason for the failure must be understood. The cause–effect diagram or fishbone diagram is a powerful tool for finding out the failure mechanism. It is a structural approach to sort ideas into categories and visualizes the path from root causes to their effect (i.e. failure). A simple fishbone diagram example is given in Figure 6.6. At the

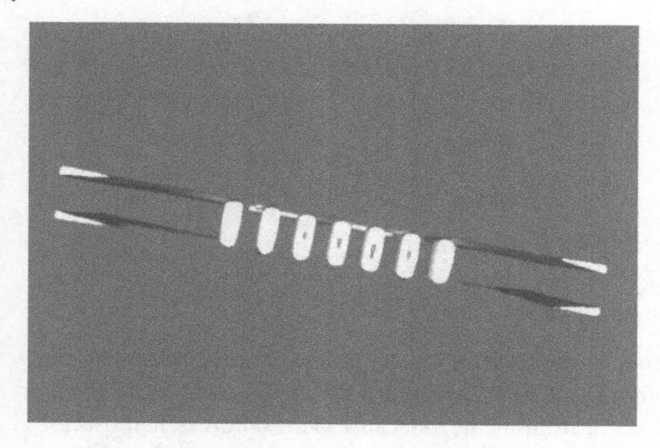

Figure 6.5 Three-dimensional X-ray imaging of voids in a thermal via of a PCB.

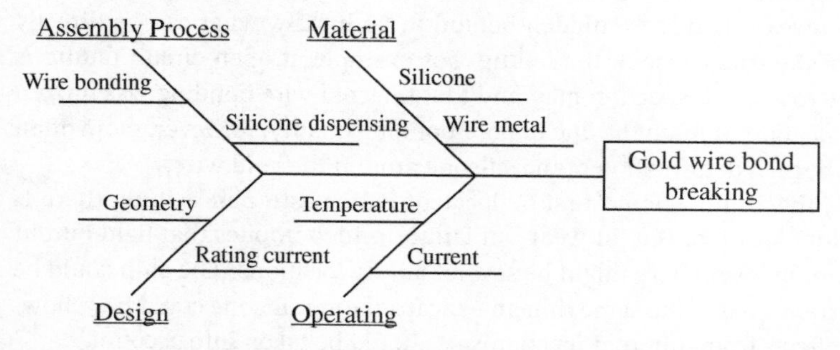

Figure 6.6 A simple fishbone example for the cause of gold wire bond breaking.

end of each branch is every possible cause of the failure. These failure causes are then summarized into a more universal cause. Such reduction should be carried out repeatedly until reaching the failure. This analysis approach enables tracing steps that ultimately produce the failure. There could be many potential causes that lead to the failure. Fish bone diagrams allow causes to be broken down into several categories, which helps finding the failure and further failure analysis.

6.2.2 Weibull Analysis and Acceleration Model for Life Assessment

The data we can get from the reliability tests include the failure time of different samples. Let the survival probability at a certain time be $R(t)$, where t is how long the reliability test has been performed or how many cycles the samples have experienced. $R(t)$ can be expressed by

$$R(t) \equiv \frac{\text{Number of samples not failed at time } t}{\text{Total sample number}} \tag{6.1}$$

Consequently, the failure probability or life distribution probability $F(t)$ is

$$F(t) = 1 - R(t) \tag{6.2}$$

Sometimes, how fast the samples are failing at a particular time is important. Failure rate, $h(t)$, is defined to describe it as

$$h(t) \equiv \lim_{\Delta t \to 0} \frac{1 - \frac{R(t+\Delta t)}{Rt}}{\Delta t} = \frac{\mathrm{d}F(t)}{\mathrm{d}t} / R(t) \tag{6.3}$$

where $\frac{\mathrm{d}F(t)}{\mathrm{d}t}$ is the failure probability density function (PDF), $f(t)$. The failure rate versus time of a common product is plotted in Figure 6.7. At the early failure period, the failure rate is high because of defects in the product. The defect samples are just like unhealthy infants who are likely to die in childhood. As the weak products keep being eliminated, the failure rate drops. Then, the products that survive the early failure period are the normal products. The failure of these products is a random event and the failure rate becomes stable at a low level. As the time extends, the product begins to wear out and then the failure rate rises again. Such a curve is named the bathtub curve, from its shape.

Differentiating the failure probability, $F(t)$, with respect to time, t, the failure PDF can be obtained. There are several different types of failure PDF. Among all these distributions, the Weibull distribution is the one most widely used in reliability analysis [6]. The Weibull PDF is expressed as

$$f(t) = \frac{\beta}{\theta}\left(\frac{t}{\theta}\right)^{\beta-1} \mathrm{e}^{-\left(\frac{t}{\theta}\right)^{\beta}} \tag{6.4}$$

where β and θ are two parameters of the Weibull PDF. Then, the Weibull cumulative distribution function or the failure probability function, $F(t)$, is

$$F(t) = 1 - \mathrm{e}^{-\left(\frac{t}{\theta}\right)^{\beta}} \tag{6.5}$$

and correspondingly, the Weibull reliability function is

$$R(t) = 1 - F(t) = \mathrm{e}^{-\left(\frac{t}{\theta}\right)^{\beta}} \tag{6.6}$$

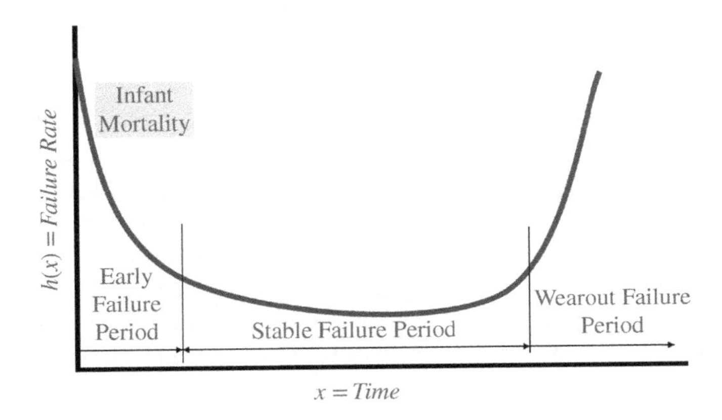

Figure 6.7 The failure rate versus time of a common product.

Table 6.1 The meaning of different β values.

Shape parameter (β)	Failure rate ($h(t)$)	Failure period
$\beta < 1$	Decreasing	Early failure
$\beta = 1$	Constant	Stable failure
$\beta > 1$	Increasing	Wear-out failure

Consequently, the Weibull failure rate, $h(t)$, is obtained as

$$h(t) = \frac{f(t)}{R(t)} = \frac{\beta}{\theta}\left(\frac{t}{\theta}\right)^{\beta-1} \tag{6.7}$$

The β is the Weibull shape parameter and θ is the characteristic life of the product. Different β values have different physical meanings, as listed in Table 6.1. If the time is equal to θ, 63.2% of the samples have failure.

From Eq. (6.6), it can be derived that

$$\ln\left(\ln\left(\frac{1}{1-F(t)}\right)\right) = \beta \ln t - \beta \ln \theta \tag{6.8}$$

Let $y = \ln\left(\ln\left(\frac{1}{1-F(t)}\right)\right)$ and $x = \ln t$, the equation becomes

$$y = \beta(x - \ln \theta) \tag{6.9}$$

As a result, the $F(t)$ and t data can be fit by linear regression to extract the shape parameter and characteristic life.

Mean time before failure (MTBF) is an arithmetic mean of the sample failure time. It is the expected time of an average sample failure. It can be calculated from the failure rate, $f(t)$, as

$$\text{MTBF} = \int_0^{\infty} tf(t)\text{d}t \tag{6.10}$$

The failure rate is the reciprocal of MTBF. The unit of failure rate is parts per million per thousand hours (10^{-6}/K or FIT). One FIT means that one failure is expected to occur in one million devices running for 1000 hours. The FIT can be calculated as

$$\text{FIT} = \frac{1 \times 10^9}{\text{MTBF}} \tag{6.11}$$

The MTBF of LED products can reach 50 000 hours, or 5.7 years. It is too long for performing a reliability test to evaluate the product's life. Therefore, the reliability tests should be performed with a harsher condition to accelerate the tests so that more samples can fail in a shorter time. A good accelerated test condition is critical for successive reliability tests. If the acceleration condition is too stressed, the failure mechanism may be changed. Meanwhile, if the condition is not stressed enough, the testing time may be too long. The acceleration factor (AF) is usually defined to quantify the difference between the life in normal operating condition, L, and the life in an accelerated condition, L_{acc}, as

$$\text{AF} = \frac{L}{L_{\text{acc}}} \tag{6.12}$$

The AF is a function of different testing parameters, such as temperature, current, voltage, and cycling amplitude, and depends on the corresponding acceleration mechanism.

The most common acceleration strategy is elevating the environment temperature. The high temperature may increase the chemical reaction rate and as a result increase degradation speed. The chemical reaction rate, k_r, follows the Arrhenius equation as

$$k_r = Ae^{-\frac{E_a}{RT}} \tag{6.13}$$

where T is the absolute temperature, A is the Arrhenius constant, E_a is the activation energy for the reaction, and R is the universal gas constant. The life should be the reciprocal of the reaction rate; therefore, the AF becomes

$$\mathrm{AF} = \frac{Ae^{\frac{E_a}{RT}}}{Ae^{\frac{E_a}{RT_{acc}}}} = e^{\frac{E_a}{R}\left(\frac{1}{T}-\frac{1}{T_{acc}}\right)} \tag{6.14}$$

E_a/R is the only parameter in the equation. Usually, a series of reliability tests with different elevated temperatures are conducted. The E_a/R can be obtained by linear regression among the life and temperature data. Consequently, the life at normal temperature can be precisely obtained. If the critical AF is forward current, the Eyring model can be implemented combining with the Arrhenius equation as

$$\mathrm{AF} = \frac{AJ^N e^{\frac{E_a}{RT}}}{AJ_{acc}^N e^{\frac{E_a}{RT_{acc}}}} = \left(\frac{J}{J_{acc}}\right)^N e^{\frac{E_a}{R}\left(\frac{1}{T_{normal}}-\frac{1}{T_{accelerated}}\right)} \tag{6.15}$$

where J and J_{acc} are the normal current and the accelerated current and N is a negative fitting parameter. Humidity sometimes is used as the acceleration parameter as well. The power law can be applied for the AF as

$$\mathrm{AF} = \frac{Ae^{BR}}{Ae^{BR_{acc}}} = e^{B(R-R_{acc})} \tag{6.16}$$

where R and R_{acc} are the relative humidity in normal condition and accelerated condition and B is a fitting parameter.

6.3 Design for Reliability

Design for reliability (DfR) is a design approach to guarantee that an LED product can fulfill specific reliability requirements, including reliability evaluation, reliability-oriented product design, and reliability tests. It includes the reliability requirement estimation, allocation of reliability margin, product design, and evaluation.

DfR is considered both a design philosophy and a rule for the entire product design. It can be a technique to eliminate potential defects and vulnerabilities so as to fulfill specific reliability requirements. It can also reduce manufacturing costs. The DfR of an LED system includes selection of components, materials, and processes; thermal management, rated performance estimation; redundancy design; and maintenance instructions.

DfR should obey the following principles:

- At the beginning of product design, the reliably requirement should be proposed beforehand with corresponding reliability test indexes, such as a lifetime in high temperature operating test or a thermal cycling test.
- At each step of an LED system's design, such as chip selection, chip bonding method, substrate, encapsulation, phosphor, or other color conversion strategy, housing, and etc., not only the LED light performance but also the reliability issues must be considered at the same time.
- DfR should include the known failure patterns so that the potential failures in the product design's life can be effectively controlled or even eliminated.
- The reliability experiences and data gleaned from past products must be carefully studied. New techniques, such as new chip types; bonding processes; and unusual material, must be attentively evaluated for their effects on the system's reliability.
- DfR should compromise the product performance, lifetime, manufacturing cost, maintenance cost, warranty, etc., so as to obtain the best design.
- The reliability margin should be allocated to different parts of the LED product according to importance, complexity, technique state, and relative failure rate.

DfR is the most crucial step in reliability engineering to realize a reliable lifespan of an LED system. Once a product design is finished, the manufacturing can achieve the potential reliability which has already been determined at the product design stage. Proper usage and maintenance will not be able to prolong the product life. Therefore, if the reliability issues are not seriously considered during the product design stage, it will bring problems such as inappropriate structure, poor thermal management, wrong selection of components and materials, and so on. These problems will not be resolved, no matter how good the manufacturing and maintenance can be. The key activities of engineering cycle mentioned in Figure 6.8 is usually carried out for one or two rounds for a product with commercial viability the key consideration. Reliability experiences can only be accumulated product by product and the reliability data must be handled with great care.

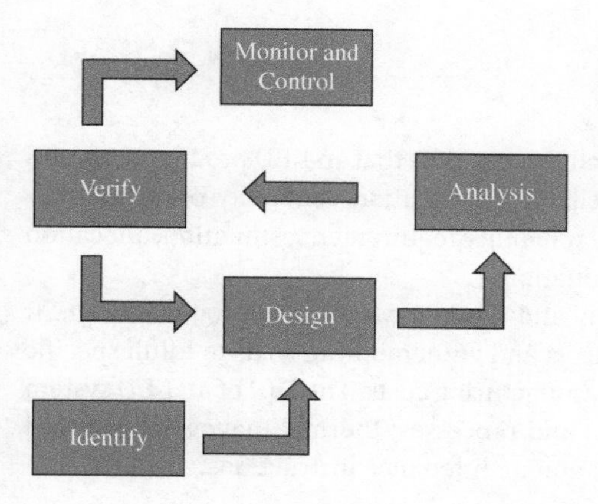

Figure 6.8 Key activity in design for reliability [7]. Source: Modified from Silverman, M., Kleyner, A. What is design for reliability and what is not? *2012 Proceedings Annual Reliability and Maintainability Symposium*, Reno, NV, USA, 2012, pp. 1–5.

References

1 IEEE (1991). *IEEE Standard Computer Dictionary: A Compilation of IEEE Standard Computer Glossaries: IEEE Standard 610*, 1–217. IEEE.

2 Luo, X., Fu, X., Chen, F. et al. (2013). Phosphor self-heating in phosphor converted light emitting diode packaging. *International Journal of Heat and Mass Transfer* 58 (1–2): 276–281.

3 Piazza, M., Dua, C., Oualli, M. et al. (2009). Degradation of TiAlNiAu as ohmic contact metal for GaN HEMTs. *Microelectronics Reliability* 49 (9–11): 1222–1225.

4 Hsu, C., Lan, W., and Wu, Y.C.S. (2003). Effect of thermal annealing of Ni/Au ohmic contact on the leakage current of GaN based light emitting diodes. *Applied Physics Letters* 83 (12): 2447–2449.

5 Electric light source acceterated aging real-time monitoring system. CN203414106U, 2014.

6 Rinne, H. (2008). *The Weibull Distribution: A Handbook*. Oxford: CRC Press.

7 Silverman, M. and Kleyner, A. (2012). What is design for reliability and what is not? In: *2012 Proceedings Annual Reliability and Maintainability Symposium*, Reno, NV, 1–5. IEEE.

7

Emerging Applications of LEDs

7.1 LEDs for Automotive Lighting

The emergence of high-brightness AlInGaP (aluminum indium gallium phosphide) and InGaN (indium gallium nitride) LED technologies has promoted LED devices replacing incandescent lamps. The luminous efficiency of these new LED technologies is equal to or higher than that of incandescent light sources, and their reliability is far higher than that of incandescent light sources. The high-brightness LED lamp meets the requirements of replacing incandescent lamps with high efficiency, reliability, and cost-effectiveness. It is widely used in traffic signals and traffic management road signs, exterior lighting of automobiles, panel lighting of commercial aircraft, night vision goggles (NVG), and compatible lighting of military aircraft [1]. Among them, high-brightness LEDs have significant advantages in automotive lighting applications, but they require power converters capable of operating over a wide range of input and output voltages and of withstanding electrical transients [2, 3].

In addition, using information technology fusion technology, such as visible light communication (VLC), the extensive functions of LEDs have been prominently applied. Vehicles based on LED headlamps and traffic lighting infrastructure (such as streetlamps, traffic lights, etc.) use very large-scale integrated circuits [4].

The main advantages of using LEDs in automotive applications are long service life, energy and space saving, shock and vibration resistance, and new modeling potential. Nowadays, more and more applications use LEDs to enter the market, such as central high-mounted stop lamps, amber and red rear combination lamps, rear combination lamps using LEDs, backup lamps, and the use of signal functions in headlamps [5].

7.1.1 Development

The first LED was created in 1969. Since then, their luminous efficiency and power consumption have been continually improved, and they have realized yellow to red, blue to green (including white) through AlInGaP and InGaN devices, respectively. Improvements in luminous efficiency and power consumption make them interesting for lighting applications in the automotive industry, where many advantages are provided from a lower time response of brake lights to a much higher working life than the vehicle itself [6, 7].

From LED to Solid State Lighting: Principles, Materials, Packaging, Characterization, and Applications, First Edition.
Shi-Wei Ricky Lee, Jeffery C. C. Lo, Mian Tao, and Huaiyu Ye.

The revolutionary victory of LEDs in the field of automobile lighting began in the late 1980s and early 1990s, when the first centrally high-mounted stop lamp came out. Today, more than 95% of lamp functions are generated by LEDs, which highlights the dynamic and advanced development and penetration of this technology. By the 2000s, rear combination lamps began to use LEDs. First, independent taillights, stop lights, and direction indicator functions were developed on the basis of LEDs to supplement the traditional incandescent bulb technology. In 2005, the first "all LED" rear combination lamp was introduced on the market. All functions, including reversing lights, are now realized based on LEDs. They clearly demonstrate the unique combination of advantages of a complete LED lamp solution: vehicle service life, reduced design space due to extremely low-profile design, and top-level energy conservation. The performance of white LEDs has been greatly improved, especially in the past few years, which has also made LED technology relevant for headlamp applications. As early as 2002, the first position light integrate white high-power LEDs into headlamps entered the market. Only one year later, a standard series of cars adopted daytime running lights. The combination of position/daytime running lights led in particular to a functional solution designed for LED technology. Here, the advantages of LEDs in terms of efficiency, dimming capability, and service life can be fully utilized. LED manufacturers predict that the first batch of main lighting functions such as fog lamps, low beam, and high beam will be realized in the next few years, thus completing the application range of LEDs in automobile exterior lighting. Various prototypes of the main lighting functions of headlamps currently displayed at auto shows around the world emphasize this development [8].

Besides, LEDs are also used in optical communications in automobiles, such as the new optical wireless communication (OWC) system equipped with an optical communication image sensor (OCI), which has been newly developed using complementary metal oxide semiconductor (CMOS) technology. In order to obtain higher transmission rate, an OCI adopts a special communication pixel that has a rapid response to light intensity changes. In addition, a new fast LED detection technology has been proposed, which is based on a one-bit logo image that only reacts to high-intensity objects. Then the communication pixel, common image pixel, and related circuits (including one-bit flag image output circuit) are integrated into the OCI.

7.1.2 Typical Structures

In the application of automotive lighting, LEDs have great potential and can replace traditional halogen or HID (high-intensity discharge) light sources. To this end, Peter Brick and Tobias Schmid explored a project of automobile headlight with a combination of low beam and high beam composed of a single LED [9]. The light source included two rows of chips arranged on a common carrier to form a compact LED. At the same time, electronic complexity was reduced by independently driving two rows. The primary optics collected emissions from two closely spaced chip rows while providing a way to separate the respective contributions. The subsequent secondary optics were based on the shape of the multisided reflector to achieve a low-beam and a high-beam pattern. Efficiency, tolerance, system size, and crosstalk were evaluated for different primary optics based on refraction, reflection, and total internal reflection.

In order to realize a LED-based lighting device in a vehicle, it is very important to ensure the functions of the non-light-source components [5].

An automotive headlamp adjusting mechanism is one of the necessary functional structures and has been standardized. The main purpose is to adjust the reflector up and down or sideways during factory commissioning and user driving. Most of them are traditional adjustment mechanisms designed to adjust a single light source. They are generally based on the structural design of the Besturn B70 LED headlamp, which aims to emulate the characteristics of a multilight source system [10].

Finally, in order to realize the successful design of automobile lighting, the requirements of the light field and the heat field need to be analyzed theoretically. First of all, it is necessary to select optimized structural parameters and appropriate packaging materials. For example, an LED package should provide a luminous flux of 1000 mm and a sharp cut-off, and the chromaticity of the package should conform to Economic Commission of Europe (ECE) regulations. In addition, when the equivalent heat transfer coefficient of the cooling system is about 5000 $W/(m^2 \cdot K)$, the package's thermal resistance can be less than 1.3 K/W. Secondly, for automotive forward lighting applications, including low-beam and high-beam functions, it is necessary to have a supporting optical system with new free-form surface multimirrors, an effective cooling system, and even a complete LED packaging module. Finally, the performance and simulation test results of the LED package module should fully comply with the provisions of ECE and national standards [11].

7.1.3 Challenges

Based on developments for white LEDs, it has become feasible to design automotive headlamps using LED light sources. While these LEDs are reaching high lumen outputs, the uniqueness of headlamp applications has not yet been sufficiently addressed [12].

7.1.3.1 Electric Driver

Since luminous flux power LEDs have experienced high growth in the past few years, they have replaced incandescent bulbs as an option for automobile signal lamps and are even an interesting option for automobile headlamps, as shown by the recent prototype, because their performance is close to that of gas discharge lamps [13].

An LED needs an intelligent driver, especially for automotive lighting applications. Compared with a bulb driver, LEDs are difficult to drive, so LED drivers require more powerful control functions. In addition, automotive applications require compact, low-cost solutions that are flexible enough to interface with digital electronic control units (ECUs) and cope with different wiring configurations and related parasitic effects. Automotive lighting systems should also be robust enough to tolerate electromagnetic interference and overheating; overcurrent and overvoltage phenomena, such as the ringing effects caused by the resonance of line inductance; and connector capacitance [14].

Lin Ying-Yan et al. studied a linear LED driver with improved efficiency. The driver works in a wide input voltage range (6.3–32 V) and can provide an output current of up to 350 mA with a 3% accuracy [15]. In order to improve driving efficiency, an operational amplifier is used to reduce the current sensing voltage, while a positive channel metal oxide semiconductor (PMOS) pass element with a fine metal layout pattern is used to reduce the voltage drop. In addition, a regulated voltage of 5 V is obtained from a wide range of input voltages to supply power to some modules in the driver, thereby further reducing the power consumption of the driver. The proposed driver is fabricated on a 0.5 μm bipolar-CMOS-DMOS (BCD) process, occupying an area of 1.4 and 1.8 mm. The postsimulation results show that, at $I_{LOAD} = 350$ mA, the proposed scheme of driving three hexagonal LEDs in series can achieve a maximum efficiency of 91.04%, which is 8.2% higher than the typical linear LED driver under the same conditions. When the load current is 350 mA, the differential pressure is 450 mV. In addition, it can recover within 99 μs [15].

Furthermore, the problem of uneven current often occurs when driving LEDs, which can lead to inconsistent luminosity. Therefore, Professor Werner proposed a new low-cost current sharing method for passive parallel LEDs [16]. The concept of current balance is based on discontinuous inductor currents. The prototype converter is established and tested. The prototype converter drives 36 high-brightness LEDs, which are composed of six parallel series and six high-brightness LEDs connected in series. The converter is designed for automotive headlamp systems that operate directly from the 14 V electric vehicle grid. The experimental results have a good correlation with theoretical prediction, which provides good current balance performance [16].

7.1.3.2 Optical Design

In recent years, automobile headlamps based on LEDs have proliferated, including various lighting functions, such as low beam, high beam, daytime running lights, and fog lights. However, in order to solve the problems of high cost and low optical efficiency of current LED automobile headlamps, a new optical design method of LED automobile headlamps has become necessary [9, 17].

By combining the light of several high-brightness LEDs with a single optical element, a high-brightness light source can be formed, and its shape and emission characteristics are suitable for almost all lighting problems. The illuminance distribution of this virtual light source is helpful to generate the required intensity pattern through its imaging projection into the far field. This projection is realized by a refraction-free and reflection-free surface, both of which are calculated simultaneously by the 3D SMS (three-dimensional simultaneous multiple surface) method [18].

Since then, some researchers have proposed an alternative scheme for automobile headlamps, which combines the low beam and the high beam in an LED. The light source includes two rows of chips arranged on a common carrier to form a compact LED. At the same time, electronic complexity is reduced by independently driving two rows. The primary optics collect emissions from two closely spaced chip rows while providing a way to separate the respective contributions. The subsequent secondary optics are based on the shape of the multisided reflector to achieve a low-beam and a high-beam pattern. Efficiency, tolerance, system size, and crosstalk will be evaluated for different primary optical systems based on refraction, reflection, and total internal reflection [9].

Owing to the rapid development of LED technology, the application of LEDs in automobile lighting is becoming more and more common. The testing methods for LEDs used in vehicle lighting are also very limited, and other tests are required, such as those regarding photometric characteristics, glare, and visibility. The experimental results are compared with those of halogen and high-intensity discharge headlamps. It is found that LED headlamps are still inferior to traditional headlamps in glare. However, many properties of LED headlamps (luminous intensity, beam width, energy efficiency, and design) are comparable with halogen and high-intensity gas discharge headlamps [19].

7.1.3.3 Thermal Optimization

High-brightness white light LEDs have shown great promise for many lighting applications, such as outdoor lighting, task and decorative lighting, and aircraft and automotive lighting, including, as we have seen, automotive headlights. Temperature and current can affect the luminous efficiency of an LED. When the temperature rises above a certain limit, luminous efficiency will drop rapidly. The increase of current will cause nonequilibrium electrons to diffuse out of the potential wall and will increase the temperature, thus reducing the luminous efficiency and the life of the LED. Strengthening the thermal management of LEDs will improve their luminous efficiency. Thermal design is checked by thermal analysis. A constant current power supply is used to drive the circuit of the LED to keep the temperature of the LED more constant [20].

Thermal design, from equipment to circuit board to system level, is required, and optimization work is needed to find the best thermal performance [3].

In order to develop the thermal management of high-power LEDs, an automatic cooling device is first integrated with the microcontroller, heat pipe, and fan. The experimental results show that the substrate temperature of the high-power LED can be automatically controlled and kept in a relatively low range to protect the LED, thus improving the performance and service life of the LED. In order to develop the thermal management of high-power LEDs, Chengdi Xiao et al. integrated an automatic cooling device, and established a numerical model of the cooling system [21]. The experimental results verified the validity of the model. The simulation results show that the junction temperature of an LED can be kept within a suitable range. The heat dissipation performance was simulated and analyzed by the constant finite volume method and the experimental measurement method. Simulation and experimental results show that natural convection cooling is not an effective cooling method for LED headlamps. The radiator combined with heat pipe technology can greatly improve the heat dissipation performance. When the liquid filling rate is 10%, a heat pipe with an evaporator length of 30 mm, an insulation length of 40 mm, and a condenser length of 50 mm has been shown to have the best cooling performance. Cooling devices with distributed heat pipes make junction temperature lower than cooling devices with heat pipes equidistantly placed on the same plane. The liquid filling rate of the heat pipe has a significant effect on the equivalent heat transfer coefficient, and the best liquid filling rate is 30% [21].

Although many researchers have accurately quantified junction temperature, factors such as electrothermomechanical boundary conditions or the microstructure of the die attach layer in automotive LEDs remain uncertain. The propagation of these variables reduces the accuracy of junction temperature quantization [22]. Studies have described two methods for directly and indirectly determining the junction temperature of LEDs or LED arrays. Both methods are based on temperature measurement, but indirect methods also require thermal resistance specified by the manufacturer. The direct method uses the inherent characteristics of junction voltage drop at low current (<0.1 mA). The computer model of typical plate-fin radiator design for high-power LED automobile lamps has been calibrated experimentally. Experimental analysis and design were carried out combined with the active radiator surface area, the convection coefficient related to airflow, and the ambient temperature. A quadratic model was generated using the results of these factors and the important interaction with the response [23].

7.1.4 Conclusion

Compared with traditional light sources, LEDs have many advantages that support their continued market penetration. The emergence of high-brightness AlInGaP and InGaN LED technology has promoted the new application of LED devices to replace incandescent lamps. The luminous efficiency of these new LED technologies is equal to or higher than that of incandescent light sources, and their reliability is much higher than that of incandescent light sources. The high-brightness LED lamp meets the requirements of replacing incandescent lamps with efficient, reliable and, cost-effective performance. The main advantages of using LEDs in automotive applications are long service life, saving energy and space, and resisting shock and vibration.

However, there are still many difficulties when LEDs are applied to automobile lighting. Compared with a bulb driver, an LED driver is difficult to drive, especially for automotive lighting applications. In the aspect of optical design, the problems of the high cost and low optical efficiency of current LED automobile headlamps need to be solved. And there is also a lack of testing methods for LEDs used for vehicle lighting. As the power of LEDs increases, the heat dissipation of automotive LEDs also becomes a very significant problem.

Although there are some difficulties, automotive LEDs still have a great application prospect. The first full-LED headlamp, including white high-power LEDs, is expected in the next few years. The timescale for LEDs to replace traditional automobile light sources mainly depends on the development of cost-per-unit luminous flux. Here, automotive applications, especially white LEDs, will benefit from the use of LEDs in general lighting applications. In addition, OWC systems based on LED transmitters and camera receivers have been developed for use in the automotive field. Automobile OWC systems need more than 1 MB/s data rate and the ability to quickly detect LEDs from images.

7.2 Micro- and Mini-LED Display

LEDs provide very high brightness levels, greater than 50 000 000 cd/m^2, which enables them to have mature performance in high environment (such as augmented or virtual reality or high-brightness TV) display applications. LEDs also provide some of the highest efficiency in converting electrical energy into light energy. Owing to these advantages and small solid-state form factors, LEDs can be a solution for display applications of various sizes.

In today's display applications, LEDs are most commonly used as illumination sources for liquid crystal displays (LCDs) of almost all sizes, including 100 in. A 0.5 in TV mini-display needs a peak brightness of 10 000 cd/m^2 to meet future high-definition content, while a micro-display needs a peak brightness of 100 000 cd/m^2 to support the brightness requirements of AR/MR (augmented/mixed reality) glasses. LEDs can easily meet these requirements, with brightness of up to 50 million cd/m^2 [13].

Mini-LEDs are defined as devices with a light-emitting area of less than 50 μm × 50 μm or 0.0025 mm^2 per pixel. The micro-LED array forms a micro-LED display, with sizes ranging from 1 to 70 in. The micro-LED display utilizes the abnormal brightness of the LED by spreading the generated photons over an area larger than the area occupied by the micro-LED itself, or by spatially distributing LED elements or by optically dispersing light [24].

7.2.1 Development

Mini-LEDs, with chip sizes ranging from 100 to 200 μm, have been commercialized and applied to backlights in consumer electronics applications. The local dimming that can be achieved can greatly increase the contrast at relatively low levels of energy consumption. The micro-LED, whose chip size is less than 100 μm, is still at the research stage. Recently, mini-LED and micro-LED displays have attracted extensive attention for their high dynamic range (HDR), high ambient contrast ratio (ACR), thin profile, and low power consumption. According to Research and Markets, a market research institution, the global micro-LED display market is predicted to soar from $0.6 billion in 2019 to $20.5 billion in 2025, with a compound annual growth rate of about 80% (Figure 7.1) [25]. The main reason for the market outbreak is the sharp increase in demand for brighter and more energy-efficient display panels, which are needed for surging devices such as smartwatches, mobile phones, TVs, laptops, augmented reality (AR), and virtual reality (VR).

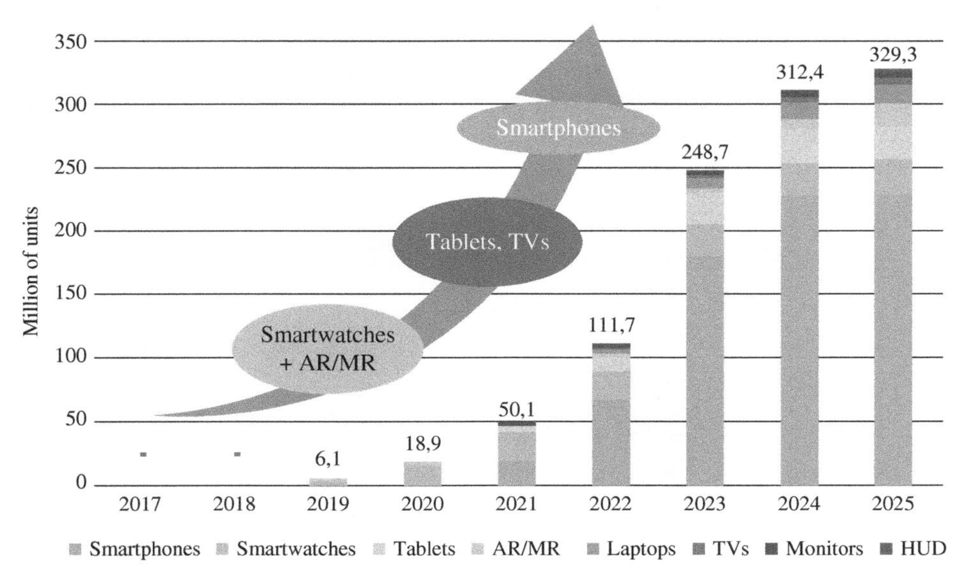

Figure 7.1 Forecast about the development of micro-LED displays [25]. Source: Wu, T.Z., Sher, C.W., Lin, Y., et al. Mini-led and micro-led: promising candidates for the next generation display technology. *Applied Sciences*, 2018, 8(9): 1557. Figure 1. Licensed under CC BY-4.0.

7.2.1.1 Development of Mini-LED Display

Mini-LED technology is considered a transitional technology between traditional LEDs and micro-LEDs. Mini-LEDs generally refer to LEDs with a grain size of 50–100 μm. Compared with micro-LEDs, mini-LEDs are easier to make. It is easier to mass produce them, and they can be quickly brought to market. The high luminance and excellent dark state, low power consumption, and thin profile, long lifetime, and free-form factors will make mini-LED displays the next-generation HDR solution.

Mini-LEDs have been introduced as a promising solution for HDR displays. Tan et al. discussed the system modeling and performance evaluation of LCDs with a mini-LED backlight [26]. First, a model LCD system with a direct-lit mini-LED backlight is set up for simulation, as shown in Figure 7.2. The backlight unit consists of a square-shaped mini-LED array. A diffuser plate is utilized to widen both spatial and angular distributions, and a liquid crystal (LC) panel is applied to control the output light. The parameters of the model, such as p, s, H_1, and H_2, are based on the device configuration reported in [27]. To validate the model, four patterns are used to simulate the dynamic contrast ratio (CR) of the model, and the simulated results agree with the measured data from [27] reasonably well.

Next, an exemplary candle picture, as shown in Figure 7.3, can be used to illustrate the light modulation process of mini-LED backlit LCDs. Here, the backlight consists of 12 × 24 local dimming zones and each zone contains 6 × 6 mini-LEDs in order to achieve the desired luminance. According to the image content, the mini-LEDs in each dimming zone are predetermined to show different gray levels, as Figure 7.3a depicts. After passing through the diffuser, the outgoing light spreads out uniformly before reaching the LCD panel, as shown in Figure 7.3b. The gray level of each LCD pixel is controlled by a thin film transistor (TFT), and each consecutive frame (CF) only transmits the designated color. Finally, a full-color image as the one shown in Figure 7.3c is generated.

Although the above simulations and experiments are based on small-size smartphone displays with a viewing distance at 25 cm, the analysis and conclusion can also be applied to display devices with different sizes and resolutions via converting the results from the spatial to the angular domain (Figure 7.4). In summary, Tan et al. demonstrated the

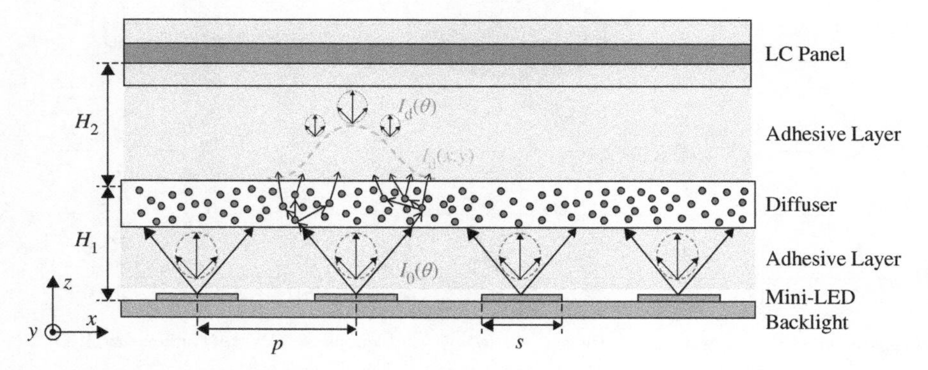

Figure 7.2 Schematic diagram of the model for the LCD with a mini-LED backlight [26]. Source: Tan, G.J., Huang, Y.G., Li, M.C., et al. High dynamic range liquid crystal displays with a mini-led backlight. *Optics Express*, 2018, 26(13): 16572–16584. Fig. 1.

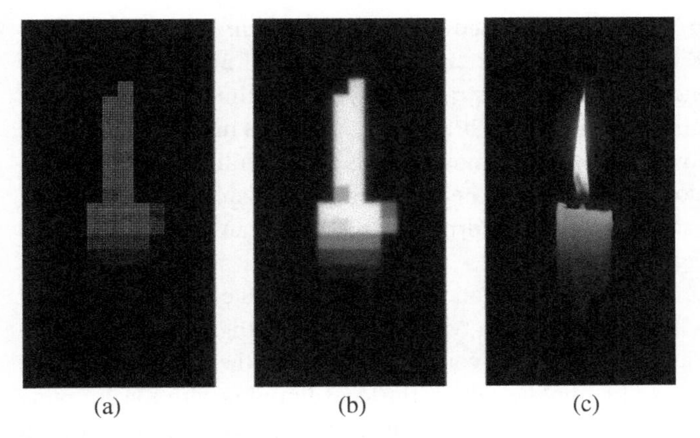

Figure 7.3 Light modulation of mini-LED backlit liquid crystal display (LCD): (a) mini-LED backlight modulation; (b) luminance distribution of the light incident on the liquid crystal (LC) layer; and (c) displayed image after LCD modulation [26]. Source: Tan, G.J., Huang, Y.G., Li, M.C., et al. High dynamic range liquid crystal displays with a mini-led backlight. *Optics Express*, 2018, 26(13): 16572–16584.

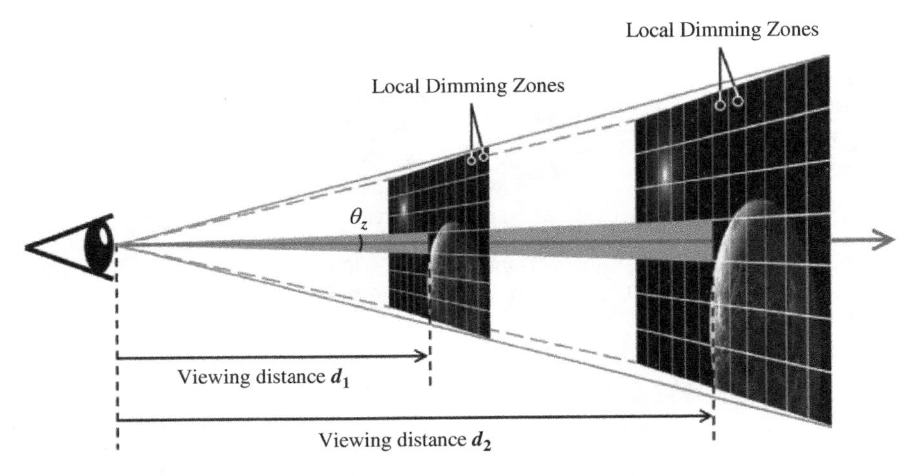

Figure 7.4 Conceptual diagram of scaling up display size based on same angular size [26]. Source: Tan G J, Huang Y G, Li M C, et al. High dynamic range liquid crystal displays with a mini-LED backlight. Optics Express, 2018, 26(13): 16572-16584. Fig. 10.

required local dimming zone number to exhibit comparable HDR performance with organic light-emitting diode (OLED), and the HDR performance could not be achieved by conventional segmented LED backlights [26]. The simulation model can provide a useful guideline to theoretically optimizing the mini-LED backlit LCDs to achieve an excellent HDR display.

The mini-LED, an ideal backlight candidate for local dimming LCDs, is currently ready to be produced in volume. Many companies have sought to develop the mini-LED backlight technology to replace the traditional LCD backlight.

AU Optronics Corporation (AUO) demonstrated several high-end mini-LED backlit LCD displays, including a 27 “4 K 144 Hz gaming monitor and a 1000 PPI 2” LTPS VR display (Figure 7.5) [25]. The gaming monitor used a straight down type mini-LED backlight to provide accurate local dimming with ultra-high brightness, creating a more realistic visual enjoyment for customers. However, the cost of mini-LEDs is still several times higher than traditional backlight technology at present. For head-mounted VR displays, AUO shows a 2 in panel equipped with an active matrix (AM) driver circuit which can achieve 1024 local dimming zones for vivid images.

JDI, a Japanese manufacturer, demonstrated an automotive central control panel based on a 16.7 in curved screen with a straight down type mini-LED backlight at the 2018 SID Display Week, as shown in Figure 7.6a [25]. The contrast and color of the screen can be presented perfectly even in complex situations because of the local dimming with 104 dimming zones in the screen. BOE, one of the largest display makers in China, exhibited a 27 in ultra-high-definition (UHD) panel, which used a mini-LED backlight with 1000 local dimming zones. Its brightness is up to 600 units and its contrast is up to 1 000 000 : 1. In addition, BOE also showed a 5.9 in mobile phone panel which was only 1.4 mm thick, and is shown in Figure 7.6b [25].

Figure 7.5 The 27 “gaming monitor and the 2” VR display of AUO [25]. Source: Wu, T.Z., Sher, C.W., Lin, Y., et al. Mini-LED and micro-LED: promising candidates for the next generation display technology. *Applied Sciences*, 2018, 8(9): 1557.

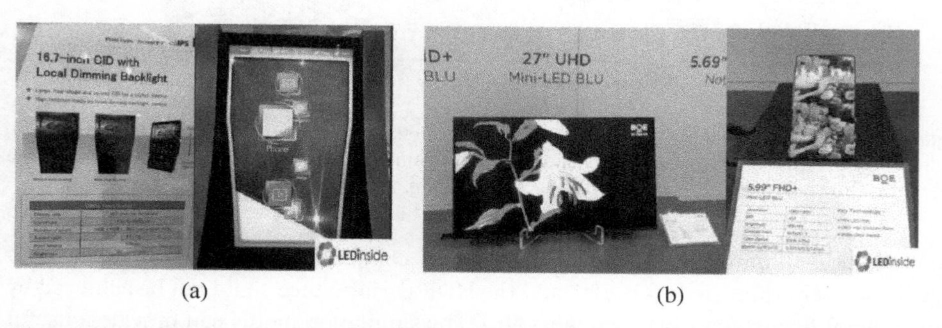

Figure 7.6 (a) The 16.7 in curved automotive display with direct backlight solution of JDI. (b) Ultra-high contrast UHD display of BOE [25]. Source: Wu, T.Z., Sher, C.W., Lin, Y., et al. Mini-LED and micro-LED: promising candidates for the next generation display technology. *Applied Sciences*, 2018, 8(9): 1557.

7.2.1.2 Development of Micro-LED Displays

At present, there are two major trends in the development of micro-LED displays. One is Sony's, whose offering focuses on: small spacing, large size, and high-resolution indoor/outdoor display screen. The other is the wearable device (such as the Apple Watch). The display part of this kind of device requires high resolution, strong portability, low power consumption, and high brightness, and these are the advantages offered by a micro-LED.

Micro-LED displays have been developed for more than 10 years, during which time several project teams in the world released results and promoted the further development of related technologies. For example, in 2001, Jiang et al. fabricated blue micro-displays [28], as shown in Figure 7.7. The array adopts a passive driving mode, and uses wiring to connect pixels and driving circuits, and three LED chips of red, green, and blue are placed on the same silicon reflector to realize colorization through the red–green–blue (RGB) mode. Although the array has achieved initial results, it also has some shortcomings that cannot be ignored. Its resolution and reliability are still very low, and the forward conduction voltages of different LEDs are quite different.

Another outstanding achievement was announced by the Hong Kong University of Science and Technology team in 2006. Passive driving is also adopted, and the micro-LED array was integrated by flip-chip bonding technology. However, the forward conduction voltages of pixels in the same row were also quite different, and when the number of pixels on the column was different, the brightness of the pixels was also affected, and the brightness uniformity was not good enough.

In 2008, the Fan team announced another passively driven 120 × 120 pixel microarray, with a chip size of 3.2 mm × 3.2 mm, a pixel size of 20 μm × 12 μm, and a pixel spacing of 22 μm [29]. The dimensions have been obviously optimized, but a lot of wiring is still needed and the layout is still very complicated. In the same year, the microarrays announced by Gong's team were still driven by a passive matrix and integrated by flip-chip technology [30]. The team has made blue (470 nm) micro-LED arrays and ultraviolet (UV) micro-LED

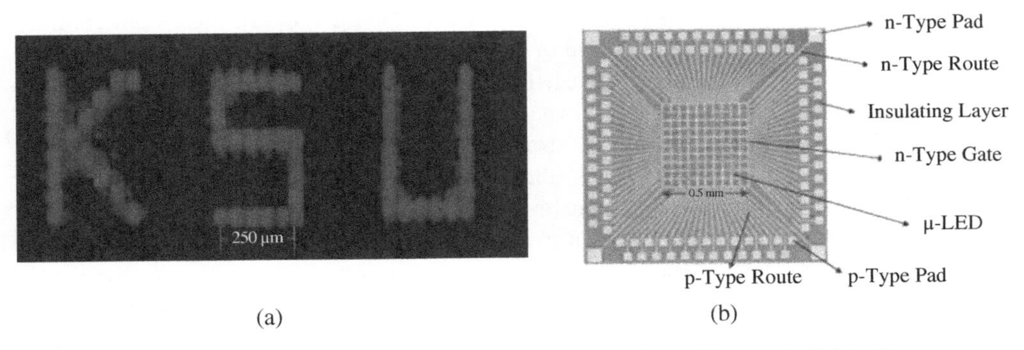

Figure 7.7 Optical microscope image (a) of a micro-display fabricated from an individually addressed μ-disk LED array [28]. Source: Jiang, H.X., Jin, S.X., Li, J., et al. III-nitride blue microdisplays. *Applied Physics Letters*, 2001, 78(9): 1303–1305.

Figure 7.8 Representative micro-displays programmed onto the flip-chip blue micro-LED array [30]. Source: Gong, Z., Gu, E., Jin, S.R. et al. Efficient flip-chip InGaN micro-pixellated light-emitting diode arrays: promising candidates for micro-displays and colour conversion. *Journal of Physics D Applied Physics*, 2008, 41(9): 094002.

(370 nm) arrays, and successfully excited green and red quantum dots through UV LED arrays to prove the feasibility of quantum dot colorization, as shown in Figure 7.8.

In addition, in 2009, the Rae team successfully integrated the Si-CMOS circuit, which can provide appropriate electrical pulse signals for UV LEDs and integrated the SPAD (single-photo avalanche diode) detector, which is mainly applied to portable fluorescence lifetime readers [31]. However, its driving capability was relatively weak and its working voltage was very high.

Also in 2009, Liu's team of Hong Kong University of Science and Technology used UV micro-LED array to excite red, green, and blue phosphors to obtain a full-color micro-LED display chip [32]. In 2010, the team made 360 PPI micro-LED display chips using red, green, and blue LED epitaxial wafers, respectively, and integrated the three chips together to realize the world's first full-color micro-LED projector without a backlight source. These are only some of the more important achievements in the developmental history of micro-LEDs. Since then, the exploration of micro-LEDs has been deepened and more progress has been announced, including further reductions in size and improvements to brightness uniformity.

7.2.2 Typical Structures

The structure of the mini-LED display is shown in Figure 7.9 [33]. Each light-emitting diode display unit (LDU) is composed of 96 × 96 LED pixels, and each pixel includes red, green, and blue LEDs with an integrated, three-in-one packaging structure that significantly reduces the pixel size, thus resulting in a seamlessly combined LDU.

Figure 7.10 illustrates the device structures of color-converted micro-LED displays without (device A) and with funnel-tube array (device B), respectively [34]. For device A (Figure 7.10a), an array of monochromatic blue micro-LEDs with an LED driving backplane is formed on the bottom substrate. Above the micro-LED array, a layer of yellow phosphor is coated to obtain a white light source. On top of the phosphor, a color filter array is used to form RGB subpixels. This system usually suffers from optical color crosstalk. To overcome this issue, some researchers propose a funnel-tube array, as illustrated in device B (Figure 7.10b). It is formed above a micro-LED layer with the tube region aligned with each subpixel. The inner surface of the funnel-tube can be either absorptive or reflective. The phosphors are filled inside the funnel-tube to obtain white light. On top of the funnel-tube

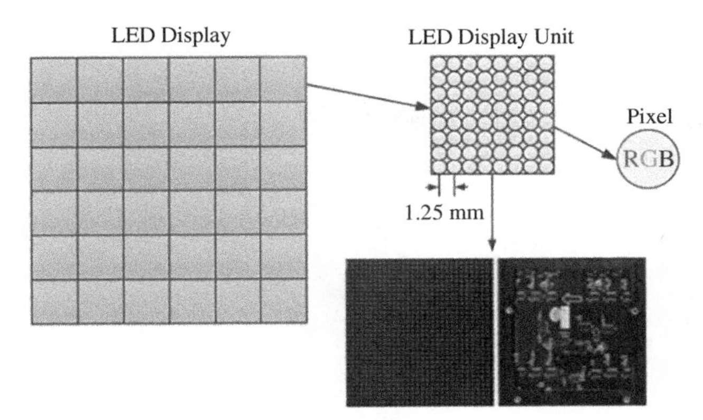

Figure 7.9 Model of LED display unit [33]. Source: Wu, W., Wang, S., Zhong, C., et al. Integral imaging with full parallax based on mini-LED display unit. *IEEE Access*. 2019, 7: 32030–32036.

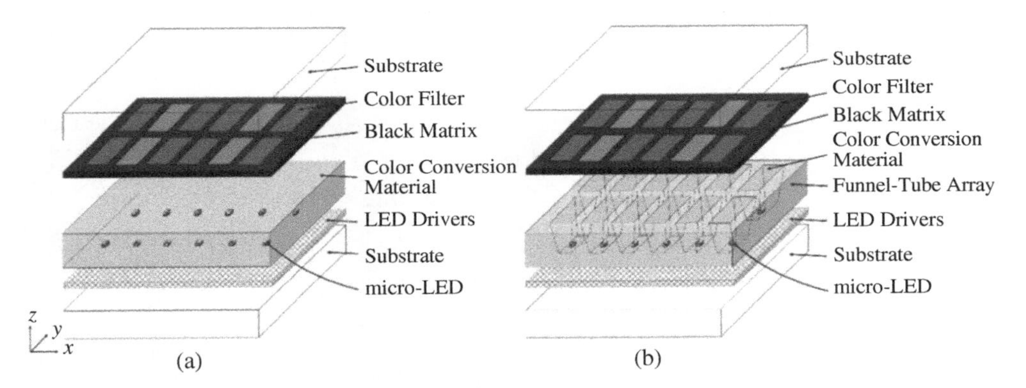

Figure 7.10 Schematic diagram for configuration of full color micro-LED display (a) device A: without funnel-tube array and (b) device B: with funnel tube array [34]. Source: Gou, F.W., Hsiang, E.L., Tan, G.J. Tripling the optical efficiency of color-converted micro-LED displays with funnel-tube array. *Crystals*, 2019. Figure 1. Licensed under CC BY-4.0.

array, the color filters with RGB subpixels are aligned with each tube region. In the system, the phosphors for each subpixel region are designed to be totally isolated. Thus, the color crosstalk will be eliminated.

Micro-LED displays take advantage of the exceptional luminance of LEDs by spreading the generated photons over a larger area than the area occupied by the micro-LEDs themselves, either by distributing the LED elements spatially or by dispersing light optically [35]. This is illustrated in Figure 7.11. There are wide differences between these two technologies, despite confusing nomenclature that refers to both as "micro-LEDs". The former technology, shown in Figure 7.11a, distributes LED elements spatially and can be used to build displays ranging from 3 to 70 in. These are referred to as direct-view micro-LED displays. The latter technology, shown in Figure 7.11b, disperses light optically and is used to build displays <2 in, which are referred to as micro-LED micro-displays.

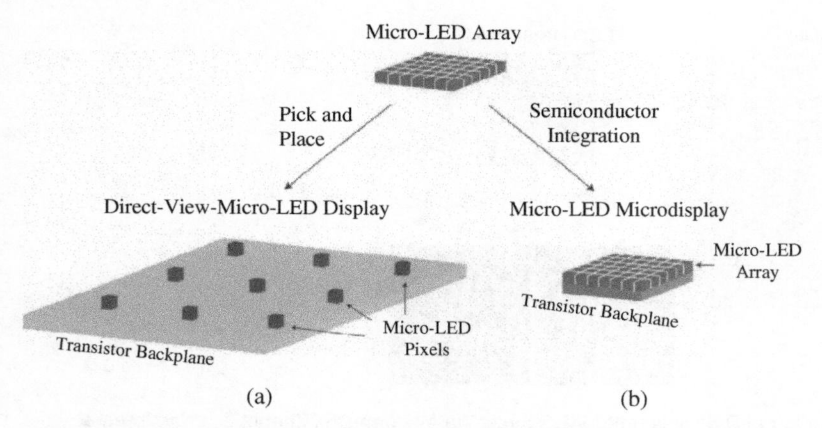

Figure 7.11 Two approaches used to build micro-LED displays. Both methods start with a micro-LED array but use either (a) a pick-and-place technology for direct-view displays or (b) semiconductor integration for micro-displays [35]. Source: Tull, B.R., Twu, N., Hsu, Y.J., et al. 19-1: Invited paper: micro-LED microdisplays by integration of III-V LEDs with silicon thin film transistors. *SID Symposium Digest of Technical Papers*, 2017,48(1): 246–248. Figure 1.

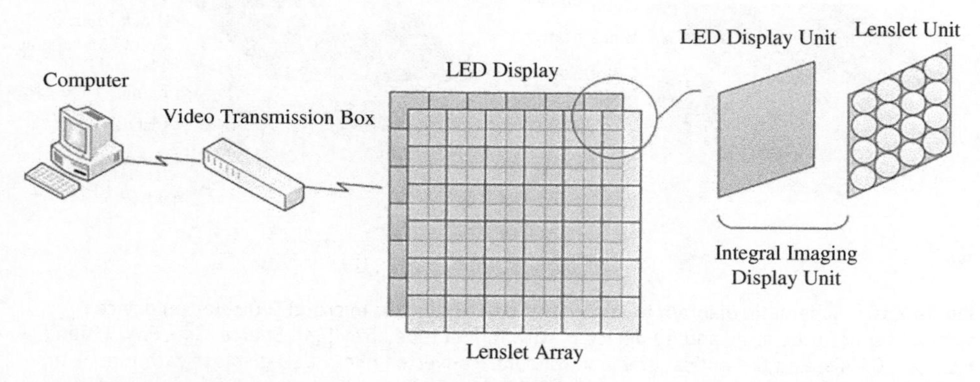

Figure 7.12 Integral imaging based on a mini-LED [33]. Source: Wu, W., Wang, S., Zhong, C., et al. Integral imaging with full parallax based on mini-LED display unit. *IEEE Access*. 2019, 7: 32030–32036.

The integral imaging system based on the mini-LED is composed of a computer, video transmission box, mini-LED, and lenslet array, as shown in Figure 7.12 [33]. The video signal is transformed into a network signal via a video transmission box, which is controlled by the computer and sent to the LED display. The 3D image is reconstructed around the central depth plane, at which the lights emitted from the LED are refracted by the lenslet array and converge.

7.2.3 Challenges

Compared with LCDs, LED displays have better contrast and higher response speeds, which makes self-luminous technologies such as LEDs an ideal choice for making display panels. Although significant progress has been made in micro-LED micro-displays, for micro-LED backlit LCDs, the influence of device contrast, the number of local dimming areas, and the local light distribution on the image quality are still in the quantitative evaluation stage. For emissive micro/micro-LED displays, the challenges of environmental contrast and size-dependent power efficiency have been analyzed. At present, the main technical challenges are in three aspects: manufacturing yield and cost due to quality transfer, ACR due to strong internal reflection, and IQE (internal quantum efficiency) decline due to chip size reduction [36].

Another challenge for micro-LED micro-displays is the scalability of integrated semiconductor processes. This method has one unique advantage: the flip-chip LED epitaxial transfer to CMOS or thin film transistor. In general, these factors will drive related costs and thus the market.

What is more, there is still much work to be done before commercial products can enter the market. The integration of color-generating materials is a challenge, although some work has been done in optical combination using three micro-LED micro-displays, integration of phosphor materials, and a stack of epitaxial layers of red, green, and blue LEDs. In addition to color material integration, the mixing of silicon integrated circuits and compound semiconductor device arrays is also crucial to the fabrication of functional hybrid chips, which have been found to have great applications in many fields. Although flip-chip technology is widely used to manufacture hybrid chips, it has several limitations that are difficult to overcome, especially when it is necessary to manufacture functional hybrid chips with higher device array density without sacrificing the chip footprint.

7.2.4 Conclusion

Mini-LED and micro-LED displays have gained wide attention due to their HDR, high ACR, thin appearance, and low power consumption. Mini-LEDs are a transitional technology between traditional LEDs and micro-LEDs. The mini-LED chip does not present production complexity. Since they are an improved version of traditional LED backlights, they can be manufactured in existing manufacturing facilities with minimal reconfiguration. Micro-LEDs do offer high luminous efficiency, brightness, contrast and reliability, and short response times. However, at the initial mass production stage, they can cost more than three times the cost of traditional LED displays. And micro-LEDs are typically made of GaN-based LED materials that produce brighter displays (many times brighter than OLEDs) and are much more efficient than traditional LEDs. They are a new generation of display technology and are a miniaturized LED with a matrix. There has been significant progress in mini/micro-LED displays, but challenges still remain for full commercialization. One of the keys to the success of micro-LEDs is their overall manufacturability. Each technology faces unique and specific challenges, such as manufacturing yield and cost, the scalability of integrated semiconductor processes, and the integration of color-generating materials.

7.3 LED for Visible Light Communication

Visible light communication (or VLC) is the name given to an OWC system that carries information by modulating light in the visible spectrum (400–700 nm) that is principally used for illumination. The communications signal is encoded on top of the illumination light. Interest in VLC has grown rapidly with the growth of high-power LEDs in the visible spectrum. The motivation to use the illumination light for communication is to save energy by exploiting the illumination to carry information and, at the same time, to use technology that is "green" in comparison to radio frequency (RF) technology, while using the existing infrastructure of the lighting system. The necessity to develop an additional wireless communication technology is the result of the almost exponential growth in the demand for high-speed wireless connectivity. Emerging applications that use VLC include: (i) indoor communication where it augments Wi-Fi and cellular wireless communications which follow the smart city concept, (ii) communication wireless links for the Internet of things (IoT), (iii) communication systems as part of intelligent transport systems (ITSs), (iv) wireless communication systems in hospitals, (v) toys and theme park entertainment, and (vi) provision of dynamic advertising information through a smartphone camera.

VLC to augment Wi-Fi and cellular wireless communication in indoor applications has become a necessity, with the result that many people carry more than one wireless device at any time, such as a smartphone, tablet, smartwatch, and smart glasses, and a wearable computer, and at the same time the required data rate from each device is growing exponentially. It is also becoming increasingly clear that in urban surroundings human beings spend most of their time indoors, so the practicality of VLC technology is self-evident. It would be extremely easy to add extra capacity to existing infrastructure by installing a VLC system in offices or residential premises.

The downlink includes illumination LED, Ethernet power line communication (PLC) modem, and LED driver, which receives a signal by a dedicated or dongle receiver as part of the device. The uplink configuration could be based, for example, on: (i) a Wi-Fi link, (ii) an IRDA (Infrared Data Association) link, or (iii) a modulated retro reflector. A modulated retro reflector is an optical device that retro reflects incident light. The amplitude of the retro reflected light is controlled by an electronic signal. As a result, modulation of the light can be achieved.

7.3.1 Development

In 2000, at the Nakagawa Laboratory at Keio University, Japan, transmission of data was carried out using LEDs [37]. As a visible light source can be used both for illumination and communication, it saves the extra power that is required in RF communication. VLC uses visible light of a spectrum between 400 THz (780 nm) and 800 THz (375 nm). It uses optical carriers in this range for data transmission and illumination purposes with LEDs at the transmit ends and photodetectors at the receive ends. Since then, the applications of VLC are rapidly developed including light fidelity (Li-Fi), vehicle-to-vehicle communication, robots in hospitals, underwater communication, information displayed on sign boards, etc.

In 2003, Tanaka et al. investigated trichromatic LEDs system for optical wireless (OW) communication and reported the simulation results for data rates up to 400 MB/s [38]. In 2010, the maximum measured data rate for a VLC system using blue-chip LEDs achieved data rates of higher than 500 MB/s with a modified version of the classical orthogonal frequency division multiplexing (OFDM) modulation technique.

In 2011, Harald Haas was the first to coin the term Li-Fi [39], which is a high-speed bi-directional fully connected, visible light wireless communication system. Li-Fi is one of the many applications of VLC which is similar to the many applications of wireless communication, i.e. Wi-Fi, GSM (global system for mobile communications), satellite, LTE (long-term evolution), etc. The term Li-Fi was inspired by its potential to compete with conventional Wi-Fi. Wi-Fi uses 2.4–5 GHz radio frequencies for data transmission and its bandwidth is limited to 50–100 MB/s. The Li-Fi uses visible light for communication to provide high-speed Internet up to 10 GB/s [40]. Importantly, VLC technology uses the visible light part of the light spectrum, whereas Li-Fi technology uses any possible light spectrum for communication, for example, infrared light is being tested in Li-Fi for communication. Speeds of about 40 GB/s have been achieved.

7.3.2 Typical Applications

7.3.2.1 Li-Fi

Li-Fi has the potential to replace Wi-Fi in many areas, especially for those applications whose hypersensitivity to RFs is a problem or when radio waves cannot be used for communication or data transfer.

- Medical and healthcare: there is no RF allowed around the magnetic resonance imaging (MRI) scanner and in the operating room; pregnant woman will have concerns over radiation.
- Sensitive areas: Li-Fi can provide safe connectivity throughout the power plant and hazardous environment.
- Giga speed technology: current solutions offer effective transmission rates of up to 10 GB/s, allowing a two-hour high-definition TV film to be transferred in less than 30 seconds. This will be extended to several 100 GB/s in future versions.
- Indoor wireless communication: it is very suitable for communication and data transmission, such as indoor positioning and broadcasting in large supermarkets.

7.3.2.2 Traffic Communication

One simple case for traffic communication is for the automotive use shown in Figure 7.13 [41]. In one-way positioning, the light sources transmit a positioning signal and the receiver processes this signal so as to determine the relative position. LED headlights and taillights are being implemented for different cars. Traffic signals, signs,

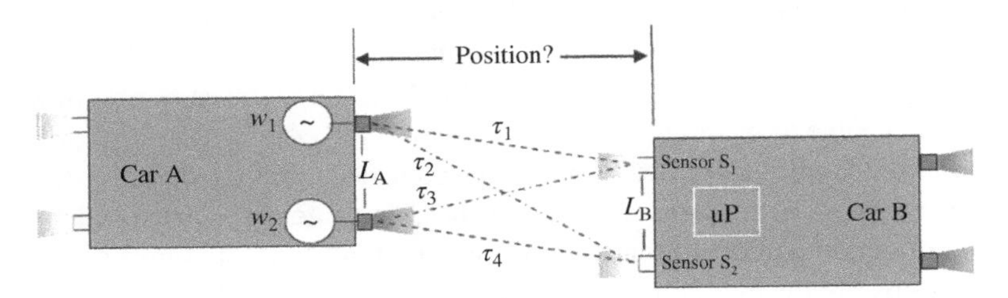

Figure 7.13 Typical automotive application [43]. Source: Roberts, R., Gopalakrishnan, P., Rathi, S. Visible light positioning: automotive use case. *Vehicular Networking Conference*, IEEE, 2011.

and streetlamps are all also transitioning to LED. With these LED lights in place, Li-Fi can be used for effective vehicle-to-vehicle as well as vehicle-to-signal communications. This could of course lead to enhanced traffic management and safety. Since the light is widely used as a traffic light, other applications are explored as well:

- Airlines and aviation: Wi-Fi is often prohibited in aircraft. However, since aircraft already contain multiple lights, Li-Fi can be used for data transmission.
- Underwater exploration and communications: radio waves cannot be used in water, owing to strong signal absorption. Acoustic waves have a low bandwidth and disrupt marine life. VLC offers a solution for conducting short-range underwater communications.

7.3.2.3 Smart City

Today's cities are complex systems characterized by massive numbers of interconnected citizens, businesses, different modes of transport, communication networks, industrialization, manufacturing, entertainment and information services, and utilities [42]. Among them, communication networks play a very important part in 5G, IoT, energy networks, and the intelligence industry.

VLC technology will contribute to not only Li-Fi and traffic communication but also all other aspects of communication. Reusing light sources will reduce the need for massive replacement of infrastructure, such as streetlights, display screens, and so on. Smart lighting can be used to achieve smart homes (classrooms/offices), smart communities, smart transportation, smart factories, and smart towns and cities with space reuse. VLC can act as an alternative in regions with high-density wireless communication where 500 or more users may be contending for Wi-Fi. This would lead to low access speeds for users. Li-Fi can be used to share some of the load of Wi-Fi. Besides, VLC will be able to serve indoor environment communications in future 5G systems with high efficiency [43].

7.3.3 Challenges

7.3.3.1 Multiple Input and Multiple Output (MIMO)

A single-input and single-output (SISO) system is a simple single variable control system with one input and one output, which refers to one LED and one receiver in the VLC system. In order to achieve a high data rate in optical communications, multiple separate LED arrays are usually utilized to provide higher data rates by means of spatial multiplexing. As a result, the multiple-input multiple-output (MIMO) technique is a natural progression which includes multiple channels between the transmitters and the receivers [44, 45]. Besides, MIMO VLC systems can solve the problem of SISO VLC systems suffering link interruptions caused by people or moving objects.

In 2004, Komine et al. found that when shadowing often occurred at 800 MB/s the performance of the outage call duration rate and the blocking rate were improved by using three LED lightings compared with one or two LED lightings [46]. McKendry et al. demonstrated a data transmission at bit rates of up to 512 MB/s using on/off keying (OOK) using nonreturn-to-zero modulation with 8×8 arrays of III-nitride-based micro-pixelated LEDs [47]. In 2015, the MIMO VLC system that uses the mobile phone camera as an optical receiver (Rx) to receive MIMO signals from the $n \times n$ RGB LED array was designed [48]. Recently, researchers used a micro-LED for the MIMO VLC system, which indicated that many more light sources can be used in one system [49, 50].

Equally importantly, different modulation schemes for the SISO and MIMO VLC systems have been developed since the invention of VLC technology, including the OOK, variable pulse-position modulation (VPPM), color-shift keying (CSK), subcarrier inverse pulse-position modulation (SCIPPM), and subcarrier index modulation orthogonal frequency division multiplexing (SIM-OFDM) [39]. The driver circuit also plays a key role. Tanaka et al. proposed and applied the carrier drawing-out circuit to the driver circuit that drives a blue LED. An optical measurement shows that the fall time decreases from 56 to 10 ns (i.e. it decreases to less than one-fifth its original size), and the horizontal eye-opening ratio has been improved from 29% to 61% (i.e. an increase of 113%) for 50 MB/s PRBS (pseudorandom binary sequence) [51]. New types of digital-to-analog converters (DACs) and other LED drivers can optimize the performance of OWC [52–56]. To achieve high efficiency, the diode-connected GaN power transistor is utilized to replace the traditional ultrafast recovery diode used in switching type LED drivers. It has been proposed that the GaN power device is to replace the traditional silicon power device of switching LED drivers to increase the switching frequency of the converter, thereby increasing the bandwidth of data transmission. The experimental results also show that system efficiency of 80.8% can be achieved at a data rate of 1 MB/s [57].

7.3.3.2 Reliability and Network Coverage

Reliability and network coverage are the major issues to be considered by companies providing VLC services. There will be interference to communication from external light sources like sunlight, normal bulbs, and opaque materials in the path of transmission. Finding ways of dealing with these problems lies at the heart of successful development [58].

7.3.3.3 Other Challenges

The high installation costs of the systems can be complemented by the large-scale implementation of VLC, and adopting this technology will reduce further operating costs, like electricity charges, maintenance charges, etc.

Unlike broadcasting applications, serving user requests and performing adaptive modulation relying on channel information requires a realization of an uplink channel, which is a challenging design task. A proposal is foreseen through wavelength-division duplexing (WDD) using different infrared (IR) wavelengths. Lighting equipment only has LEDs. Therefore, it is necessary to install a photodetector to allow bidirectional communication [59].

In optical systems, the LED is a major source of nonlinearity. This nonlinear behavior is particularly important when an analog OFDM modulating signal is used. Another consideration in the overall design of VLC using lighting LEDs is how to include a dimming function. It is desirable that LED lights can be dimmed, but the dimming techniques used for conventional incandescent lights are not compatible with LED driver circuits. Most LED dimming systems switch the LEDs on for only a percentage of the time: the lower the percentage, the dimmer the light.

7.3.4 Conclusion

VLC technology has many attractive characteristics to support a wide range of demands for wireless communication, such as Li-Fi, traffic communication, smart city, etc. Compared to using radio waves, it is an attractive technique for future communication in electromagnetically sensitive areas and is nonhazardous to health. Besides, Li-Fi is 250 times faster than its analogous Wi-Fi, which uses RF for communication. This means VLC can be considered a potential access option for 5G wireless communications, since its strengths lie in its energy efficiency and ultrawide bandwidth. However, VLC also has weakness in many aspects. One of the most important limitations is that the line of sight between transmitter and receiver must be maintained to avoid obstacles in transmission path. Therefore, most current VLC applications focus on fixed point-to-point links. Also, the lack of mobility is due to the feature of a narrow beam width of light sources. Secondly, the VLC suffers from solar and artificial visible light interference, which, like obstacles blocking line of sight, degrades the performance of the system. In short, we definitely still need Wi-Fi and we still need RF cellular systems, but VLC deserves to be considered for a wide array of applications.

References

1 Evans, D.L. (1997). High-luminance LEDs replace incandescent lamps in new applications. *Light-Emitting Diodes: Research, Manufacturing, & Applications. International Society for Optics and Photonics* 3002: 142–154.

2 Mednik, A. (2005). Automotive LED lighting needs special drivers. *Power Electronics Technology Magazine* August.

3 Lai, Y. and Cordero, N.S. (2006). Thermal management of bright LEDs for automotive applications. In: *International Conference on Thermal, Mechanical and Multiphysics Simulation and Experiments in Micro-Electronics and Micro-Systems*, EuroSimE 2006, 1–5. IEEE.

4 Yoo, J.H., Lee, R., Oh, J.K. et al. (2016). Demonstration of vehicular visible light communication based on LED headlamp. *International Conference on Ubiquitous & Future Networks* 17 (2): 347–352.

5 Eichhorn, K. (2006). LEDs in automotive lighting. *Proceedings of SPIE - The International Society for Optical Engineering* 6134: 40–45.

6 Gacio Vaquero, D., Cardesín, J., Corominas, E.L. et al. (2008). *Comparison among Power LEDs for Automotive Lighting Applications*, 1–5. Industry Applications Society.

7 Ortega, A.V. and Silva, I.N. (2008). Neural network model for designing automotive devices using SMD LED. *International Journal of Automotive Technology* 9 (2): 203–210.

8 Takai, I., Ito, S., Yasutomi, K. et al. (2013). LED and CMOS image sensor based optical wireless communication system for automotive applications. *IEEE Photonics Journal* 5 (5): 1–1.

9 Brick, P. and Schmid, T. (2011). Automotive headlamp concepts with low-beam and high-beam out of a single LED. In: *Proceedings of SPIE - The International Society for Optical Engineering*, vol. 8170. https://doi.org/10.1117/12.895215.

10 Yanghongqiao, B. (2011). Research on Multi-LED headlamp lighting system adjustment mechanism design: based on Besturn B70 LED headlamp design. In: *2011 International Conference on Mechatronic Science, Electric Engineering and Computer (MEC), IEEE*. https://doi.org/10.1109/MEC.2011.6025558.

11 Chen, F., Wang, K., Zhao, M.X. et al. (2011). Design of LED packaging module for automotive forward-lighting application. In: *Proc. 13th Electronics Packaging Technology Conference (EPTC)*, Singapore (7–9 December 2009), 213–217. IEEE.

12 Jiao, J. and Wang, B. (2004). Etendue concerns for automotive headlamps using white LEDs. *Proceedings of SPIE – The International Society for Optical Engineering* 5187: 234–242.

13 Lee, V.W., Twu, N., and Kymissis, I. (2016). Micro-LED technologies and applications. *Information Display* 32 (6): 16–23.

14 Saponara, S., Pasetti, G., Costantino, N. et al. (2012). A flexible LED driver for automotive lighting applications: IC design and experimental characterization. *IEEE Transactions on Power Electronics* 27 (3): 1071–1075.

15 Ying-Yan, L., Jing, Z., Xue-Cheng, Z. et al. (2008). An efficiency-enhanced low dropout linear HB LED driver for automotive application. In: *International Conference on Electron Devices and Solid-State Circuits*, 1–4. IEEE.

16 Thomas, W., Pforr, J. (2009). A novel low-cost current-sharing method for automotive LED-lighting systems. In: *13th European Conference on Power Electronics and Applications*, Barcelona, Spain (8–10 September 2009).

17 Zhu, X., Zhu, Q., Wu, H. et al. (2013). Optical design of LED-based automotive headlamps. *Optics and Laser Technology* 45: 262–266.

18 Cvetkovic, A., Dross, O., Chaves, J. et al. (2006). Etendue-preserving mixing and projection optics for high-luminance LEDs, applied to automotive headlamps. *Optics Express* 14 (26): 13014–13020.

19 Kang, B., Yong, B., and Park, K. (2010). Performance evaluations of led headlamps. *International Journal of Automotive Technology* 11 (5): 737–742.

20 Kwok, K.F., Divakar, B.P., and Cheng, K.W.E. (2009). Design of an LED thermal system for automotive systems. In: *3rd International Conference on Power Electronics Systems & Applications*, 1–4. *IEEE*.

21 Xiao, C., Liao, H., Wang, Y. et al. (2017). A novel automated heat-pipe cooling device for high-power LEDs. *Applied Thermal Engineering* 111: 1320–1329.

22 Zhou, J., Long, X., He, J. et al. (2018). Uncertainty quantification for junction temperature of automotive LED with die-attach layer microstructure. *IEEE Transactions on Device & Materials Reliability* 18 (1): 86–96.

23 Bielecki, J., Jwania, A.S., Khatib, F.E. et al. (2007). Thermal Considerations for LED Components in an Automotive Lamp. In: *IEEE Semiconductor Thermal Measurement & Management Symposium*, 37–43. IEEE.

24 Sun, C.W., Chao, C.H., Chen, H.Y. et al. (2011). Development of micro-pixellated GaN LED array micro-display system. *SID Symposium Digest of Technical Papers, Session 71* 42 (1): 1042–1045.

25 Wu, T.Z., Sher, C.W., Lin, Y. et al. (2018). Mini-LED and micro-LED: promising candidates for the next generation display technology. *Applied Sciences* 8 (9): 1557.

26 Tan, G.J., Huang, Y.G., Li, M.C. et al. (2018). High dynamic range liquid crystal displays with a mini-LED backlight. *Optics Express* 26 (13): 16572–16584.

27 Deng, Z., Zheng, B.Y., Zheng, J.P. et al. (2018). High dynamic range incell LCD with excellent performance. *SID Symposium Digest of Technical Papers, Book 2: Session 74* 49 (1): 996–998.

28 Jiang, H.X., Jin, S.X., Li, J. et al. (2001). III-nitride blue microdisplays. *Applied Physics Letters* 78 (9): 1303–1305.

29 Fan, Z.Y., Lin, J.Y., and Jiang, H.X. (2008). III-nitride micro-emitter arrays: development and applications. *Journal of Physics D Applied Physics* 41 (9): 94001–94012.

30 Gong, Z., Gu, E., Jin, S.R. et al. (2008). Efficient flip-chip InGaN micro-pixellated light-emitting diode arrays: promising candidates for micro-displays and colour conversion. *Journal of Physics D Applied Physics* 41 (9): 094002.

31 McKendry, J., Rae, B.R., Zheng, G. et al. (2009). Individually addressable AlInGaN micro-LED arrays with CMOS control and subnanosecond output pulses. *IEEE Photonics Technology Letters* 21 (12): 811–813.

32 Liu, Z.J., Wong, K.M., Keung, C.W. et al. (2009). Monolithic LED microdisplay on active matrix substrate using flip-chip technology. *IEEE Journal of Selected Topics in Quantum Electronics* 15 (4): 1298–1302.

33 Wu, W., Wang, S., Zhong, C. et al. (2019). Integral imaging with full parallax based on mini LED display unit. *IEEE Access* 7: 32030–32036.

34 Gou, F.W., Hsiang, E.L., and Tan, G.J. (2019). Tripling the optical efficiency of color-converted micro-LED displays with funnel-tube array. *Crystals*.

35 Tull, B.R., Twu, N., Hsu, Y.J. et al. (2017). 19-1: Invited paper: micro-LED microdisplays by integration of III-V LEDs with silicon thin film transistors. *SID Symposium Digest of Technical Papers* 48 (1): 246–248.

36 Huang, Y., Tan, G., Gou, F. et al. (2019). Prospects and challenges of mini-LED and micro-LED displays. *Journal of the Society for Information Display* 27 (7): 387–401.

37 Tanaka, Y., Haruyama, S., and Nakagawa, M. (2000). Wireless optical transmissions with white colored LED for wireless home links. In: *11th IEEE International Symposium on Personal Indoor and Mobile Radio Communications*, IEEE. https://doi.org/10.1109/PIMRC.2000.881634.

38 Tanaka, Y., Komine, T., Haruyama, S. et al. (2003). Indoor visible light data transmission system utilizing white led lights. *IEICE Transactions on Communications* 86 (8): 2440–2454.

39 Sarkar, A., Agarwal, S., and Nath, A. (2015). Li-Fi technology: data transmission through visible light. *IJARCSMS* 3 (6): 1–10.

40 Khan, L.U. (2017). Visible light communication: applications, architecture, standardization and research challenges. *Digital Communications & Networks* 3 (2): 78–88.

41 Roberts, R., Gopalakrishnan, P., and Rathi, S. (2011). Visible light positioning: Automotive use case. In: *2010 IEEE Vehicular Networking Conference*, Jersey City, NJ, 309–314. IEEE.

42 Neirotti, P., Marco, A.D., Cagliano, A.C. et al. (2014). Current trends in smart city initiatives: some stylised facts. *Cities* 38 (5): 25–36.

43 Wu, S., Wang, H., and Youn, C.H. (2014). Visible light communications for 5G wireless networking systems: from fixed to mobile communications. *IEEE Network* 28 (6): 41–45.

44 Zeng, L., O'Brien, D.C., Minh, H.L. et al. (2009). High data rate multiple input multiple output (MIMO) optical wireless communications using white LED lighting. *IEEE Journal on Selected Areas in Communications* 27 (9): 1654–1662.

45 Wang, Q., Wang, Z.C., and Dai, L.L. (2015). Multiuser MIMO-OFDM for visible light communications. *IEEE Photonics Journal* 7 (6): 1–11.

46 Komine, T., Haruyama, S., and Nakagawa, M. (2004). A study of shadowing on indoor visible-light wireless communication utilizing plural white LED lightings. In: *International Symposium on Wireless Communication Systems*, vol. 34, 211–225. IEEE.

47 McKendry, J.J.D., Massoubre, D., Zhang, S. et al. (2012). Visible-light communications using a CMOS-controlled micro-light-emitting-diode array. *Journal of Lightwave Technology* 30 (1): 61–67.

48 Chen, S.H. and Chow, C.W. (2015). Hierarchical scheme for detecting the rotating MIMO transmission of the in-door RGB-LED visible light wireless communications using mobile-phone camera. *Optics Communications* 335: 189–193.

49 Santos, J.M.M., Jones, B.E., Schlosser, P.J. et al. (2015). Hybrid GaN LED with capillary-bonded II–VI MQW color-converting membrane for visible light communications. *Semiconductor Science and Technology* 30 (3).

50 Griffiths, A.D., Islim, M.S., Herrnsdorf, J. et al. (2017). CMOS-integrated GaN LED array for discrete power level stepping in visible light communications. *Optics Express*: 1094–4087.

51 Tanaka, H., Umeda, Y., and Takyu, O. (2011). High-speed LED driver for visible light communications with drawing-out of remaining carrier. In: *2011 IEEE Radio and Wireless Week, RWW 2011 – 2011 IEEE Radio and Wireless Symposium*, 295–298. IEEE.

52 Armstrong, J. (2013). Optical domain digital-to-analog converter for visible light communications using led arrays. *Photonics Research* 1 (2): 92–95.

53 Zhang, M. (2014). An experiment demonstration of a LED driver based on a 2nd order pre-emphasis circuit for visible light communications. In: *2014 23rd Wireless and Optical Communication Conference (WOCC)*, IEEE. https://doi.org/10.1109/WOCC.2014.6839943.

54 Jalajakumari, A.V.N., Tsonev, D., Cameron, K. et al. (2015). An energy efficient high-speed digital LED driver for visible light communications. In: *2015 IEEE International Conference on Communications (ICC)*, IEEE. https://doi.org/10.1109/ICC.2015.7249125.

55 Modepalli, K. and Parsa, L. (2015). Dual-purpose offline LED driver for illumination and visible light communication. *IEEE Transactions on Industry Applications* 51 (1): 406–419.

56 Gao, Y., Li, L., and Mok, P.K.T. (2018). An ac input inductor-less LED driver for efficient lighting and visible light communication. *IEEE Journal of Solid State Circuits*: 1–13.

57 Gong, C.S.A., Lee, Y.C., Lai, J.L. et al. (2016). The high-efficiency LED driver for visible light communication applications. *Science Reports* 6: 30991.

58 Chen, S.H. and Chow, C.W. (2015). Differential signaling spread-spectrum modulation of the LED visible light wireless communications using a mobile-phone camera. *Optics Communications* 336: 240–242.

59 Suzuki, K., Asahi, K., and Watanabe, A. (2015). Basic study on receiving light signal by LED for bidirectional visible light communications. *Electronics and Communications in Japan* 98 (2): 1–9.

8

LEDs Beyond Visible Light

8.1 Applications of UV-LED

The ultraviolet (UV) spectrum is commonly divided into UVA (wavelengths of 315–400 nm), UVB (280–315 nm), UVC or short-wave UV (200–280 nm), and vacuum UV (10–200 nm) (Figure 8.1) [1]. Group III-nitride materials (GaN, AlN, and InN) have tunable band-gap energies, which are 3.43 eV, 6.04 eV, and 0.65 eV, respectively [2]. Through the selection of various compositions of ternary compounds, a wide range of emission wavelengths can be obtained. As listed in Table 8.1, systems with higher mole fractions of Al (e.g. $Al_xGa_{1-x}N$) have higher band gaps and emit at shorter wavelengths while high mole fractions of In (e.g. $In_xGa_{1-x}N$) emit at longer wavelengths due to the smaller band gap.

Mercury vapor lamps, widely used in the majority of UV disinfection systems, have many disadvantages: large size, low resistance to shock, and high energy consumption. However, these lamps have a short lifespan of approximately 4000–10 000 hours and contain mercury. In comparison to UV lamps, ultraviolet-light-emitting diodes (UV-LEDs) not only are mercury free but also have a higher operating efficiency (more electrical input converted to light and lower heat generation), a longer lifetime (exceeds 100 000 hours), and more constant light intensity (as listed in Table 8.2) [3]. Notably, the different wavelength determines the different function. For instance, UVA LEDs have steadily replaced mercury arc and microwave lamps in clinical treatment, whose efficiencies are almost up to 70% at peak wavelengths of 385–405 nm [4]. For UVB LEDs, the most promising applications are in the field of medical skin therapy and novel concepts of horticulture and plant growth (irradiation of plants for the generation of phytamines or to reduce hormone-like mixtures). UVC applications focus on the disinfection of air, surfaces, and water at 265 or 280 nm [5]. Apparently, the advantage of LED lamps is to be compact light sources and to produce UV light at a single wavelength, which allows UV disinfection systems with optimized numbers and wavelength LEDs as a function of microorganisms to be inactivated to be designed. For all these reasons, UV-LEDs are expected to be more widely used in the future [3].

8.1.1 Structures

8.1.1.1 UV-LED Chip

Figure 8.2 shows the typical structures of the common UV-LED chip. Light is extracted through the UV-transparent AlGaN base layer and the sapphire substrate. Figure 8.3 shows

From LED to Solid State Lighting: Principles, Materials, Packaging, Characterization, and Applications, First Edition.
Shi-Wei Ricky Lee, Jeffery C. C. Lo, Mian Tao, and Huaiyu Ye.

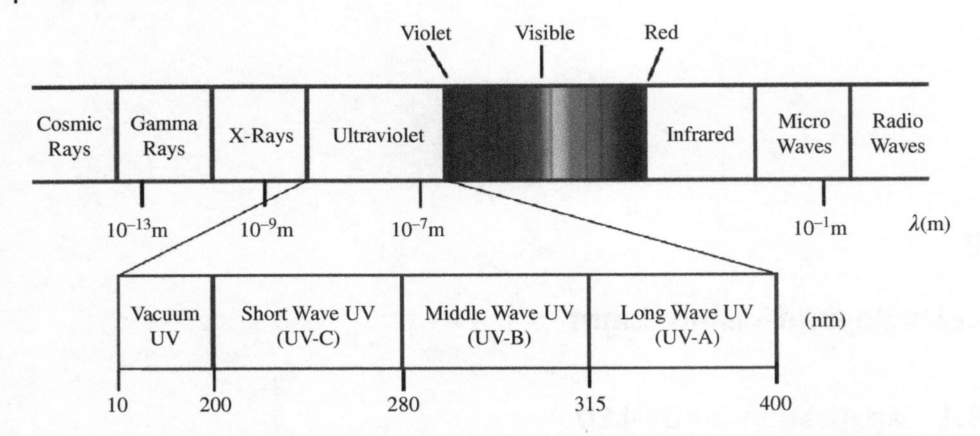

Figure 8.1 The electromagnetic spectrum of light [1]. Source: Based on Taniyasu, Y., Kasu, M., Makimoto, T. An aluminium nitride light-emitting diode with a wavelength of 210 nanometres. *Nature*, 2006, 441(7091): 325–328.

Table 8.1 Band gap energy and emission wavelength ranges that can be achieved with the compound semiconductors.

Compounds	E_g (eV) range	λ (nm) range
$In_xAl_{1-x}N$	0.7–6.2	200–1771
$In_xGa_{1-x}N$	0.7–3.47	357–1771
$Al_xGa_{1-x}N$	3.47–6.2	200–357

emission spectra of different LED devices with AlGaN, InAlGaN, and InGaN multiple quantum well (QW) active regions. Depending on the Al and In mole fractions, the emission wavelength can be tuned between 382 and 289 nm.

The performance of an LED is characterized by its external quantum efficiency (EQE), also known as its wall-plug efficiency (WPE), which is the ratio of optical power output to electrical power input. It is a function of internal quantum efficiency (IQE), luminous efficiency (LEE), and extraction efficiency (EE). LEE is the number of photons produced by each injected electron in the active region, and EE is the number of photons emitted by each photon produced in the active region from the LED. The WPE decides the heat power dissipates from die to heat slug and finally to the outside of UV-LEDs. The WPE is closely related to the EE and EQE of UV-LEDs based on Eqs. (8.1) and (8.2), which means that the EQE of UV-LED chip plays a decisive role in the thermal performance of the device. Table 8.3 lists the performance of blue LEDs and UV-LEDs, the efficiency of blue LEDs is much better than UV-LEDs.

$$\eta_{\text{EQE}} = \eta_{\text{IQE}}\,\eta_{\text{LEE}} \tag{8.1}$$

$$\eta_{\text{WPE}} = \eta_{\text{EQE}}\,\eta_{\text{EE}} \tag{8.2}$$

Table 8.2 Feature of mercury lamps and UV-LED lamps.

	Mercury lamps	UV-LED lamps
Structure	Bulky	Compact flexible
Emission spectrum	Fixed, limited	Arbitrary spectral power distribution from 210 nm up
Power consumption	High	Low
Standby time	Long	Short
Environmental burden	Toxic	Benign
Heat generation	High	Low

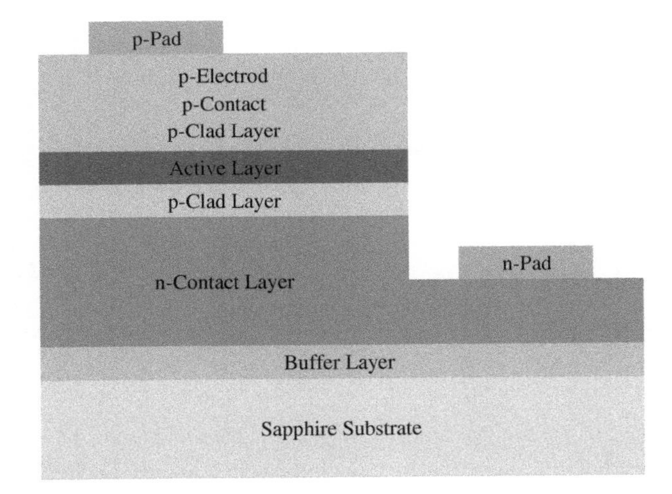

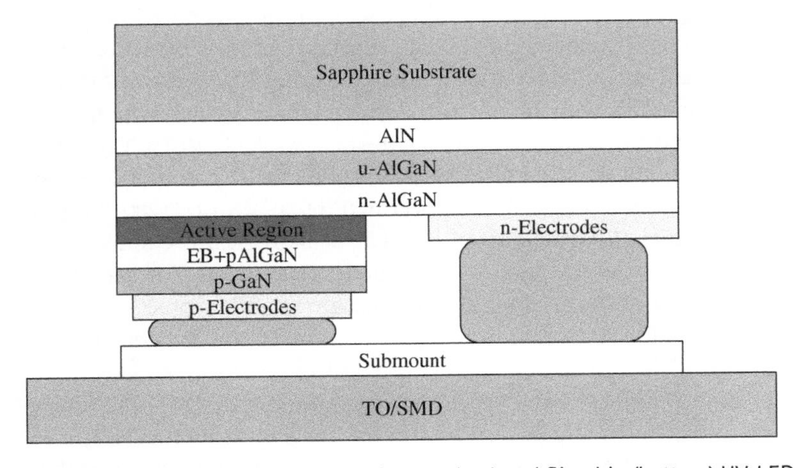

Figure 8.2 Structural diagram of a face-up (top) and flip-chip (bottom) UV-LED chip [3]. Source: Based on Muramoto, Y., Kimura, M., Nouda, S. Development and future of ultraviolet light-emitting diodes UV-LED will replace UV lamp. IEEE Photonics Society Summer Topical Meeting Series (SUM). Semiconductor Science and Technology, 2015, 084004.

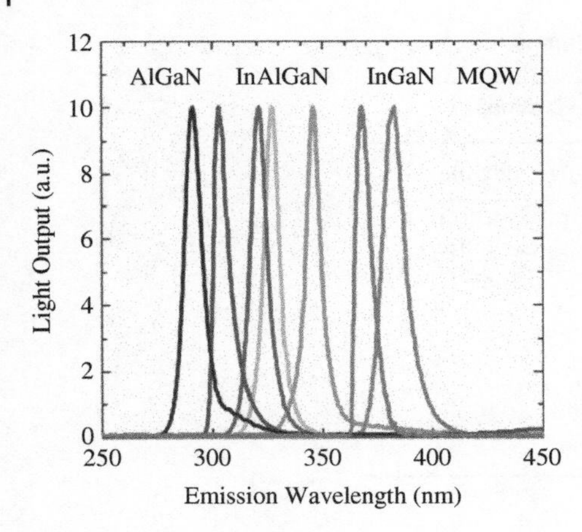

Figure 8.3 Emission spectra of AlGaN, InAlGaN, and InGaN more quantum well (MQW) UV-LEDs [6]. Source: Kneissl, M., Kolbe, T., Chua, C., et al. Advances in group III-nitride-based deep UV light-emitting diode technology. *Semiconductor Science and Technology*, 2010, 26(1): 014036. Figure 4.

Table 8.3 The performance of Blue LEDs and UV-LEDs.

Item	Blue LED	UV LED
IQE	>80%	<60%
LEE	>80%	<20%
EQE	>80%	<10%
EE	>70%	<70%
WPE	>50%	<5%
Max power	>5 W	<100 mW

In 1998, Han et al. reported on the growth and characterization of UV-LED with wavelengths shorter than 360 nm using $Al_{0.2}Ga_{0.8}N$/GaN multiple QWs for the first time [7]. The emission wavelength was 353.6 nm at 20 mA with an output power of 13 μW, and the EQE is less than 1%. Thereafter, UVA, UVB, and UVC LEDs have successively been developed. In 2006, Taniyasu et al. reported a homojunction LED based on AlN PIN (p-type/intrinsic/n-type) materials with an emission wavelength of 210 nm, which is the shortest reported to date for nitride LEDs [1]. However, the EQE was only 10^{-6}%. Currently, the EQE of deep ultraviolet-light-emitting diodes (DUV-LEDs) in UVB and UVC bands is mostly in the 10% or even less than 5%. The highest EQE of DUV-LEDs (280–300 nm) was obtained by Hirayama et al. [8], which reached a peak of 14.3% at 2 mA and 10.5% at 20 mA. The EQE of near-ultraviolet ray light-emitting diodes (NUV-LEDs), in particular, has greatly improved because of developments in crystal growth, chip processing, and packaging technologies, reaching 30% at a wavelength of 365 nm, 50% at 385 nm, and 60% at 405 nm.

Since 2010, the emission wavelength and output power of the nitride DUV-LEDs have made considerable progress, which largely depends on the development of AlGaN fabrication technology. Besides, the EQE of DUV-LEDs has much room for improvement than NUV-LEDs and blue LEDs, owing to the epitaxial and doping level of AlGaN materials.

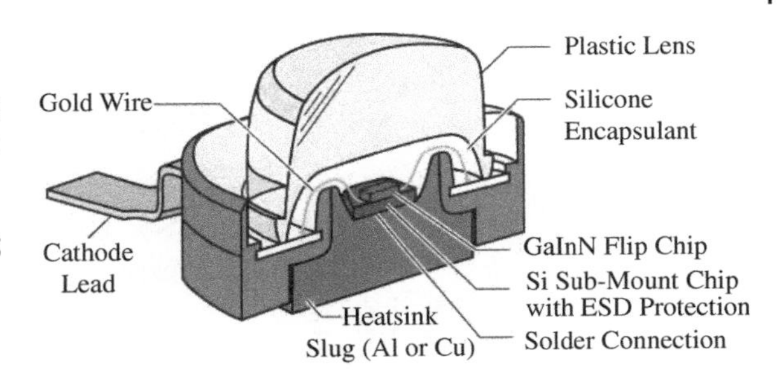

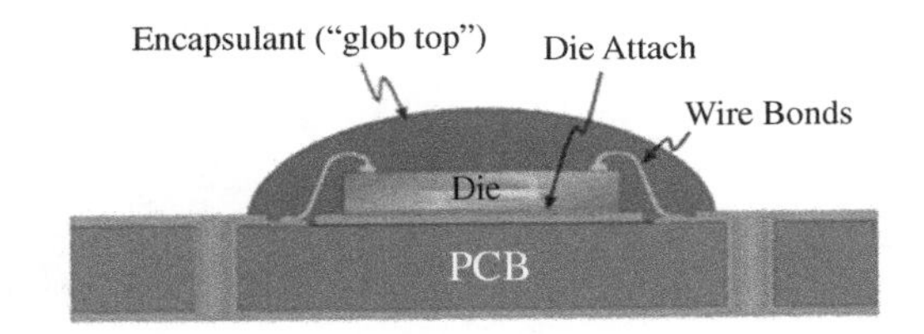

Figure 8.4 Cross-sectional diagram of lead-frame-based plastic (LFP) packaging (up) [10] and chip-on-board (COB) packaging (down) [11]. Sources: Schubert, E.F., Gessmann, T., Kim, J.K. *Light Emitting Diodes*. New York: Cambridge University Press, 2006 (up); Habtemichael Y T. Packaging Designs for Ultraviolet Light Emitting Diodes. G.W. Woodruff School of Mechanical Engineering, Georgia, 2012 (down).

8.1.1.2 UV-LED Package

The typical lateral area of UV-LED chips is 1 mm^2 with an input power of just 1 W [9]. The enormously high heat-flux (as high as 100 W/cm^2) is loaded in the chip and then transported down through minimal contact pads inside the conventional UV-LED package (Figure 8.4). For this thermal bottleneck before the underlying package can spread the heat, increasing the contact cross-section area between the device and submount is an efficient way to reduce the thermal resistance. In addition, precautions in device geometry optimization should be taken to avoid lateral current crowding, as it may well result in an increase of the device resistance and hence a reduction of quantum efficiency [12, 13].

Figure 8.5 shows the total system resistance of leadframe plastic (LFP) package and chip-on-board (COB) package. Note that four interface resistance remains the main bottleneck for efficient heat conduction in LFP packaging. The COB packaging is a different solution, whereby the chip is directly mounted onto a printed circuit board (PCB) and can be directly packaged on a heat sink. This leads to a more compact package and allows for a higher packaging density of chips on the same board [9, 14]. COB packaging has several advantages compared to LFP packages for high-power LEDs. A large thermal resistance is removed because of the simplified structure of COB packages and assembly parts, and overall manufacturing processes is reduced efficiently. However, light intensity control of a large surface light source and the thermal dissipation of the high density heat flux are still the main concerns for COB packaging.

Furthermore, novel packaging configurations have been developed. As shown in Figure 8.6, Jiangsu XGL Tech. Corp. [15] developed and manufactured a packaging-free UV-LED solidification light source module. The measured the heat resistance coefficient of the module as lower than 3 °C/W, and the maximum driving power of the single LED chip can reach 5 W.

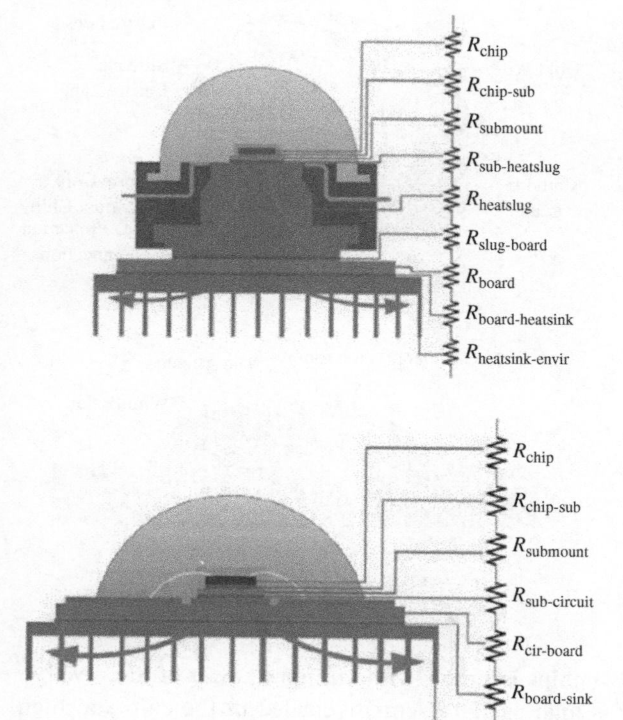

Figure 8.5 Thermal resistance network for LFP (up) and COB packaging (down) [14]. Source: Liu, Z., Liu, S., Wang, K., et al. Status and prospects for phosphor-based white led packaging. *Frontiers of Optoelectronics in China*, 2009, 2(002): 119–140. Fig. 10, Fig. 11.

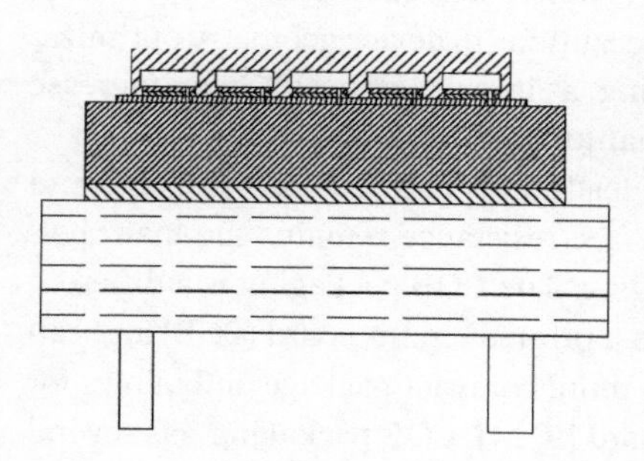

Figure 8.6 Packaging-free UV-LED solidification light source module [15]. Source: Huang, H.S., Guo, W.P., Ke, Z.J., et al. Packaging-free UV-LED solidification light source module, China Patent, CN103794603 A, 2014.

8.1.2 Applications

UV light in the 250–350 nm range has many useful and attractive functions. Some potential applications include air and water purification, UV photolithography, in situ activation of drugs through optical stimulus, solid state lighting, polymer curing, and laser surgery for III-nitride-based UV light sources [2]. The combined markets are fairly large, and the water purification market alone is estimated to be worth over \$5 billion [16]. In fact, the UV-LEDs industry reached \$233 million in 2017. It is expected to reach \$1.2 billion by 2022. The UV-LED market grew at a CAGR of 33% (Figure 8.7) [17].

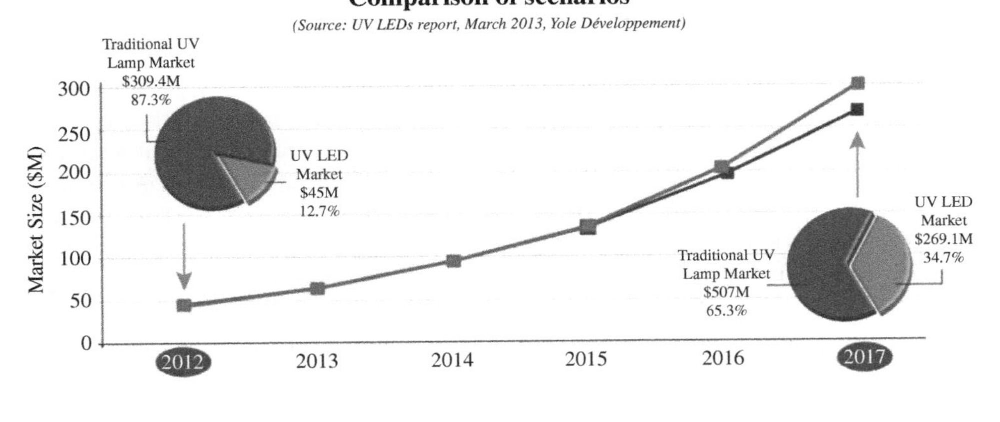

Figure 8.7 "UV-LEDs: Technology & Application Trends", a report from Yole Développement [17]. Source: Mukish, P. The UV-LED market is booming, announces Yole Développement. *Solid State Technology*, 2013.

8.1.2.1 Sterilization and Disinfection

UV light sources can be used for the purification of river water, industrial wastewater, and atmospheric gases. The entire UV spectrum can kill/inactivate many microorganisms, such as the log-inactivation of *Escherichia coli* [18, 19], and the inactivation of *Pseudomonas aeruginosa* [20] is enhanced remarkably using UV-LED. Legionella seems to be very sensitive to UVC irradiation and to visible violet light. But UVC energy, with 265 nm being the optimum wavelength, provides the most germicidal effect [21]. UVC light can destroy the deoxyribonucleic acid (DNA) and ribonucleic acid (RNA) molecular structures of bacteria, viruses, spores, and other pathogens in a short time, owing to its short wavelength and high energy [22]. The bacteria and viruses lose the ability to self-replicate, thereby being effectively killed. It was also found that UV-LEDs emitting at 253.7 nm were effective as disinfectants in water and bactericides in fresh meat [23, 24]. Besides, ozone can be produced by V-UV (185 nm) bands in the air and dissolved in water. It can destroy the cell walls of microorganisms by oxidation. Eskandarian et al. studied the effect of different organic chemicals using UVA-, UVB-, UVC-LEDs, and titanium oxide (TiO_2), respectively [25]. UV wavelength was found to be a more important parameter for decomposition than light intensity. Chevremont et al. tested the efficiency of UVA- and UVC-LED radiations, used alone or coupled, on bacteria in the wastewater and chemical indicators [26, 27]. It appears that coupling UVA/UVC oxidizes up to 37% of creatinine and phenol, which is comparable to that commonly obtained with photoreactants such as TiO_2. A novel UV fluorescence sensor was designed to monitor the dissolved organic matter (DOM) online based on the fact that 280 nm UV light could excite both protein-like and humic-like fluorescence. The unique aspects of UV-LEDs improve inactivation effectiveness by applying LED special features, such as multiple wavelengths and pulsed illumination; however, more studies are needed to investigate the influencing factors and mechanisms [28].

8.1.2.2 Curing

UV curing is one of the main directions of the UV-LED application industry. Anyaogu et al. found that UV-LED sources successfully matched the performance of conventional light sources in both the polymerization dynamics occurring within a cured layer and the final properties of the coatings [29]. UV-LEDs can be used as the light source in the microlithography system, and the potential usage of UV-LEDs to exposure photo-sensitized PCBs was evaluated by Huang and Sung [30]. A Khalifa University research team developed a low-cost portable UV-LED lithography system by using 172 dominant wavelength of 380 nm InGaN UV-LED arrays package in the region of $3 \times 3\,in^2$ [31]. Owing to the reduction in system cost and complexity, UV-LED lithography can be rendered as a perfect candidate for microlithography with large process windows typically suitable for microelectromechanical systems (MEMS) and microfluidics applications. Suzhou Institute of Biomedical Engineering and Technology has designed a system of UV-LED curing automated industrial production lines. It discovered that illumination intensity and uniformity are both superior to conventional UV curing systems [32] (Figure 8.8). These technical and market developments present both opportunities and challenges for the development of UV-LED-based curing systems. In other areas, under the radiation of UV-LED, gel nail polishes can be cured better, to enjoy greater durability and more sustainable development. These novel gel nail polishes are greener alternatives to the current products on the market, and promise greater consumer acceptance [33].

8.1.2.3 UV Communications Technology

The UVC band is solar blind (280 nm and below) at the ground level within the entire UV spectrum 4–400 nm. It means that solar radiation is negligible in this band, because of ozone absorption in the upper atmosphere. Thus covert or non-line-of-sight (NLOS)

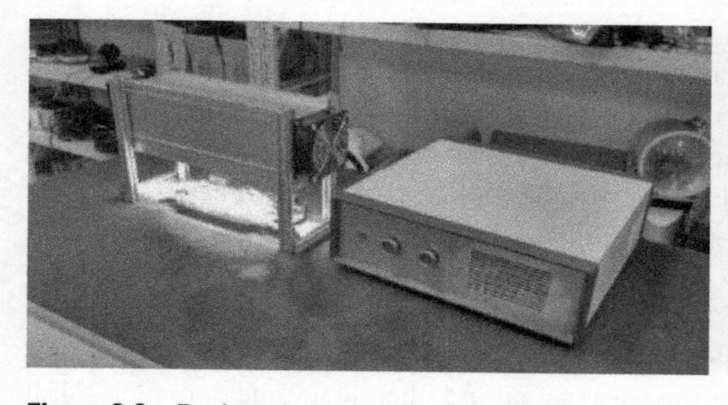
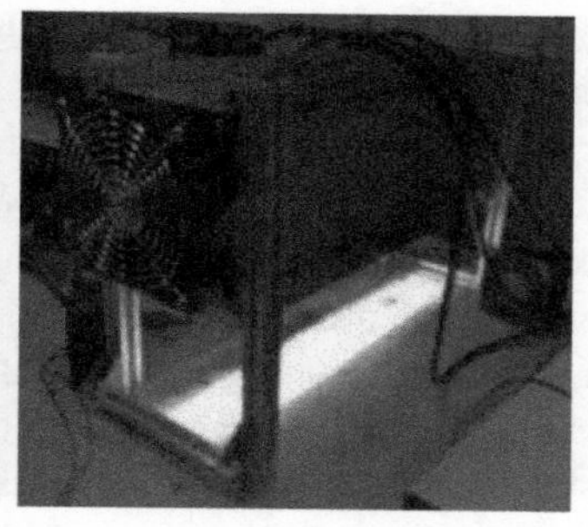

Figure 8.8 Engineering prototype of UV-curing system [32]. Source: Wang, P., Jin, Z., Xiong, D. Optimal optical design of UV-LED curing system with high illumination and luminance uniformity. 2018 15th *China International Forum on Solid State Lighting: International Forum on Wide Bandgap Semiconductors China (SSL China: IFWS)*. IEEE 2018. Fig. 8.

communication can be realized in the "solar blind" region [34–36]. By operating in the UVC band, a ground-based photodetector with low dark noise can exploit the low solar radiation and approach quantum noise-limited photon-counting detection. Besides, there are many advantages for UV communication technology, such as a huge unlicensed spectrum, low-power and miniaturized transceivers, higher-power densities, and high capacity [37]. Encouraged by these characteristics, signal transmission by UV communication technology develops rapidly in complex environments. UV communication technology has potential military applications, as well, including unattended ground sensor (UGS) networks and small unit communications, flame sensing, biological fluorescence detection, missile or artillery fire detection, ground–air communications, optical tag identification, and covert networking [38].

8.1.2.4 White Light Generation

There are currently three different approaches to generate white light based on LEDs. The most commercial way is by using a blue-emitting GaN LED chip (mostly at 450–470 nm) covered by a yellowish phosphor coating; by mixing reds, greens, and blues, i.e. red–green–blue (RGB) LEDs; or by using a UV-LED excited light for RGB phosphors [39–41]. The white light produced by the UV technique only depends on phosphors. Thus, UV-LEDs have the potential to make a big impact in the white light LED market in the near future.

8.1.2.5 Medical Treatment

UV light therapy, like skin treatment, is a typical application of narrow band UVB. The UV wavelength of 310 nm has a strong effect on speeding up the metabolism of skin and preventing the production of melanin precipitation [43]. Thus UVB light can effectively treat dermatitis, such as vitiligo, pityriasis rosea, polymorphous light eruption, chronic actinic dermatitis, light prurigo, and others [44–46]. Stronger and faster medications can be guaranteed, owing to UV-LED's pure spectrum compared to traditional UV light sources. Moreover, studies have shown that UVB accelerates certain polyphenols, produced by leafy vegetable (i.e. red lettuce), which have been declared to have anticancer, antiproliferation, and anticancer mutation properties [47, 48]. Inada et al. irradiated Jurkat tumor cells with UV-LED (365 nm) and found that the numbers of apoptotic and necrotic cells were approximately the same as the number of those cells irradiated by a conventional lamp system [49]. UV-LEDs have the possibility of realizing a new UV light source for phototherapy, which can replace heavy, short-lived, large traditional energy ultraviolet fluorescent lighting. Guo adjusted the concentration of Sr^{2+} or Mg^{2+} to change the wavelength of UV-LEDs, indicating their potential application in the phototherapy of infant jaundice [50]. Besides, dentists can diagnose caries easily at the early stage of tooth demineralization by way of fluorescence imaging using a UV-LED [51, 52]. The Technical University of Denmark has developed a UV-LED system to measure cardiac marker troponin I with a concentration of 200 ng/L in immunoassay, which will likely soon replace xenon flash lamp excitation in immunoassays time-resolved detection systems in medical treatment [52].

Table 8.4 Crystal lattice mismatch and thermal mismatch between AlN and various substrates.

Substrate	Lattice constant (Å)	CTE (10^{-6}/K)	Crystal lattice mismatch (%)	Thermal mismatch (%)
AlN	$a = 3.112, c = 4.982$	$\Delta a/a = 4.2, \Delta c/c = 5.3$	—	—
GaN	$a = 3.189, c = 5.186$	$\Delta a/a = 5.59, \Delta c/c = 3.17$	−0.24	−0.25
Sapphire	$a = 4.758, c = 12.991$	$\Delta a/a = 7.5, \Delta c/c = 8.57$	−34.6, 13.3	−44
6H-SiC	$a = 3.0817, c = 15.1123$	$\Delta a/a = 4.2, \Delta c/c = 4.68$	0.98	0
Si(111)	$a = 5.4301, c = 3.843$	$\Delta a/a = 3.59$	42.7, −19	17

8.1.3 Challenges

8.1.3.1 Heteroepitaxy of DUV-LEDs

Owing to the large difference between the lattice parameters of AlInGaN alloys and sapphire, it is very difficult to grow high-quality epilayers (heteroepitaxially grown layers) directly on sapphire [53]. Theoretically, the sapphire substrate is not an ideal choice for UV-LEDs because there is considerable lattice mismatch and thermal mismatch with AlGaN materials and low thermal conductivity. However, the transmittance of sapphire in the band gap of 200–400 nm is nearly transparent and the fabrication process of epitaxy sapphire substrate is being perfected, which has led to systematic experimental study in this area. Besides, different types of heterogeneity substrates, such as sapphire, silicon carbide (SiC), silicon (Si), and homogeneity substrate (e.g. GaN and AlN) are also studied for their crystal lattice qualities (as listed in Table 8.4).

The crystal mismatch of 6H-SiC substrate and AlGaN is only about 1%, which is more suitable for AlGaN epitaxial; the SiC substrates can achieve good electrical conductivity, which can be implemented in the vertical structure of the device. Since SiC absorbs UV light with the wavelengths shorter than 360 nm, most research of UV-LEDs focuses on the SiC substrate UVA band [54, 55]. Although cracking of epitaxial AlN in Si substrate can be alleviated by lateral epitaxial overgrowth (LEO)/epitaxial lateral overgrowth (ELO) technology, the stress control of and quality improvement to the Al(Ga)N epitaxial layer is still a bottleneck for Si-based UV-LED research [56, 57].

Because it is very difficult to grow heteroepitaxy layers directly on sapphire, a low-temperature and thin buffer layer of GaN or AlN is usually grown on sapphire to initiate nucleation, relax the strain, and aid in the growth of subsequent high-quality layers [2]. However, with the increasing lattice mismatch at high Al mole fractions in the growing of AlN/AlGaN films on sapphire, the growth of DUV-LEDs on sapphire becomes more challenging in the UVC range because of high threading dislocation densities (TDDs). Moreover, the issues of doping challenges of $Al_xGa_{1-x}N$ layers at high Al mole fractions, biaxial tensile strain in AlGaN films, and cracking have emerged.

8.1.3.2 Doping of AlGaN UV-LEDs and External Quantum Efficiency

Due to the low ionization efficiency of silicon (n-type dopant) and magnesium (p-type dopant), the n-type and p-type doping efficiency of GaN is relatively low, especially for AlGaN with a high aluminum content [54]. High "activation energy" of dopants, low active efficiency, and concentration of a carrier can be achieved by enhancing the Al concentration of AlGaN. However, increasing the concentration of dopants induces the deterioration of crystal quality and increases the number of defects (including vacancies, impurities, dislocations, etc.). Furthermore, it leads to a dope in sheet resistance in the AlGaN epitaxial layer, which results in severe current crowding and low electrical injection and luminous efficiency of UV-LED.

Furthermore, as the Al content increases, the quantum conned stark effect (QCSE) becomes serious. QCSE can increase a carrier's life, reduce radiative recombination efficiency and IQE, and induce redshift of the emission wavelength. Besides, UV light is absorbed or totally reflected by the epitaxial layer, which leads to a fairly low EQE of UV-LEDs. Three possible causes related to crystal growth for this decrease in efficiency are:

- A decrease in LEE due to a high dislocation density in the active layer.
- The hole concentration of p-AlGaN is low, resulting in low injection efficiency (IE).
- A decrease in the light-extraction efficiency due to UV absorption by the GaN buffer layer and total internal reflection and polarization.

8.1.3.3 Thermal Degradation

Generally, DUV-LEDs exhibit both catastrophic and gradual output power degradation under dc bias. This catastrophic degradation is mostly caused by the rough surface morphology and current crowd of the p-GaN contact layer [58–60]. The V-defects in AlGaN and p-GaN layers work as leakage paths, as confirmed by the electron beam's induced current measurements [61]. The current density in the vicinity of these defects is increased by long-term stress, which leads to local overheating followed by the p-metal.

The low WPE and poor thermal conductivity of sapphire (35 W/(m · K)) of DUV-LEDs induces excessive joule heating or self-heating in sapphire-based UV-LEDs under high operating voltages, and leads to a reduction in device lifetime and a spectral shift in emission. Shatalov et al. compared the optical power decay with time under dc and pulse constant current stress and found that the degradation rate varies noticeably with the junction temperature at the same pump current value [42]. Besides, they took out the comparison of *I–V* curves of the DUV-LED devices before and after degradation, and concluded that the increase of the operating voltage corresponds to the change of differential resistance but not to the buildup of the turn-on voltage.

8.1.3.4 Structures of UV-LED Packages

AlGaN/GaN-based DUV-LEDs with emission wavelengths between 200 and 280 nm produce a lot of undesired heat because of the low WPE of UV-LEDs. The thermal resistance of differently packaged UV-LEDs can differ largely, which is closely related to the thermal performance and lifetime of the device. In view of this, effective thermal management to decrease the package's junction-to-air thermal resistance with cost-effective solutions is critical to the packaging design of DUV-LEDs. Habtemichael et al. investigated both the thermal and stress response of AlGaN/GaN-based UV-LED packages through finite element analysis (FEA) and experimental analysis [11]. They found that the largest bottleneck for improved thermal performance is the thermal interface material layer and the submount. Shatalov et al. performed aging studies on 10 × 10 micropixel 280 nm emission

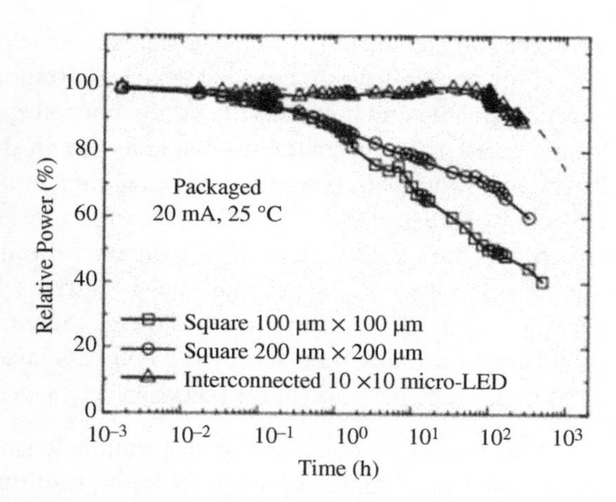

Figure 8.9 The power degradation of packaged deep UV-LED at 20 mA dc and 25 °C heatsink temperature [42]. Source: Shatalov, M., Streubel, K.P., Yao, H.W., et al. Reliability of AlGaN-based deep UV-LEDs on sapphire. *Proceedings of SPIE: The International Society for Optical Engineering*, 2006, 6134: 1–11. Figure 6.

DUV-LEDs mounted onto TO-66 headers for thermal management [42]. The power degradation of a packaged DUV-LED for a 200 μm × 200 μm square device is lower than the standard 100 μm × 100 μm square device at 20 mA dc current. As presented in Figure 8.9, a larger junction area leads to superior reliability performance of micropixel LEDs with a projected operational lifetime for 50% power reduction in excess of 1000 hours.

8.1.3.5 Materials of UV-LED Packages

Exposure to an aggressive environment (i.e. UV radiation, thermal exposure, oxidative atmosphere) induces changes in polymer materials' physical, chemical, and mechanical characteristics [62–65]. UV light is able to age the package materials and induces yellowing, carbonization, and nebulization. Most polymers absorb solar UV radiation and can be degraded by photo- and thermo-oxidative reactions [66, 67]. Epoxy resins are versatile as the standard choice for the encapsulation of indicator LEDs [68]. However, they suffer from yellowing under the influence of heat and UV radiation [69]. Silicone resins are excellent in thermal stability and UV resistance compared with epoxy resin [70], but UV light causes a scission of C–H and Si–C bonds and forms byproducts such as hydrocarbons [71]. Moreover, most encapsulants have low transmittance of UV/DUV light, which reduces the WPE and increases the joule heat of UV-LED package. However, there is still a lack of quantitative analysis regarding the impact of UV light on the components of a UV-LED package. From the aspect of personal safety, too much UV radiation can damage the DNA in skin cells, and even lead to skin cancer. Thus, protection design is also an essential part of UV-LED application.

8.1.4 Conclusion

There are many challenges to overcome when fabricating UV-LEDs. It is important to:

- Manage the strain to enable the growth of crack-free, thick, doped AlGaN epitaxial layers.

- Realize n-type doping in AlGaN with a high aluminum composition and reduce the concentration of nonradiative recombination centers in AlGaN that are responsible for low IQE.
- Cater for the strong polarization effects in the AlGaN-based QW active region of DUV-LEDs.
- Realize a sufficiently high p-type doping of the AlGaN electron blocking or cladding layer.

This should mean that reasonable structures of UV-LED chips can be obtained.

For the packaging of UV-LEDs, the device's structure and materials are the essential issues for thermal management. Thus, the thermal resistance of UV-LEDs can be reduced efficiently by optimizing the design of the packaging structure. Besides, the luminous efficiency of the device can be improved by using high-transmission encapsulant materials and high-reflectance reflectors.

For the UV-LED module/system, the cooling system should be designed to keep the cell temperature low and uniform, but still be simple and reliable.

Finally, the intellectual property of technological development of UV-LEDs is analyzed by one patent map. Even though the optical performance and service life of UV-LEDs have been significantly improved by advanced technology, there is still great room for the development of thermal solutions at the chip level and the package level.

8.2 Applications of IR-LEDs

Infrared optoelectronic technology is an important branch of modern optoelectronic technology, and its development affects all fields of human life and social development. There exists a great demand for efficient, broad spectrum infrared (IR) light sources for various military and commercial sensor applications [72]. The IR spectrum is divided into near infrared (NIR) (wavelength of 0.75–1.4 μm), short- (SWIR) (1.4–3 μm), medium- (MWIR) (3–8 μm), and long-wavelength infrared (LWIR) (8–15 μm), and far infrared (FIR) (50–1000 μm) (Figure 8.10). Similarly, different wavelengths of IR light determine different properties of themselves. IR-A DIN(NIR) is commonly used in fiber optic communication, owing to the low rate of decay in silica glass, as well as because it is sensitive to image enhancement. IR-B DIN (SWIR) is generally used in long-distance communication. IR-C DIN (MWIR) is also known as mini-IR and is usually used in guided missile technology for passive IR heat-seeking technology. As for IR-C DIN (LWIR), it can obtain a complete passive image of heat emissions which do not require additional light or external heat sources.

Commonly used IR optoelectronic materials include III–V, II–VI, and IV semiconductors. However, owing to the different types of materials, the band gap, lattice constant, and cut-off wavelength of these materials are also different. At present, the latest technology used in the large-scale chip integration foundry is Si's complementary metal oxide semiconductor (CMOS) or bipolar complementary metal oxide semiconductor (BiCMOS) process. Therefore, the development of Si-based IR optoelectronic devices compatible with CMOS or BiCMOS processes and the realization of the single-chip integration of photonic devices and electronic devices has become a consistent goal of many researchers. And the key is how to integrate the IR band of active devices into silicon-based circuits.

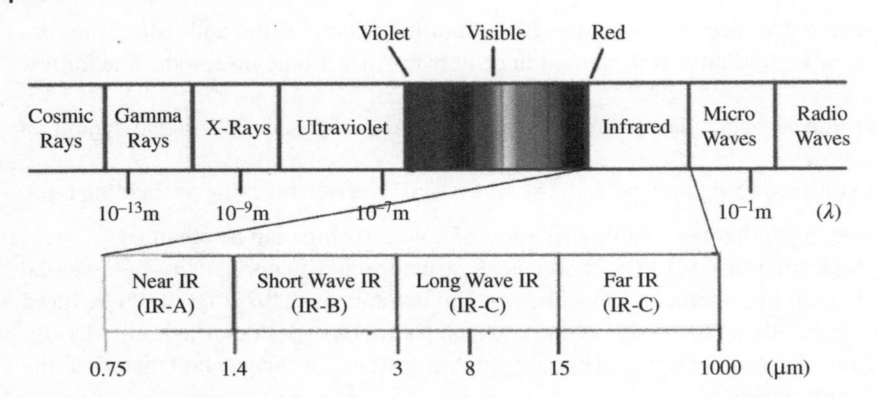

Figure 8.10 The electromagnetic spectrum of light.

Table 8.5 Some performance parameters of the IR-LED [75].

Parameters	Symbol	Value	Unit
Power consumption	Pad	100	mW
Forward current	I_F	100	mA
Peak forward current	I_{FP}	1	A
Pulse duration	D	100	μs
Duty ratio	P	1%	—
Reverse voltage	V_R	5	V
Operating temperature	T	−40 to 85	°C

8.2.1 Structures

Most IR-LEDs are plastic encapsulated, but some are metal or ceramic based and glass or resin encapsulated. The basic shape of the plastic packaging of the IR diode and visible light LED are the same. In addition to a transparent white shell, there are other colors such as light blue and black resin packaging in IR-LEDs. Like regular diodes, the IR chip is mounted on a concave reflector and sealed with epoxy resin. The p–n junction chip of LED is made of a III-group semiconductor material, which has a higher refractive index than air and will reflect at the interface of air. To prevent reflection, an anti-reflection film can coat the surface or a condenser lens can be used to improve the brightness. The top of the diode is encapsulated with a lens-type structure, so that the beam can be focused, as far as possible, along the direction of the central axis.

The kernel of the IR-LED, which is essentially a chip of the p–n junction, attaches on a holder and connects to a cathode directly while anode bonds to a positive pillar via a gold wire. The reflector is added to enhance the lighting effects, and the whole chip is encapsulated with a kind of resin. Some of its performance parameters from the product specification are listed in Table 8.5.

A kind of structure of IR-LED was proposed by Davis and his team to improve the photoluminescence quantum yields [73]. Quantum dot (QDs) nanocrystals, specifically lead chalcogenides, are candidate LED materials since they exhibit tunable luminescence across the whole near-IR region. It has been demonstrated that the luminescence of a donor acceptor QD system can be significantly improved by embedding low-energy QDs with high photoluminescence quantum efficiency (PLQE) into a matrix of higher-energy QDs with lower PLQE and then incorporating a relatively low concentration of high-quality PbS/CdS (lead sulfide/cadmium sulfide) QDs into a densely packed crosslinked film of PbS QDs. Shao et al. also analyzed the emissive power decrease of LEDs during their lifetime and identified that preliminary irradiation increased the reliability and operating time of LEDs [74].

8.2.2 Applications

8.2.2.1 Sensors and Detecting

One of the applications of the IR-LED is as a gas sensor. In 1997, Matveev et al. studied mid-IR-LEDs based on InGaAs (indium gallium arsenic), InAsSb(P) (indium arsenic antimony (phosphorus)), and InGaAsSb (indium gallium arsenide antimonide) alloys. Its transmission band is 0.3–0.5 km. Its output power is 10–50 μW. And its long-term working time can reach 20 000 hours [75]. The LED-based analyzer prototypes with PbSe (lead selenide) photoresistors as detecting elements have been made and characterized with respect to CO, CO_2, and H_2O detection, which can be employed in online monitoring applications and multicomponent analysis. Just as they did other gas sensors, Massie et al. investigated a low-cost portable optical sensor for methane detection based on near-IR-LED [76]. The sensor was not only very sensitive to methane within 1% which decrease the explosive level, but also operated in harsh environments with temperature variations between 20 and 50 °C. Experiments were carried out with a photodiode (PD) as the detector (Figure 8.11a), where it can be seen that there is a different absorption between the standard detector and the lensed detector. The knowledge gained from this investigation provides a good basis for the structure of portable sensors for other hydrocarbon gases and for the development of portable laser-based systems for higher sensitivity.

Subsequently, Kuusela et al. combined LED technology with enhanced cantilever trace gas detection with mid-IR 37 mm wavelength [77]. They tested the sensitivity of the PD to methane (CH_4), propane (C_3H_8), carbon dioxide (CO_2), and sulfur dioxide (SO_2) gases and made a preliminary evaluation. It is a remarkable improvement that the detection limit for methane is 250 times smaller than previously reported. However, the current status of LED technology for the longwave range (710 mm) prevents creating sensitive instruments for such gases as SO_2.

Another fact is that a new type of light-collecting structure that can greatly improve the signal-to-noise ratio for gas testing was designed on the basis of the mid-IR absorption spectrum principle postulated by Zhang et al. [78]. Theoretical analysis showed that this structure could perfectly enhance the gains of signals by analyzing the absorption of methane gas. Based on this design, a small system has been demonstrated with an ultimate sensitivity of 5×10^{-5} (50×10^{-6}) and the precision is approximately 3%.

In addition, as shown in Figure 8.12, a novel mid-IR-LED and PD light source/detector combination, used within a nondispersive infrared (NDIR) carbon dioxide gas sensor,

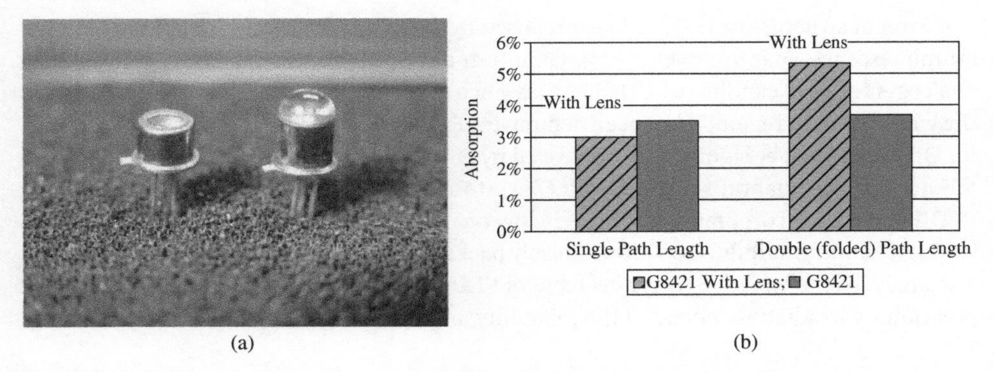

Figure 8.11 (a) Photograph of a standard detector and the lensed detector; (b) comparison of absorption from 100% methane using lensed and unlensed detectors for single and double path length configurations [42]. Source: Shatalov, M., Streubel, K.P., Yao, H.W., et al. Reliability of AlGaN-based deep UV-LEDs on sapphire. *Proceedings of SPIE: The International Society for Optical Engineering*, 2006, 6134: 1–11.

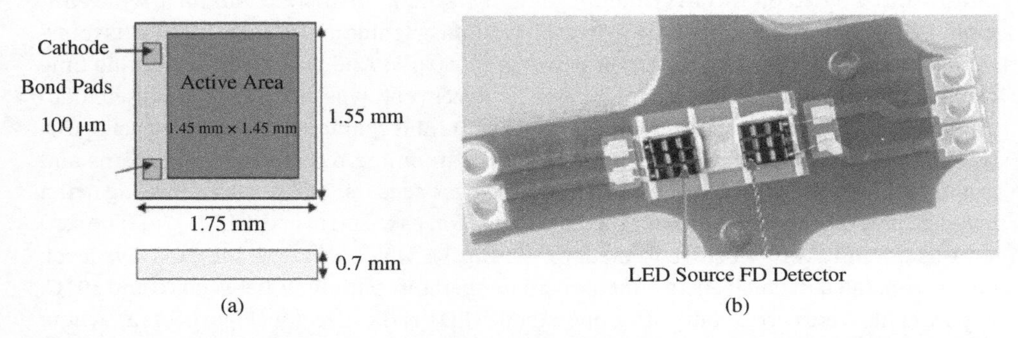

Figure 8.12 (a) LED and PD detector structure; (b) LED and PD detector mounted on Bridgeboard [79]. Source: Gibson, D., MacGregor, C. A novel solid state non-dispersive infrared CO_2 gas sensor compatible with wireless and portable deployment. *Sensors (Basel)*, 2013, 13(6): 7079–7103.

was developed in 2013. It can be widely used in buildings, transport systems, and horticultural greenhouses, and for portable deployment in safety, industrial, and medical applications [79].

8.2.2.2 Imaging Techniques

IR imaging technology is widely used in many areas, including charge coupled devices (CCDs), image positioning systems, and night vision equipment. And it has been widely used in the field of imaging.

The American Institute of Physics has presented experimental results which support a proposed scheme for IR imaging through the combined use of a photon frequency up-conversion device with a CCD camera [80]. The structure is shown in Figure 8.13 the IR-LED is used for imaging onto an Si CCD camera.

In addition, IR-LEDs are also used in the production of intelligent remote control (Figure 8.14). Yunjung Park and Minho Lee presented a new low-cost intelligent remote control that consists of a camera and IR-LED to control various electronic devices, such as smart TVs, air conditioning, etc. [81]. Users can easily operate the invisible IR-LED

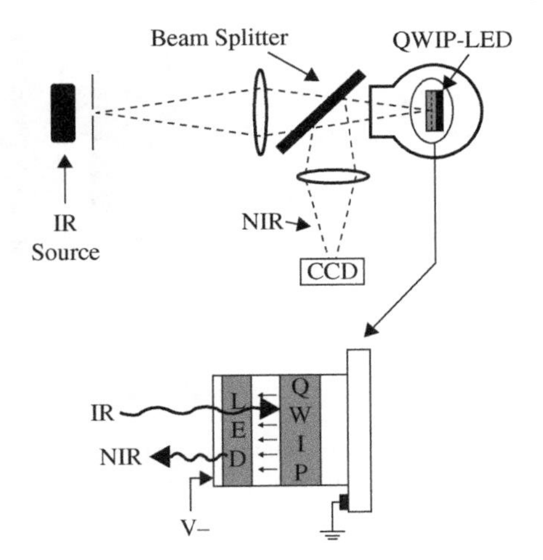

Figure 8.13 Experimental setup employed to assess the quantum well infrared photoconductor (QWIP) LED integrated device's functionality as an imaging detector [80]. Source: Allard, L.B., Liu, H.C., Buchanan, M., et al. Pixelless infrared imaging utilizing a p-type quantum well infrared photodetector integrated with a light emitting diode. *Applied Physics Letters*, 1997, 70(21): 2784–2786. Fig. 3.

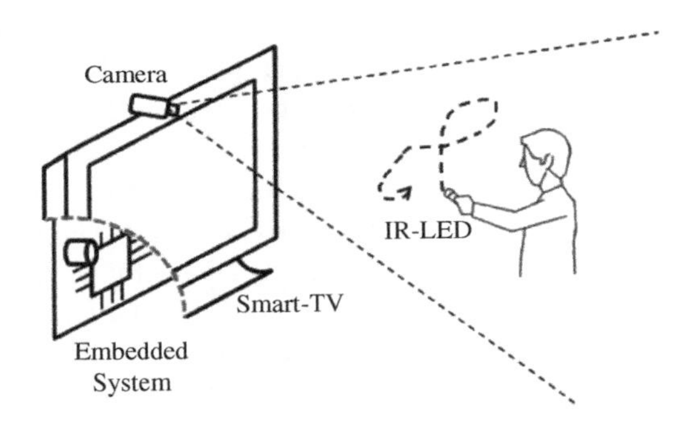

Figure 8.14 Concept of proposed system [81]. Source: Park, Y., Lee, M. Cost effective smart remote controller based on invisible IR-LED using image processing. *2013 IEEE International Conference on Consumer Electronics (ICCE)*, Las Vegas, NV, (2013), pp. 434–435. Fig. 1.

and make specific commands by simply blinking the IR-LED. On this basis, using the embedded processor of the digital device for image processing, it can easily analyze the state of the IR-LED and understand the human intention. The system can directly understand and generate specific command signals to control digital instruments. And it has the advantages of simple operation and successful execution of human intention instructions.

Even a small mouse cursor can be positioned. Edwin Walsh et al. proposed a head-mounted pointing device (i.e. the Head-mouse) [82]. It consisted of an optical sensor based on a small CMOS camera with an IR bandpass filter and an IR-LED (850 nm wavelength). The advantage of the sensor was that the proposed sensor resulted in an absolute positioning of the mouse cursor and did not require laborious recalibration as the user changed their position.

This kind of IR-LED is also applied in traffic systems (Figure 8.15). America has developed a new IR-LED stereo camera, which can help detect pedestrians during the day and evening [83]. Using the image differencing and denoising, once pedestrians are detected, traffic signals at street intersections change phases to alert the drivers of approaching vehicles. The initial test results using images collected at a street intersection show that the system can detect pedestrians in near real time.

Figure 8.15 An example of a street intersection and traffic controller [83]. Source: Ling, B., Zeifman, M.I., Gibson, D.R.P. Multiple pedestrian detection using IR-LED stereo camera. Intelligent Robots and Computer Vision XXV: Algorithms, Techniques, and Active Vision, 2007.

Das et al. reported a 300% increase in optical power from LWIR-LED devices by thinning the bottom substrate till each pixel was isolated from the others [84]. Further, they optimized the chemical etching procedure. The lower input power requirement for an etched device compared to an unetched device will be useful in designing large-format IR-LED arrays. IR-LEDs are widely employed in Internet protocol cameras to increase the imaging effects on very dark scenes. Hence, the production yield of IR-LEDs significantly affects the performance of the displaying effects for IP cameras. It has been proposed as a very efficient way to estimate the IR-LED production yield based on image processing techniques. The results show that the method effectively improves the working efficiency of the operators on the production line [85]. In addition, Xiong et al. described an experimental setup and the characterization results of various flight model visible LEDs for development of on-board calibration unit of advanced wide field sensor (AWiFS) camera [86]. The AWiFS camera caters to high temporal resolution requirement, which is composed of three types of light: visible, NIR, and SWIR. Surface mountable wide view angle LEDs may be used in the future for a more compact calibration unit development of miniaturized payloads.

8.2.2.3 Biomedical Engineering

LED therapy is becoming increasingly important in several health treatment areas, such as biomedical engineering, deontology, physical therapy, and dermatology. Biomedical engineering, for example, has been experiencing major advances in the research into the application of LEDs in the treatment of premalignant or even malignant skin lesions, rejuvenation treatments, acne, hair loss, skin injuries, and postoperative incision recovery, and in the psychological improvement of patients where LED lights can be applied [87].

Progress in the field of neuroscience is critical in our rapidly aging society as the fatality rate due to brain degenerative diseases is rising. Recently, functional near-infrared spectroscopy (fNIRS) has established its application in measuring the blood chromophores' concentration changes in an active state. Yaqub et al. developed a lab-developed fNIRS system capable of computing the concentration changes in oxyhemoglobin and deoxyhemoglobin from optical NIR light signals [88]. Blood oxygen saturation (SpO_2) of the human

Figure 8.16 Electronic adjustable current circuit driver for LED and IR/RED separation [89]. Source: Yahya, M.A.M., Gan, K.B., Sameen, A.Z. Two channel dynamic optical phantom for transabdominal and fetal oxygen saturation with heart rate measurement. *2017 International Conference on Robotics, Automation and Sciences (ICORAS)*, 2017. IEEE 1–4. Fig 3.

physical structure is also very important, and it's a technique related to photoplethysmography (PPG), which is a noninvasive way to work out our blood oxygen continuously. A two-channel dynamic adjustable phantom of oximetry instrumentation has been developed, using a 735–890 nm IR-LED, for several biomedical applications, especially for transabdominal and fetal oximetry models [89]. It can be seen from Figure 8.16 that the circuit makes the separation for both IR and RED detection by using a four-channel electronic switch integrated circuit (IC). An advanced input/output USB (universal serial bus) multifunction data acquisition (DAQ) is used to control the circuit and for data acquisition.

The IR-LED can also be used in phototherapy on mast cells on the dorsum of the tongue of rodents based on the effects of laser light on tissue. The irradiation of IR-LED can induce the release of mediators responsible for vasodilatation; both red and IR-LED light causes increased mast cell degranulation and IR-LED light results in a greater number of mast cells [90].

8.2.3 Challenges

8.2.3.1 Monolithic Integration

Since the middle of the twentieth century, with the rapid development of the Si CMOS integration process and the increasing scale of integrated circuits, the chip integration of optoelectronic devices has become an inevitable trend in the development of modern optoelectronic technology. Optoelectronic chip integration can provide newer features, higher performance, and greater reliability with ever smaller sizes, lighter weights, and reduced power consumption. In addition, its large-scale commercial production can greatly reduce its development costs. Currently, the latest technology used in large-scale chip integration foundries is the Si CMOS or BiCMOS process. Therefore, the development of Si-based mid-IR optoelectronic devices compatible with CMOS or BiCMOS processes to achieve a monolithic integration of photonic devices and electronic devices has become the universal goal of many researchers. The integration of active devices in the band in silicon-based integrated circuits has led to the development of Si-based photonic devices that cover the mid-IR band, are inexpensive, environmentally friendly, and easy to integrate, and have become the focus of research around the world.

8.2.3.2 Driving Currents and Efficiency

IR-LEDs emit IR rays which are invisible to the human eye. That means we cannot judge an IR-LED's performance by observation.

The light intensity of an IR-LED is related to the forward currents driving the IR-LED to excite IR rays. During the rated forward current range, light intensity increases with the increase of forward currents. However, when the forward current exceeds the rated value, it will decrease with the increase of forward currents. So the forward current is a significant influencing factor for driving an IR-LED.

8.2.3.3 Thermal Management

The waste heat generated by LEDs must be removed through a cold plate or a cryogenic cold finger attached to the backside of the driver array. This heat must travel across the LED array–driver interface and through the driver array. The thermal resistance of these components can be significant. The time of the temperature drop can be significantly reduced by changes in the LED layout. These proposed guidelines to minimize thermal issues in LEDs should result in better performing and more-reliable IR-LEDs [91]. Generally speaking, whether the IR-LED works stably and its quality is good or bad is crucial to the heat dissipation of the lamp body itself. The heat dissipation of high-brightness IR-LED lamps in the market often adopts natural heat dissipation, and the effect is not ideal. IR-LED lamps made by an LED light source are composed of IR-LED, heat dissipation structure, driver, and lens. Therefore, heat dissipation is also an important part of IR-LED lamps. If IR-LED lamps cannot dissipate heat well, their life will be affected.

8.2.4 Conclusion

Nowadays, IR-LEDs are widely used in various fields, not only in special-purpose cameras but also for image positioning and in other fields. And people are realizing their importance. According to the characteristics of an IR-LED chip, IR-LED products with different wavelengths can be applied to different fields, such as remote control, medical instruments, space optical communication, IR lighting, the pump source of solid lasers, automatic card swipe systems, camera, monitoring, building interphones, burglar alarms, etc. With the acceleration of the development and application of the high-power IR ray, IR-LED manufacturers have tended to focus on the research and development for remote control, mouse, communication, and other applications with low added value, to mobile phones, vehicles, security control products IR monitoring, IR medical, and other high added-value market applications.

And by combining the sensing device with recognition technology, it can be applied to iris recognition, face recognition, and other special applications. Or it could be applied to biosensors in wearable devices, which can quantify the physiological state of the human body and become a new tool for health management. The Asia-Pacific region will become the world's largest market segment for IR-LEDs, given the surge in penetration of smartphones, cars, surveillance systems, and other IR-LED applications.

IR-LEDs used to be used in a small number of devices, and the power was not great. They were only used as sensors originally. However, being applied to night lighting and unmanned vehicle automated driving will spur the growth of the IR-LED. In future, the demand for better-quality mobile images, the growth of the Internet of things (IoT), and the ubiquity of portable devices generally will continue to encourage the development of IR-LEDs.

8.3 Future Outlook and Other Technology Trends

8.3.1 Better Light Sources

8.3.1.1 Display Technology

Wide gamut LED technology is defined as having a >90% NTSC (National Television Standards Committee) display gamut. Based on this technology, the liquid crystal display (LCD) can accurately present images and enrich colors. At present, the industrialization level of an LCD color gamut based on the new LED backlight source exceeds 90% of NTSC, and developing new phosphor powder and LED backlight sources, and further improve the LCD color gamut to 110% NTSC [92] have become an imperative.

The organic light-emitting diode (OLED) has many advantages, such as active luminescence, high luminescence efficiency, good luminescence color purity, bright color, and low power consumption, is an ultra-thin device, flexible, etc., which is beneficial to panchromatic display and has a good development prospect in the display field, and is favored by the industry. Recently, the first mobile phone with a flexible screen exhibited the rapid development of OLED.

QD materials have excellent luminescent properties, such as high quantum efficiency, continuously adjustable luminescence wavelength, half-peak width, and narrowness [93]. The traditional phosphor powder is replaced by QDs, and the color range of the display screen is improved to 110% NTSC.

8.3.1.2 High-Quality Light

With the development of white LEDs, the market is demanding higher-quality light sources, especially for indoor applications. The requirements for white LED light sources have been converted from the original pure pursuit of "high brightness" to the "high quality" of color rendering performance such as color rendering index, color temperature, and even the full spectrum of sunlight.

Owing to the phenomenon of LED "efficiency plunge" (i.e. when working in high current density), the IQE will decrease sharply. At present, scientists around the world are looking for a new generation of high-quality light sources. In the near future, LED technology will eventually be replaced by laser diodes because of the physical limits of its luminous efficiency. Compared with LEDs, lasers can achieve a higher efficiency. Semiconductor lasers are considered the most promising high-end and high-quality light source for display after LEDs, and will become a development trend in the future.

8.3.1.3 Special Light Sources

With the rapid development of NIR detectors, their application in facial recognition, iris recognition, security monitoring, LIDAR (light detection and ranging), health detection, 3D sensing, and other fields has become the focus of international research. IR detection is an important part of communication and IoT systems, and it requires NIR (especially 780–1600 nm) emitted in high-efficiency narrow band or special wide band. In addition, the blue-ray chip and NIR phosphor powder are combined and encapsulated, with the advantages of simple preparation process, low cost, and high luminous efficiency, which has attracted wide attention around the world. New kinds of NIR phosphor to meet diverse application needs are urgently needed.

Artificial sunlight (LEDs) can keep plants growing at high speed for a long time without considering the weather and seasons. Using LED-source light to precisely control yields, farm lettuce, kale, basil, and chives could be harvested 20–25 times a year, saving 85% of energy. LED light sources are more energy efficient than traditional light sources, which will effectively reduce the cost of power enterprises. At the same time, LED light sources produce less heat, which means that enterprises will need to use less air conditioning power. Philips says that indoor planting, which is controlled by LED lights, can cut the growth cycle of plants by up to 50% compared with traditional methods. This means it is possible for consumers to enjoy freshly grown local crops all year round. Therefore, the development of new luminescent materials matching plants and the research on the ratio of blue, red, and far-red light are the key directions of the development of biological agriculture.

8.3.2 Interconnection

Visible light is the electromagnetic spectrum region that can be used for communication in a new generation of lighting equipment, and is the future developmental direction of white LED. The rapid spread of indoor and outdoor solid-state lighting will provide a powerful platform for a new data transmission method, visible light communication (VLC), or Li-Fi. Conceptually, the operating mechanism of VLC or Li-Fi is reminiscent of a signal light. Data are piped through the power line to an LED lighting system equipped with a signal processing system. Light is directed at the user, and their electronic device is able to transmit data quickly through brightness modulation so that it is unrecognizable to the human eye [94]. An optical sensor on the user device detects fluctuations in light intensity and converts them to a digital signal. Li-Fi is more secure than Wi-Fi, because visible light cannot penetrate walls or other entities, so hackers must be in the same place to steal data.

The ultimate goal is to achieve information interaction between different devices under LED lighting and build an information smart city, which makes our lives easier, faster, and safer. Possible application scenarios include:

- In commercial applications, a system of LED lights in stores connect to consumers' smartphones. Through Li-Fi, customers can enjoy special offers based on their location in the store. People can find items on their shopping lists or get coupons when they pass the goods in the aisle. Retailers can send targeted information, such as recipes and coupons, to consumers based on their precise location within the store [95].
- LED headlights and taillights on vehicles allow vehicles to communicate with each other, and traffic signals to reduce red lights, collisions, and deaths.
- Medical applications can interfere with cell phone and wireless network signals. A Li-Fi network can provide secure and localized communication for patients, healthcare workers, and visitors, while low-data-based systems can track people's locations and mobile devices in the hospital.

8.3.3 Interaction with Humans

Lighting is an important environmental factor for human health, industrial production, and daily life. It is well known that short-wavelength ("blue") light affects the "nonvisual" effects of light in humans, and these effects go beyond mere "visual" functions to affect human health and performance centers [96]. Light perceived by ganglion cells in the retina can affect melatonin release, sleep, and circadian rhythms. For example, adjusting the LED spectrum to reduce clinicians' sleepiness and workload may shorten the execution time of clinical operations, as well as the occurrence of medical errors, while improving clinicians' health. At the same time, appropriate LEDs can delay the visual fatigue of primary and middle school students, improve learning efficiency, and help students to read and write. All of these indicate a bright future for the health industry.

Another potential application for LEDs is phototherapy. Phototherapy refers to the treatment of a patient's skin with light [94]. The use of UV light to treat skin diseases is a common phototherapy method. Its principle is the bactericidal properties of UV light. At the same time, visible light has also been studied for the direct treatment of dermatitis and muscle analgesia, as well as the elimination of bacteria *in vitro* [97]. Studies have shown that blue LEDs are more effective than quartz tungsten halogen lamps in inhibiting the proliferation of human gingival fibroblasts by producing intracellular reactive oxygen species [98]. In particular, a study on the phototoxicity and bactericidal effects of blue LED illumination showed that LED irradiation reduces the initial growth rate of melanoma cells by activating mitochondria-mediated pathways, thereby inducing apoptosis, indicating that the potential use of LEDs will be far greater than expected [99]. Oh observed that blue LED irradiation reduced the survival rate of mouse A20 and human RAMOS b-cell lymphoma cells, resulting in apoptosis [100]. Although it is necessary to further study the autophagy mechanism by which blue light irradiation induces lymphocyte apoptosis, LEDs certainly have many potential opportunities in new medical applications.

8.3.4 Light on Demand

LEDs can be integrated in many substrates like glass, textile, wood, and even biological materials. Since 1992, flexible LEDs made from soluble conducting polymers have been tested [101]. And much research has been conducted into the applications of soft material LEDs, and especially OLEDs [102, 103]. Besides, other materials are used as substrates. Choi et al. reported on the enhanced electroluminescence (EL) of GaN LEDs on glass substrates [104]. And Hu et al. reported that the textile-based device emits light when a voltage of 200 V AC at a fixed 400 Hz frequency is applied [105]. Besides, OLED materials were successfully manufactured on flexible, low-CTE (coefficient of thermal expansion), and optically transparent wood–cellulose nanocomposites [106]. And even electronics can be applied in the human body [107].

Light only where it is needed and only as long as it is required: LEDs have a great chance of achieving this goal for various applications, from illumination, perception and imaging, interconnection, and intelligence to biological requirements, and beyond. In future, LEDs can be arranged as required, in glass, wood, textile, etc., for illumination, sterilization, healthcare, data transmission, etc., which can be hidden and integrated.

References

1 Taniyasu, Y., Kasu, M., and Makimoto, T. (2006). An aluminium nitride light-emitting diode with a wavelength of 210 nanometres. *Nature* 441 (7091): 325–328.
2 Razeghi, M. and Henini, M. (2004). *Optoelectronic Devices: III-Nitrides*. Oxford: Elsevier.

3 Muramoto, Y., Kimura, M., and Nouda, S. (2015). Development and future of ultraviolet light-emitting diodes UV-LED will replace UV lamp. *Semiconductor Science and Technology* 29: 084004.

4 Kumar, M. (2019). Advances in UV-A and UV-C LEDs and the applications they enable. In: *Light-Emitting Devices, Materials, and Applications*. SPIE.

5 Nieland, S., Mitrenga, D., Schaedel, M. et al. (2019). Discussion on reliability issues for UV-B and UV-C LED-based devices. In: *Light-Emitting Devices, Materials, and Applications*. SPIE.

6 Kneissl, M., Kolbe, T., Chua, C. et al. (2010). Advances in group III-nitride-based deep UV light-emitting diode technology. *Semiconductor Science and Technology* 26 (1): 014036.

7 Han, J., Crawford, M.H., Shul, R.J. et al. (1998). AlGaN/GaN quantum well ultraviolet light emitting diodes. *Applied Physics Letters* 73 (12): 1688–1690.

8 Hirayama, H., Maeda, N., Fujikawa, S. et al. (2014). Recent progress and future prospects of AlGaN-based high-efficiency deep-ultraviolet light-emitting diodes. *Japanese Journal of Applied Physics* 53 (10): 100209.

9 Ha, M.S. (2009). *Thermal Analysis of High Power LED Arrays*. Atlanta, GA: School of Mechanical Engineering, Georgia Institute of Technology.

10 Schubert, E.F., Gessmann, T., and Kim, J.K. (2006). *Light Emitting Diodes*. New York: Cambridge University Press.

11 Habtemichael, Y.T. (2012). *Packaging Designs for Ultraviolet Light Emitting Diodes*. Atlanta, GA: G.W. Woodruff School of Mechanical Engineering.

12 Shatalov, M., Chitnis, A., Yadav, P. et al. (2005). Thermal analysis of flip-chip packaged 280 nm nitride-based deep ultraviolet light-emitting diodes. *Applied Physics Letters* 86 (20): 4762–4768.

13 Fischer, A.J., Allerman, A.A., Crawford, M.H. et al. (2004). Room-temperature direct current operation of 290 nm light-emitting diodes with milliwatt power levels. *Applied Physics Letters* 84 (17): 3394–3396.

14 Liu, Z., Liu, S., Wang, K. et al. (2009). Status and prospects for phosphor-based white led packaging. *Frontiers of Optoelectronics in China* 2 (002): 119–140.

15 Huang, H.S., Guo, W.P., Ke, Z.J. et al. (2014). Packaging-free UV-LED solidification light source module, China. Patent CN103794603 A. Issued 14 May 2014.

16 Khan, A., Fareed, Q., and Adivarahan, V. (2010). Lamps boost output in the deep ultraviolet. *Compound Semiconductor* June.

17 Mukish P. The UV-LED market is booming, announces YOLE Developpement. *Solid State Technology,* 2013.

18 Zou, X.Y., Lin, Y.L., Xu, B. et al. (2019). Enhanced inactivation of *E. coli* by pulsed UV-LED irradiation during water disinfection. *Science of the Total Environment* 650 (1(1–834)): 210–215.

19 Li, G.Q., Wang, W.L., Huo, Z.Y. et al. (2017). Comparison of UV-LED and low pressure UV for water disinfection: photoreactivation and dark repair of *Escherichia coli*. *Water Research* 126: 134–143.

20 Gora, S.L., Rauch, K.D., Ontiveros, C.C. et al. (2019). Inactivation of biofilm-bound *Pseudomonas aeruginosa* bacteria using UVC light emitting diodes (UVC LEDS). *Water Research* 151: 193–202.

21 Lau, J. (2008). Ultraviolet germicidal irradiation current best practices. *ASHRAE Journal* 50 (8): 28–36.

22 Sinha, R.P. and Häder, D.P. (2002). UV-induced DNA damage and repair: a review. *Photochemical & Photobiological Sciences* 1 (4): 225–236.

23 Meulemans, C. (1987). The basic principles of UV-disinfection of water. *Ozone: Science & Engineering* 9: 299–313.

24 Stermer, R.A., Margaret, L.S., and Brasington, C.F. (1987). Ultraviolet radiation: an effective bactericide for fresh meat. *Journal of Food Protection* 50 (2): 108–111.

25 Eskandarian, M.R., Choi, H., Fazli, M. et al. (2016). Effect of UV-LED wavelengths on direct photolytic and TiO_2 photocatalytic degradation of emerging contaminants in water. *Chemical Engineering Journal* 300: 414–422.

26 Chevremont, A.C., Farnet, A.M., Coulomb, B. et al. (2012). Effect of coupled UV-A and UV-C LEDs on both microbiological and chemical pollution of urban wastewaters. *Science of the Total Environment* 426: 304–310.

27 Chevremont, A.C., Farnet, A.M., Sergent, M. et al. (2012). Multivariate optimization of fecal bioindicator inactivation by coupling UV-A and UV-C LEDs. *Desalination* 285: 219–225.

28 Song, K., Mohseni, M., and Taghipour, F. (2016). Application of ultraviolet light-emitting diodes (UV-LEDs) for water disinfection: a review. *Water Research* 94: 341–349.

29 Anyaogu, K.C., Ermoshkin, A.A., Neckers, D.C. et al. (2010). Performance of the light emitting diodes versus conventional light sources in the UV light cured formulations. *Journal of Applied Polymer Science* 105 (2): 803–808.

30 Huang, C.K. and Sung, J.G. (2009). The Application of UV-LEDs to Microlithography. In: *2008 Second International Conference on Integration and Commercialization of Micro & Nanosystems*, 579–582. American Society of Mechanical Engineers Digital Collection.

31 Lai, K., Erdmann, A., Yapici, M.K. et al. (2014). UV-LED exposure system for low-cost photolithography. *Proceedings of SPIE: The International Society for Optical Engineering* 9052: 120–125.

32 Wang, P., Jin, Z., and Xiong, D. (2018). Optimal Optical Design of UV-LED Curing System with High Illumination and Luminance Uniformity. In: *2018 15th China International Forum on Solid State Lighting: International Forum on Wide Bandgap Semiconductors China (SSLChina: IFWS)*. IEEE.

33 Forough, Z. and Vijay, M. (2018). "Green" UV-LED gel nail polishes from bio-based materials. *International Journal of Cosmetic Science* 40 (6): 555–564.

34 Brown, R.G.W. (1989). Optical channels. Fibers, clouds, water and the atmosphere. *Optica Acta: International Journal of Optics* 36 (4): 552–552.

35 Xu, Z. and Sadler, B.M. (2008). Ultraviolet communications: potential and state-of-the-art. *IEEE Communications Magazine* 46 (5): 67–73.

36 Kedar, D. and Arnon, S. (2006). Non-line-of-sight optical wireless sensor network operating in multiscattering channel. *Applied Optics* 45 (33): 8454–8461.

37 Gagliardi, R.M. and Karp, S. (1976). *Optical Communications*. New York: Wiley-Interscience.

38 Schreiber, P., Dang, T., Pickenpaugh, T. et al. (1999). Solar-blind UV region and UV detector development objectives. *Proceedings of SPIE: The International Society for Optical Engineering* 3629: 230–248.

39 Steigerwald, D.A., Bhat, J.C., Collins, D. et al. (2002). Illumination with solid state lighting technology. *IEEE Journal of Selected Topics in Quantum Electronics* 8 (2): 310–320.

40 Ye, S., Xiao, F., Pan, Y.X. et al. (2011). Phosphors in phosphor-converted white light-emitting diodes: recent advances in materials, techniques and properties. *Materials Science & Engineering R: Reports* 71 (1): 1–34.

41 Adivarahan, V., Wu, S., Zhang, J.P. et al. (2004). High-efficiency 269 nm emission deep ultraviolet light-emitting diodes. *Applied Physics Letters* 84 (23): 4762–4764.

42 Shatalov, M., Streubel, K.P., Yao, H.W. et al. (2006). Reliability of AlGaN-based deep UV LEDs on sapphire. *Proceedings of SPIE: The International Society for Optical Engineering* 6134: 1–11.

43 Krutmann, J., Czech, W., Diepgen, T. et al. (1992). High-dose UVA1 therapy in the treatment of patients with atopic dermatitis. *Journal of the American Academy of Dermatology* 26 (2): 225–230.

44 Westerhof, W. and Nieuweboer-Krobotova, L. (1997). Treatment of vitiligo with UV-B radiation *vs.* topical psoralen plus UV-A. *Archives of Dermatology* 133 (12): 1525–1528.

45 Lennox C D, Beaudet S P. (1995).Medical treatment of deeply seated tissue using optical radiation. Patent: US5454807. Issued 30 December 1993.

46 Leenutaphong, V. and Jiamton, S. (1995). UVB phototherapy for pityriasis rosea: a bilateral comparison study. *Journal of the American Academy of Dermatology* 33 (6): 996–999.

47 Tsormpatsidis, E., Henbest, R.G.C., Battey, N.H. et al. (2010). The influence of ultraviolet radiation on growth, photosynthesis and phenolic levels of green and red lettuce: potential for exploiting effects of ultraviolet radiation in a production system. *Annals of Applied Biology* 156 (3): 357–366.

48 Tsormpatsidis, E., Henbest, R.G.C., Davis, F.J. et al. (2008). UV irradiance as a major influence on growth, development and secondary products of commercial importance in Lollo Rosso lettuce "revolution" grown under polyethylene films. *Environmental and Experimental Botany* 63 (1–3): 232–239.

49 Inada, S.A., Streubel, K.P., Jeon, H. et al. (2009). Effect of UV irradiation on the apoptosis and necrosis of Jurkat cells using UV LEDs. *Light-emitting Diodes: Materials, Devices, & Applications for Solid State Lighting XIII, International Society for Optics and Photonics* 7231: 72310J.

50 Guo, Z. (2019). $(Ca_{0.8}Mg_{0.2}Cl_2/SiO_2):Eu^{2+}$: a violet-blue emitting phosphor with a low UV content for UV-LED based phototherapy illuminators. *New Journal of Chemistry* 43: 3921–3926.

51 Gu, Y. (2018). Smart dental detector. *Optics in Health Care and Biomedical Optics VIII. SPIE - International Society for Optics and Photonics* https://doi.org/10.1117/12.2500799.

52 Rodenko, O., Fodgaard, H., Lichtenberg, P.T. et al. (2016). 340 nm pulsed UV LED system for europium-based time-resolved fluorescence detection of immunoassays. *Optics Express* 24 (19): 22135–22143.

53 Khan, A., Balakrishnan, K., and Katona, T. (2008). Ultraviolet light-emitting diodes based on group three nitrides. *Nature Photonics* 2 (2): 77–84.

54 Edmond, J. (2004). High efficiency GaN-based LEDs and lasers on SiC. *Journal of Crystal Growth* 272 (1): 242–250.

55 Nishida, T., Saito, H., and Kobayashi, N. (2001). Milliwatt operation of AlGaN-based single-quantum-well light emitting diode in the ultraviolet region. *Applied Physics Letters* 78 (25): 3927–3928.

56 Mino, T., Hirayama, H., Takano, T. et al. (2011). Realization of 256–278 nm AlGaN-based deep-ultraviolet light-emitting diodes on Si substrates using epitaxial lateral overgrowth AlN templates. *Applied Physics Express* 4 (9): 092104.

57 Mino, T. (2012). Characteristics of epitaxial lateral overgrowth AlN templates on (111)Si substrates for AlGaN deep-UV LEDs fabricated on different direction stripe patterns. *Physica Status Solidi* 9 (3–4): 802–805.

58 Jain, R., Sun, W., Yang, J. et al. (2008). Migration enhanced lateral epitaxial overgrowth of AlN and AlGaN for high reliability deep ultraviolet light emitting diodes. *Applied Physics Letters* 93 (5): 5532.

59 Gong, Z., Gaevski, M., Adivarahan, V. et al. (2006). Optical power degradation mechanisms in AlGaN-based 280 nm deep ultraviolet light-emitting diodes on sapphire. *Applied Physics Letters* 88 (12): L1419.

60 Pinos, A., Marcinkevicius, S., Yang, J. et al. (2009). Aging of AlGaN quantum well light emitting diode studied by scanning near-field optical spectroscopy. *Applied Physics Letters* 95 (18): 2969.

61 Zhang, J.P., Wang, H.M., Sun, W.H. et al. (2003). High-quality AlGaN layers over pulsed atomic-layer epitaxially grown AlN templates for deep ultraviolet light-emitting diodes. *Journal of Electronic Materials* 32 (5): 364–370.

62 Valko, E.I. and Chiklis, C.K. (2010). Effects of thermal exposure on the physicochemical properties of polyamides. *Journal of Applied Polymer Science* 9 (8): 2855–2877.

63 Murty, E.M. and Yehl, T.W. (1990). Adaptation of photoacoustic Fourier transform infrared spectroscopy for studying the thermal oxidation of nylon 66 at 150 °C correlated to mechanical properties. *Polymer Engineering & Science* 30 (24): 1595–1598.

64 Scaffaro, R., Dintcheva, N.T., and Mantia, F.P.L. (2008). A new equipment to measure the combined effects of humidity, temperature, mechanical stress and UV exposure on the creep behaviour of polymers. *Polymer Testing* 27 (1): 49–54.

65 Ludwick, A., Aglan, H., Abdalla, M.O. et al. (2008). Degradation behavior of an ultraviolet and hygrothermally aged polyurethane elastomer: Fourier transform infrared and differential scanning calorimetry studies. *Journal of Applied Polymer Science* 110 (2): 712–718.

66 Diepens, M. and Gijsman, P. (2008). Photo-oxidative degradation of bisphenol a polycarbonate and its possible initiation processes. *Polymer Degradation and Stability* 93 (7): 1383–1388.

67 Das, P.K., DesLauriers, P.J., Wood, F.K. et al. (1995). Photodegradation and photostabilization of poly(p-phenylene sulfide). Part 2. UV induced physicochemical changes. *Polymer Degradation and Stability* https://vdocuments.mx/photodegradation-and-photostabilization-of-polyp-phenylene-sulfide-part.html (accessed 17 March 2021).

68 Huang, J.C., Chu, Y.P., Wei, M. et al. (2010). Comparison of epoxy resins for applications in light-emitting diodes. *Advances in Polymer Technology* 23 (4): 298–306.

69 Narendran, N., Gu, Y., Freyssinier, J. et al. (2004). Solid-state lighting: failure analysis of white LEDs. *Journal of Crystal Growth* 268 (3–4): 449–456.

70 Shimomura, K. (2004). Light emitting device having a silicone resin. US Patent US6710377 B2. Issued 23 March 2004.

71 Yoshimura, N., Kumagai, S., and Nishimura, S. (1999). Electrical and environmental aging of silicone rubber used in outdoor insulation. *IEEE Transactions on Dielectrics and Electrical Insulation* 6 (5): 632–650.

72 Das, N.C. and Towner, F. (2010). Two color IR LED array. *Proceedings of SPIE – The International Society for Optical Engineering* 7663 https://doi.org/10.1117/12.852343.

73 Davis, N.J.L.K. et al. (2019). Improving the photoluminescence quantum yields of quantum dot films through a donor/acceptor system for near-IR LEDs. *Materials Horizons* 6: 137–114.

74 Shao, C., Zhang, S.N., and Liu, Z.H. (2014). Primary research on accelerated degradation tests for IR-LEDs under high pulse currents. In: *Proceedings of 2014 Prognostics and System Health Management Conference*, 664–668. IEEE.

75 Matveev, B.A., Gavrilov, G.A., Evstropov, V.V. et al. (1997). Mid-infrared (3–5 μm) LEDs as sources for gas and liquid sensors. *Sensors & Actuators B: Chemical* 39 (1–3): 339–343.

76 Massie, C., Stewart, G., Mcgregor, G. et al. (2006). Design of a portable optical sensor for methane gas detection. *Sensors & Actuators B: Chemical* 113 (2): 830–836.

77 Kuusela, T., Peura, J., Matveev, B.A. et al. (2009). Photoacoustic gas detection using a cantilever microphone and III–V mid-IR LEDs. *Vibrational Spectroscopy* 51 (2): 289–293.

78 Zhang, Y., Gao, W., Song, Z. et al. (2010). Design of a novel gas sensor structure based on mid-infrared absorption spectrum. *Sensors & Actuators B: Chemical* 147 (1): 5–9.

79 Gibson, D. and MacGregor, C. (2013). A novel solid state non-dispersive infrared CO_2 gas sensor compatible with wireless and portable deployment. *Sensors (Basel)* 13 (6): 7079–7103.

80 Allard, L.B., Liu, H.C., Buchanan, M. et al. (1997). Pixelless infrared imaging utilizing a p-type quantum well infrared photodetector integrated with a light emitting diode. *Applied Physics Letters* 70 (21): 2784–2786.

81 Park, Y. and Lee, M. (2013). Cost effective smart remote controller based on invisible IR-LED using image processing. In: *2013 IEEE International Conference on Consumer Electronics*, 434–435. IEEE.

82 Walsh, E., Daems, W., and Steckel, J. (2015). An optical head-pose tracking sensor for pointing devices using IR-LED based markers and a low-cost camera. In: *2015 IEEE SENSORS*, IEEE. https://doi.org/10.1109/ICSENS.2015.7370112.

83 Ling, B., Zeifman, M.I., and Gibson, D.R.P. (2007). Multiple pedestrian detection using IR-LED stereo camera. *Proc. SPIE.* 674: 1–12.

84 Das, N.C., Buford, J.A., Murrer, R.L., and Ballard, G.H. (2012). Performance of bottom emitting isolated LWIR LED devices for IR scene projection. *Proc. SPIE.* 8356: 1–6.

85 Lin, P.H., Cheng, F.C., and Huang, S.C. (2015). An IR-LED production yield estimation method for IP-camera. In: *2015 IEEE International Conference on Consumer Electronics*, Taiwan, IEEE.

86 Xiong, X.J., Kuriakose, S.A., Kimura, T. et al. (2016). LED characterization for development of on-board calibration unit of CCD-based advanced wide-field sensor camera of Resourcesat-2A. *Spie Asia-pacific Remote Sensing* 9881: 98811W.

87 Moreira, M.C., Prado, R., and Campos, A. (2011). Application of high brightness LEDs in the human tissue and its therapeutic response. In: *Applied Biomedical Engineering* (ed. G. Gargiulo), InTech, https://doi.org/10.5772/19510.

88 Yaqub M A, Zafar A, Ghafoor U, et al. (2018). Development of a high density neuroimaging system using functional near-infrared spectroscopy. In: *18th International Conference on Control, Automation and Systems (Iccas)*. PyeongChang, GangWon Province, Korea (17–20 October 2018).

89 Yahya, M.A.M., Gan, K.B., and Sameen, A.Z. (2017, 2017). Two channel dynamic optical phantom for transabdominal and fetal oxygen saturation with heart rate measurement. In: *International Conference on Robotics, Automation and Sciences (ICORAS)*, 1–4. IEEE.

90 Santos, D.C.M.J., De Oliveira, S.C.P.S., Lima, d.F.F. et al. (2011). Effect of LED red and IR photobiomodulation in tongue mast cells in Wistar rats: histological study. *Photomedicine and Laser Surgery* 29 (11): 767–771.

91 Murrer, R.L. (2007). Technologies for synthetic environments: hardware-in-the-loop testing XII. In: *Proceedings of SPIE: The International Society for Optical Engineering*. SPIE.

92 Xianjichina.com. (2020). Application status and trend of white light LED light source luminescent materials. https://www.xianjichina.com/special/detail_360074.html (accessed 3 September 2020).

93 Kebin, L., Jun, X., Li, N.Q. et al. (2018). Perovskite light-emitting diodes with external quantum efficiency exceeding 20 per cent. *Nature* 562 (7726): 245.

94 Cho, J., Park, J.H., Kim, J.K. et al. (2017). White light-emitting diodes: history, progress, and future. *Laser & Photonics Reviews* 11 (2): 1600147.1–1600147.17.

95 W-O-Lifi.Blogspot. (2015). Philips Creates Shopping Assistant with LEDs and Smart Phone. https://spectrum.ieee.org/tech-talk/computing/networks/philips-creates-store-shopping-assistant-with-leds-and-smart-phone (accessed 3 September 2020).

96 Perez, O.L., Strother, C., Vincent, R. et al. (2018). Effects of 'blue-regulated' full spectrum LED lighting in clinician wellness and performance, and patient safety. In: *Congress of the International Ergonomics Association*, 667–682. Cham: Springer.

97 Lim, W., Lee, S.G., Kim, I. et al. (2007). The anti-inflammatory mechanism of 635 nm light-emitting-diode irradiation compared with existing COX inhibitors. *Lasers in Surgery and Medicine* 39 (7): 614–621.

98 Yoshida, A., Yoshino, F., Makita, T. et al. (2013). Reactive oxygen species production in mitochondria of human gingival fibroblast induced by blue light irradiation. *Journal of Photochemistry and Photobiology B: Biology* 129: 1–5.

99 Kim, S., Kim, J., Lim, W. et al. (2013). In vitro bactericidal effects of 625, 525, and 425 nm wavelength (red, green, and blue) light-emitting diode irradiation. *Photomedicine and Laser Surgery* 31 (11): 554–562.

100 Oh, P.S. (2016). Blue light emitting diode induces apoptosis in lymphoid cells by stimulating autophagy. *International Journal of Biochemistry & Cell Biology* 70: 13–22.

101 Gustafsson, G. (1993). Flexible light-emitting diodes made from soluble conducting polymers. *ChemInform* 24 (2): 477–479.

102 Wang, Z.B., Helander, M.G., Qiu, J. et al. (2011). Unlocking the full potential of organic light-emitting diodes on flexible plastic. *Nature Photonics* 5 (12): 753–757.

103 Kim, W., Kwon, S., Lee, S.M. et al. (2013). Soft fabric-based flexible organic light-emitting diodes. *Organic Electronics* 14 (11): 3007–3013.

104 Choi, J.H., Ahn, H.Y., Lee, Y.S. et al. (2012). Gan light-emitting diodes on glass substrates with enhanced electroluminescence. *Journal of Materials Chemistry* 22 (43): 22942–22948.

105 Hu, B., Li, D., Ala, O. et al. (2011). Textile-based flexible electroluminescent devices. *Advanced Functional Materials* 21 (2): 305–311.

106 Okahisa, Y., Yoshida, A., Miyaguchi, S. et al. (2009). Optically transparent wood–cellulose nanocomposite as a base substrate for flexible organic light-emitting diode displays. *Composites Science and Technology* 69 (11–12): 1958–1961.

107 Rogers, J.A. (2015). Electronics for the human body. *JAMA* 313 (6): 561–562.

Index

From LED to Solid State Lighting: Principles, Materials, Packaging, Characterization, and Applications, First Edition.
Shi-Wei Ricky Lee, Jeffery C. C. Lo, Mian Tao, and Huaiyu Ye.